Lectures on

Methods of Electronic Structure Calculations

Lectures on

Methods of Electronic Structure Calculations

Proceedings of the Miniworkshop on "Methods of Electronic Structure Calculations" and Working Group on "Disordered Alloys"

ICTP, Trieste, Italy 10 August – 4 September 1992

Editors

V Kumar
Indira Gandhi Centre for Atomic Research, Kalpakkam, India
and International Centre for Theoretical Physics, Trieste, Italy

O K Andersen
Max – Planck – Institut für Festkorperforschung, Stuttgart, Germany

A Mookerjee
S. N. Bose National Centre for Basic Sciences, Calcutta, India

World Scientific
Singapore • New Jersey • London • Hong Kong

Published by

World Scientific Publishing Co. Pte. Ltd.

5 Toh Tuck Link, Singapore 596224

USA office: 27 Warren Street, Suite 401-402, Hackensack, NJ 07601

UK office: 57 Shelton Street, Covent Garden, London WC2H 9HE

British Library Cataloguing-in-Publication Data
A catalogue record for this book is available from the British Library.

Dr. Armin Burkhardt contributed to the design of the cover of these proceedings.
It shows nodal surface of the *s* tight-binding LMTO in the bcc structure (see article by O. K. Andersen, O. Jepsen and G. Krier)

LECTURES ON METHODS OF ELECTRONIC STRUCTURE CALCULATIONS

ISBN-13 978-981-02-1485-2
ISBN-10 981-02-1485-5
ISBN-13 978-981-02-3808-7 (pbk)
ISBN-10 981-02-3808-8 (pbk)

PREFACE

The developments during the last thirty years of density functional theory, of efficient methods for calculating the electronic structure, and of computers with rapidly decreasing cost/performance ratio have made *ab-initio* studies of cohesive, magnetic, and electronic properties of real materials possible. While with the linear band structure methods one may nowadays perform accurate calculations for s-, p-, and d-electron systems containing hundreds of atoms in the unit cell, or even lacking translational symmetry, *ab-initio* molecular dynamics calculations were made possible with the pseudopotential plane-wave method by Car and Parrinello.

The present lecture notes were written for the Miniworkshop on 'Methods of Electronic Structure Calculations' and the following Working Group on 'Disordered Alloys' which took place at the International Centre for Theoretical Physics in Trieste from 10 August to 4 September in 1992. The purpose of these activities was to discuss some of the widely used electronic structure methods and to provide instructions on the use of the computer codes. Accordingly, most lecturers concentrated on methodology rather than applications, for which the reader is referred to specific papers. Each contribution is self-contained and no effort has been made to keep notation uniform throughout the book.

The first two chapters by M. P. Das and U. von Barth cover the density functional formalism including very recent developments. O.K. Andersen et al. explain the linear muffin-tin orbital (LMTO) method in a way which links it with the KKR method and which points towards future use, for instance in molecular dynamics calculations. The recursion method does not rely on translational symmetry and it allows one to treat for instance liquid and amorphous matter. The modern version of this method is presented by R. Haydock while P. Vargas explains its practical implementation with tight-binding LMTOs. The problem of the electronic structure of disordered alloys is discussed by A. Mookerjee who uses the augmented space formulation together with ideas from the recursion and TB-LMTO methods. The KKR-CPA method for treating substitutionally disordered alloys is covered by R. Prasad and by Staunton et al. who also consider magnetic alloys. Kudrnovsky et al. present their efficient self-consistent Green-function TB-LMTO-CPA method for such alloys and their surfaces. Applications of the LMTO supercell method to the study of interfaces are presented by G.P. Das. Finally, the Car-Parrinello method for density functional molecular dynamics calculations is discussed by G. Pastore, V. Kumar, and K. Laasonen who, in particular, deals with the use of supersoft pseudopotentials to treat tightly bound electrons.

As it always happens, not all the speakers responded with a manuscript; understandably either they were very busy or did not want to proliferate the literature. We would like to thank the speakers who have made their contributions available and the participants in the Miniworkshop and the Working Group for their enthusiastic response. Finally the efficient and excellent help of the Condensed Matter Secretariat, and in particular Ms. Marina de Comelli is gratefully acknowledged.

V. Kumar, O.K. Andersen, and A. Mookerjee

CONTENTS

DENSITY FUNCTIONAL THEORY: MANY-BODY EFFECTS WITHOUT TEARS

M. P. Das

Department of Theoretical Physics
Research School of Physical Sciences
The Australian National University
Canberra, ACT 2601
Australia

Abstract

Density functional theory (DFT) provides a fundamental basis for the understanding of atomic and electronic structure of condensed matters. In this paper we discuss various aspects of the DFT and highlight the successes and failures of the most commonly used approximation known as the local density approximation (LDA).

1. Introduction

In 1964 Hohenberg and Kohn[1] established rigorously that the ground state properties of a quantum many body system can be expressed in terms of single particle density, rather than many-body wave functions. Since then DFT has emerged as a powerful tool for analysing a large variety of many-body systems as diverse as atoms, molecules, bulk and surfaces of solids, liquids, dense plasmas, nuclear matter and heavy ion systems.

Although the DFT does not make a frontal attack to a many-body problem, as does the many-body perturbation theory, it does not have a small parameter problem. Unlike other many-body models characterised by idealised Hamiltonians (e.g. Hubbard model, Anderson model, Kondo model etc.) one can describe a physical system in an *ab-initio* manner within the realm of the DFT. This is one of the attractive features for which the theory has wider applicability to a large variety of many-body systems. The soundness of a theory is judged by its successes in relation to explaining observations. A correct theory makes predictions which are verifiable in future. In this respect the DFT has many proven records[2-24].

In this paper we deal only with the electronic structure aspect of condensed matter. Suffice to say that the DFT provides a strong foundation of the electronic structure theory. In Sec.2 we present basic notations followed by proofs of the Hohenberg-Kohn theorems. A number of generalisations of the formalism relevant to various physical situations are mentioned. Solution of a many-body problem requires approximation(s). In Sec.3 we discuss the most commonly used approximation the LDA by pointing out its validity. Successes and failures of a theory are judged from its applications. Apart from the vast literature[2-24] an interested reader may find many examples in this proceedings on applications of the LDA to the condensed matter systems. In the final section some concluding remarks are presented.

2. Basic Notations

A condensed matter system is constituted of electrons and nuclei. Apart from their charges and masses etc. the Hamiltonian of the system depends on their coordinates. A very useful approximation is used what is known as the adiabatic approximation that helps to separate the coordinates of the nuclei from those of the electrons. Let $\{\mathbf{R}_\alpha\}$ and $\{\mathbf{r}_i\}$ denote the positions of the nuclei of charge Z and of electrons of unit charge respectively. The electronic Hamiltonian in the adiabatic approximation is given by

$$\begin{aligned} H &= H_{ee} + H_{en} \\ &= \sum_i p_i^2/2m + \frac{1}{2}\sum_{ij}{}' (\mathbf{r}_i - \mathbf{r}_j)^{-1} - Z\sum_{i\alpha} (\mathbf{r}_i - \mathbf{R}_\alpha)^{-1} \\ &\equiv T + U + V\ . \end{aligned} \tag{1}$$

The first term is the kinetic energy operator of the electrons, the second is the electron-electron interaction, prime on summation indicates $i \neq j$ and the third term represents the attractive electron-nuclear interaction. Since the nuclei do not belong to the electronic subsystem, we shall call the effect of the last term as external to the system of electrons.
In the second quantized representation we rewrite the Hamiltonian (1) as

$$\begin{aligned} H = & (\hbar^2/2m) \sum_{\sigma} \int \nabla \psi^{\dagger}(x)\, \nabla \psi(x)\, d\mathbf{r} \\ & + (1/2) \sum_{\sigma} \iint (\mathbf{r} - \mathbf{r}')^{-1}\, \psi^{\dagger}(x)\, \psi^{\dagger}(x')\, \psi(x)\, \psi(x')\, d\mathbf{r}\, d\mathbf{r}' \\ & + \sum_{\sigma} \int v(\mathbf{r} - \mathbf{R}_{\alpha})\, \psi^{\dagger}(x)\, \psi(x)\, d\mathbf{r}\ . \end{aligned} \tag{2}$$

$\psi(x)$ and $\psi^{\dagger}(x)$ are annihilation and creation operators for the electrons. They obey equal time anti-commutation relations. x stands for the space $\mathbf{r}$ and spin σ coordinates. The electron density operator $\hat{\rho}(\mathbf{r})$ is

$$\hat{\rho}(\mathbf{r}) = \sum_{\sigma} \psi^{\dagger}(x)\, \psi(x)\ . \tag{3}$$

The expectation value of $\hat{\rho}(\mathbf{r})$, with respect to the ground state Ψ of H, gives the one particle density $n(\mathbf{r})$. The total number of particles N is given by $N = \int n(\mathbf{r})\, d\mathbf{r}$ and the ground state energy is

$$E = \langle \Psi | H | \Psi \rangle\ . \tag{4}$$

We remind that $n(\mathbf{r})$ is non-negative definite and continuous. It is said to be N-representable when an anti-symmetric N particle wave function generates it and it is V-representable if it stems from an anti-symmetric ground state of Ψ of H that contains the external potential V.

We need to mention the following three important theorems which are very useful to construct approximations and to check the consistency. These principles help us to understand clearly a quantum system through the classical concepts in an analogous manner.

(i) Raleigh-Ritz Variational Principle : In absence of the knowledge of the exact many-body eigenstate Ψ, a variational trial state Ψ(trial) is constucted and by using this trial state, the trial ground state energy , E(trial) can be calculated. E(trial) is close to E(exact), if Ψ(trial) is close to Ψ(exact). This principle , the variation of the ground state energy with respect to Ψ, leads to the basic many-body Schrödinger equation.

(ii) Hellmann-Feynman Theorem : It is a further variational property of the derivative of the energy with respect to a parameter present in the Hamiltonian. For example; $\mathbf{R}_\alpha$ is the position coordinate of a nuclei as in (1). $dE/d\mathbf{R}_\alpha = \langle\Psi\,|\,dH/d\mathbf{R}_\alpha\,|\,\Psi\rangle \equiv -\mathbf{F}_\alpha$ is the force on the nuclei at $\mathbf{R}_\alpha$. Such properties are calculated as variational consequences.
(iii) Virial Theorem: If a length scale λ is chosen and all the coordinates are scaled by λ, one obtains the virial theorem. This theorem is analogous to the Hellmann-Feynman force theorem which gives the expression for the pressure in terms of internal operators of the Hamiltonian.

2.1 Hohenberg-Kohn Theorems[1]

The density functional formalism is stated by two theorems.

Theorem 1. For a nondegenerate ground state Ψ , the external potential $v(\mathbf{r})$ is (to within an additive constant) uniquely determined by the density distribution $n(\mathbf{r})$.

The proof is by *reductio ad absurdum*. Let us consider N electrons given by the density $n(\mathbf{r})$ subject to an external potential $v(\mathbf{r})$. The Hamiltonian (1) with external potential energy V has the ground state $|\Psi\rangle$ and the ground state energy E (eq.4). Let us assume that there is another potential $v'(\mathbf{r})$ which gives rise to the same density $n(\mathbf{r})$, but now with the Hamiltonian $H'(= T + U + V')$, the ground state $|\Psi'\rangle$ and the ground state energy E'. From the minimal properties of the ground state and due to nondegeneracy,

$$\langle\Psi|H|\Psi\rangle \equiv E < \langle\Psi'|H|\Psi'\rangle = \langle\Psi'|H + V' - V'|\Psi'\rangle$$

$$= \langle\Psi'|H' - V' + V|\Psi'\rangle = E' - \int[v'(\mathbf{r}) - v(\mathbf{r})]\, n(\mathbf{r})\, d\mathbf{r} \,. \quad (5)$$

Similarly by interchanging the primed and unprimed quantities, we obtain

$$E' < E - \int[v(\mathbf{r}) - v'(\mathbf{r})]\, n(\mathbf{r})\, d\mathbf{r} \,. \quad (6)$$

By adding (5) and (6) an inconsistent result is obtained

$$E + E' < E' + E \,. \quad (7)$$

The conclusion from the above argument follows that the density $n(\mathbf{r})$ associated with the ground state cannot sustain two different potentials $v(\mathbf{r})$ and $v'(\mathbf{r})$ (except $v(\mathbf{r}) - v'(\mathbf{r})$ = an additive constant). Further it follows that since $n(\mathbf{r})$ determines the potential uniquely, it determines the ground state (assumed nondegenerate) completely. For a proof of degenerate ground state see Kohn (in ref. 21).

Theorem 2. For a given $v(\mathbf{r})$, the correct $n(\mathbf{r})$ minimizes the (nondegenerate) ground state energy, which is unique functional of $n(\mathbf{r})$.

We define the following energy functional :

$$E_v[n'] \equiv \int v(\mathbf{r})\, n'(\mathbf{r})\, d\mathbf{r} + F[n'(\mathbf{r})] \;, \tag{8}$$

where

$$F[n] = \langle \Psi | T + U | \Psi \rangle \;. \tag{9}$$

F represents the sum of kinetic and Coulomb interaction energies. In (8) $v(\mathbf{r})$ is not assumed to be functional of $n(\mathbf{r})$. From the Rayleigh-Ritz principle the energy functional $\mathcal{E}_v\,[\Psi']$ is given by

$$\mathcal{E}_v[\Psi'] \equiv \langle \Psi' | V | \Psi' \rangle + \langle \Psi' | T + U | \Psi' \rangle \tag{10}$$

and it is minimum for the correct ground state $\Psi' = \Psi$ with the same total number of particles N. Let Ψ' be the ground state for another density $n'(\mathbf{r})$ as Ψ for $n(\mathbf{r})$. Then

$$\mathcal{E}_v[\Psi'] = \int v(\mathbf{r})\, n'(\mathbf{r})\, d\mathbf{r} + F[n'(\mathbf{r})] = E_v[n']$$

$$> \mathcal{E}_v[\Psi] = \int v(\mathbf{r})\, n(\mathbf{r})\, d\mathbf{r} + F[n(\mathbf{r})] \equiv E_v[n] \;. \tag{11}$$

Eq. (11) establishes the minimal property of the energy associated with $v(\mathbf{r})$ and N.

F[n] (eq.9) is defined for those n which satisfy the full many-body equation containing some external potential $v(\mathbf{r})$, hence it is called v-representable. Now it has been possible to construct more general functionals, which are non-v-representable, by using a 'constrained search approach' as pointed out by Levy[25] and Lieb[26]. These authors defined the universal functional F [n] as :

$$F[n] = \min \langle \Psi_n | T + U | \Psi_n \rangle \;. \tag{12}$$

This functional searches all anti-symmetric wavefunctions Ψ_n (which yield the fixed n) and gives the minimum expectation value. Formally this approach replaces the stronger V-representability by a milder N-representability and displays clearly the statistics which the particles obey. Of course, F[n] is *a priori* unknown. But it incorporates all the relevant effects of the many-body dynamics.

2.2 Where is Many-body Effect?

The HK functional F[n} is defined as the ground state expectation value of the operators of kinetic and electron-electron interaction energies. The many-body effects are hidden inside this functional. It is not difficult to find out the many-body contribution to this functional.

Let us consider a system of electrons given by a charge density distribution $n(\mathbf{r})$ (Fig.1) and they are interacting via classical Coulomb force only. For simplification we ignore the kinetic energy of the electrons here. We bring a probe of unit charge to a point $\mathbf{r}$, which will experience a potential $\Phi(\mathbf{r})$, due to the system of electrons.

$$\Phi(\mathbf{r}) = \int d\mathbf{r}'\, n(\mathbf{r}')/(\mathbf{r}-\mathbf{r}') \ . \tag{13}$$

This is known as the Hartree potential and the classical Coulomb (Hartree) energy is given by

$$E_c = \frac{1}{2}\int d\mathbf{r}\, \Phi(\mathbf{r})\, n(\mathbf{r}) \tag{14}$$

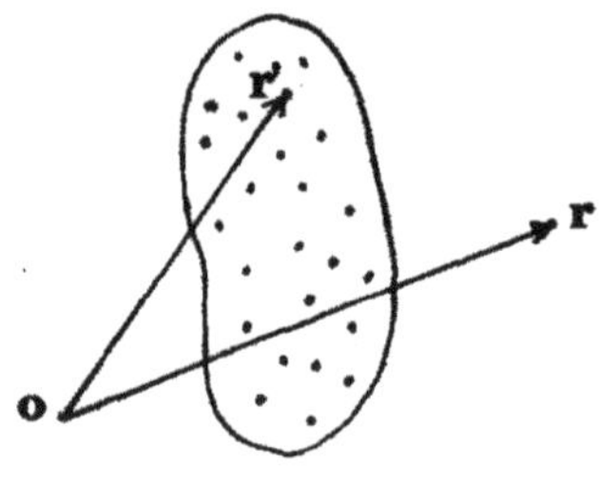

Figure 1

But, if the probe is not a spectator, it interacts with other particles as a component of the system,then the potential that it will experience is given by

$$\Phi(\mathbf{r}) = \int d\mathbf{r}'\, n(\mathbf{r},\mathbf{r}')/(\mathbf{r}-\mathbf{r}') \ . \tag{15}$$

The quantity $n(\mathbf{r},\mathbf{r}')$ is a conditional particle density at $\mathbf{r}$, when there is already a particle at $\mathbf{r}'$. $n(\mathbf{r},\mathbf{r}')$ can be further written as

$$n(\mathbf{r},\mathbf{r}') = n(\mathbf{r}')\, g(\mathbf{r},\mathbf{r}') \ , \tag{16}$$

where $g(\mathbf{r},\mathbf{r}')$ is known as pair-correlation function. In the Hartree theory $g(\mathbf{r},\mathbf{r}')$ is unity, therefore (15) reduces to (13) and we obtain instantaneous Coulomb interaction energy from (14). Since electrons are quantum particles having spins, $g(\mathbf{r},\mathbf{r}')$ contains correlations beyond Hartree approximation. One part is due to parallel spins obeying the Pauli exclusion principle, called exchange correlation and the other due to anti-parallel spins called Coulomb correlation. Generally corrections to the energy beyond the Hartree mean-field and the Dirac-Fock exchange-field is known as the correlation energy.

Pair correlation function in the many-body theory is an important and difficult subject particularly for an inhomogeneous system. During the past two decades a satisfactory understanding has been made on the homogeneous electron gas. Short and long range spatial correlations and dynamical aspects in the electron gas have been studied in detail[27]. We shall borrow several results from this in our later discussions on the approximations.

The total interaction energy U, using eqs. (15) and (16) is written as

$$U = \frac{1}{2}\iint d\mathbf{r}\, d\mathbf{r}'\, n(\mathbf{r})\, n(\mathbf{r}')\, g(\mathbf{r},\mathbf{r}') \frac{1}{|\mathbf{r}-\mathbf{r}'|}$$

$$= \frac{1}{2}\iint d\mathbf{r}\, d\mathbf{r}' \frac{1}{|\mathbf{r}-\mathbf{r}'|} n(\mathbf{r})\, n(\mathbf{r}')\, [g(\mathbf{r},\mathbf{r}') + 1 - 1]$$

$$= \frac{1}{2}\iint d\mathbf{r}\, d\mathbf{r}' \frac{1}{|\mathbf{r}-\mathbf{r}'|} n(\mathbf{r})\, n(\mathbf{r}') + \frac{1}{2}\iint d\mathbf{r}\, d\mathbf{r}' \frac{1}{|\mathbf{r}-\mathbf{r}'|} n(\mathbf{r})\, n(\mathbf{r}')\, [g(\mathbf{r},\mathbf{r}') - 1]$$

$$\equiv E_c + E_{xc}\ . \tag{17}$$

Here E_c is the classical Coulomb-Hartree energy clearly separated from the total interaction energy. The explicit many-body effect beyond the Hartree mean-field has now appeared as the exchange-correlation potential energy.

2.3 Energy Functional and Variational Equations

The energy functional F[n] can now be written as: F[n] = G[n] + E_c[n], where the functional G[n] contains kinetic and many-body effects. A useful decomposition of G[n] can be done in the following manner. Let the exchange-correlation (XC) energy functional accounts for all the many-body processes. Then by definition : $E_{xc}[n] = G[n] - T_s[n]$, where T_s is the kinetic energy functional of a noninteracting system of the same density. Note that E_{xc}[n] also contains contributions from the kinetic energy beyond the single particle approximation.

Now we use the variational principle. The energy functional E[n] attains its minimum, the *ground state energy*, if and only if, the density n(**r**) is the ground state density. This is achieved in two distinct ways.

A. Direct Method

The energy minimization is carried out by varying the density with the constraint that N is fixed (*i.e.*, $\int n(\mathbf{r})\, d\mathbf{r} = N[n]$). The variational procedure is

$$\delta\{E[n]\} = 0 \;,$$

or

$$\int d\mathbf{r}\, \delta n(\mathbf{r}) \left[\delta E[n]/\delta n(\mathbf{r}) \right] = 0 \;. \tag{18}$$

Since $\int d\mathbf{r}\;\; \delta n(\mathbf{r}) = 0$, the quantity inside the square brackets is a constant, independent of $\mathbf{r}$. This gives the famous Euler-Lagrange equation.

$$\left(\delta E[n]/\delta n(\mathbf{r}) \right) = \mu \;, \tag{19}$$

or

$$\left(\delta T[n]/\delta n(\mathbf{r}) \right) + \Phi(\mathbf{r}) + v_{xc}[n] = \mu \tag{20}$$

where

$$\Phi(\mathbf{r}) = v(\mathbf{r}) + \int d\mathbf{r}'\; n(\mathbf{r}')/|\mathbf{r}-\mathbf{r}'| \;, \tag{21}$$

and

$$v_{xc}[n(\mathbf{r})] = \delta E_{xc}[n]/\delta n(\mathbf{r}) \;. \tag{22}$$

Now the quantity, $v_{eff}[n(\mathbf{r})]$ using (21) and (22) is

$$v_{eff}(\mathbf{r}) = \Phi(\mathbf{r}) + v_{xc}[n(\mathbf{r})] \;. \tag{23}$$

v_{eff} is an effective one-body potential in which the many-body effect is accounted for by v_{xc}. If the exchange-correlation effect is dropped in eq.(20), it represents the Thomas-Fermi model.

B. Self-consistent Field Method

An exact generalization of the Hartree method can be done by deriving one-particle self-consistent Schrödinger-like equations. Let us start with a trial density n(**r**) constucted out of some fictitious one-particle normalised orbitals $\psi_i(\mathbf{r})$,

$$n(\mathbf{r}) = \sum_i |\psi_i(\mathbf{r})|^2 \;. \tag{24}$$

The summation is over the N lowest occupied states and the density integrates to the total number of particles, N. The orbitals, $\psi_i(\mathbf{r})$ minimize the total energy E[n] with the constraint of normalization and T[n] is given by the single particle kinetic energy $T_s[n]$. $T_s[n] = -(1/2)\langle \Psi | \nabla^2 | \Psi \rangle$. Now the energy variation with respect to the orbitals $\psi_i(\mathbf{r})$ is carried out.

$$\delta/\delta\psi_i^*(\mathbf{r}) \left(E[n] - \sum_j \varepsilon_j \int d\mathbf{r}\; |\psi_j(\mathbf{r})|^2 \right) = 0$$

or
$$\left(-(1/2)\nabla_i^2 + v_{eff}[n(\mathbf{r})]\right)\psi_i(\mathbf{r}) = \varepsilon_i\,\psi_i(\mathbf{r}) \ , \tag{25}$$

where $v_{eff}[n(\mathbf{r})]$ and $n(\mathbf{r})$ are defined by (23) and (24) respectively. Here ε_i's appear as Lagrange multipliers for each normalized orbitals ψ_i [41]. They are fictitious and strictly are not true physical single particle energies for the reason to be clear shortly. Note that if $v_{xc}[n(\mathbf{r})]$ in (25) is dropped, we recover the self-consistent Hartree-eqns.

To calculate $T_s[n]$, we multiply $\psi_i^*(\mathbf{r})$, integrate in $\mathbf{r}$ and sum over occupied states i in eq. (25).

$$T_s[n] = \sum_i \varepsilon_i - \int d\mathbf{r}\, v_{eff}[n(\mathbf{r})]\, n(\mathbf{r}) \ . \tag{26}$$

The total energy E[n] (eq.(8)) is

$$E[n] = \int d\mathbf{r}\, v(\mathbf{r})\, n(\mathbf{r}) + T_s[n] + E_c[n] + E_{xc}[n] \ . \tag{27}$$

Substituting (26) for $T_s[n]$ in (27)

$$E[n] = \sum_i \varepsilon_i - E_c[n] + E_{xc}[n] - \int d\mathbf{r}\, v_{xc}[n(\mathbf{r})]\, n(\mathbf{r}) \ . \tag{28}$$

It is clear that the sum of the eigenvalues (i.e. the first term on the right hand side of (28)) is not the total energy as is shown. The scheme dealing with the above procedure is the well-known Kohn-Sham method[41]. Particle density, total energy and kinetic energy are exactly obtained by this method.

Let us identify the Lagrange multiplier μ, which has occured as a parameter in the process of minimization. It can be shown as the chemical potential by using the variational eq.(19).

$$\{\partial E/\partial N\} = E[n_N] - E[n_{N-1}]$$

$$= \int d\mathbf{r} \left\{\delta E[n]/\delta n(\mathbf{r})\right\}_{n=n_N} \left\{n_N(\mathbf{r}) - n_{N-1}(\mathbf{r})\right\} = \mu \ . \tag{28}$$

The eigenvalues ε_i's of the Kohn-Sham effective Hamiltonian are not true physical energies. Physical energies of excitations are obtainable from the non-local Dyson eq. of the many-body theory,

$$-(1/2)\nabla_i^2\, \psi_i(\mathbf{r}) + \int \Sigma\, (\mathbf{r},\mathbf{r}'; \tilde{\varepsilon}_i)\, \psi_i(\mathbf{r}')\, d\mathbf{r}' = \tilde{\varepsilon}_i\, \psi_i(\mathbf{r}) \ . \tag{29}$$

By comparing (25) with (29), one can immediately see that the difference between v_{eff} and Σ lies in the exchange-correlation potential, which is non-local, complex and energy-dependent in the Dyson equation (29), whereas it is local and real in the Kohn-Sham eq. (25). The energies $\tilde{\varepsilon}_i$

are complex with finite life-time of the states, whereas ε_i are real. For an infinite system it can be shown from the density functional analogue of the Koopmans theorem that the highest occupied state of Kohn-Sham eq.(25) corresponds to the same energy, $\varepsilon_i = \tilde{\varepsilon}_i = \mu$. Hence the Fermi energy, E_F(μ at T=0) of the system is exactly determined in the Kohn-Sham theory. We shall return to further discussion of the excitation energies later.

2.4 Generalization of HK Theorems

In the original paper Hohenberg and Kohn[1] considered the external potential as a local, scalar and nonrelativistic one and the system of particles is spinless, one component, nondegenerate in the ground state. A number of generalizations have been made by relaxing the above restrictions. We shall not prove these theorems here and only refer to the original papers. An interested reader may consult ref. 9, where most of the theorems are discussed.

A. DFT at Non- Zero Temperature

Mermin[28] made the finite temperature extension of the Hohenberg-Kohn theorems. He proved that the equilibrium density $n(\mathbf{r},\tau)$ at a given temperature, τ (measured in units of Boltzmann constant) and a chemical potential, μ in the grand canonical ensemble, is determined by the external potential $v(\mathbf{r})$. Further he showed that the correct equilibrium density subject to $v(\mathbf{r})$, minimizes the Gibbs thermodynamic grand potential.

This finite temperature version does not require the assumption of nondegeneracy and the function, $(v(\mathbf{r}) - \mu)$ is uniquely determined by the $n(\mathbf{r},\tau)$ (irrespective of an additive constant). Now in this generalization it will be useful to present self-consistent Kohn-Sham equations for $\tau \neq 0$, which are required for realistic calculations.

The grand potential is written as

$$\Omega_{v-\mu}[n] = \int d\mathbf{r} \left\{v(\mathbf{r}) - \mu\right\} n(\mathbf{r}) + F[n] \ , \tag{30}$$

where

$$F[n] = F_c[n] + G_s[n] + F_{xc}[n] \ . \tag{31}$$

Here $F_c[n]$, $G_s[n]$ and $F_{xc}[n]$ are the Hartree, noninteracting and exchange-correlation part of the Helmholtz free energies respectively. $G_s[n] = T_s[n] - \tau\, S_s[n]$, where $S_s[n]$ is the noninteracting entropy defined as,

$$S_s[n] = -\sum_i \left\{ f_i \ln f_i + (1-f_i)\ln(1-f_i) \right\} , \tag{32}$$

and f_i is the Fermi function

$$f_i = 1/\left\{ \exp(\varepsilon_i - \mu)/\tau + 1 \right\} . \tag{33}$$

Analogous to the ground state problem, the self-consistent Kohn-Sham eqs. (at $\tau \neq 0$) are given by

$$\left\{ -(1/2)\nabla^2 + v_{eff}[n(\mathbf{r})] - \mu \right\} \psi_i(\mathbf{r}) = \varepsilon_i \, \psi_i(\mathbf{r}) , \tag{34}$$

$$n(\mathbf{r}) = \sum_i |\psi_i(\mathbf{r})|^2 f_i , \tag{35}$$

where $v_{eff}[n(\mathbf{r})]$ is the same effective potential with $v_{xc}[n(\mathbf{r})]$ defined as : $v_{xc}[n(\mathbf{r})] = \delta F_{xc}[n] / \delta n(\mathbf{r})$.

B. Other Generalizations

As mentioned before a number of generalizations of the DFT has been made to meet different physical situations. These include: spin[29,30], multi-component[31], degenerate ground state[32], velocity-dependent[33], nonlocal[34], relativistic[35], excited states[36], and time-dependent[37] cases.

In view of these developments the DFT has a much broader appeal to attack a variety of many-body problems. The spin density functional theory and its local spin density (LSD) version have immense utility to spin polarized and open shell atomic systems. The relativistic and time-dependent formulations have made contact with the relativistic quantum field theory and the quantum hydrodynamics respectively at a very basic level.

2.5 Electronic Structure Studies

Understanding of the electronic band structure is fundamental to many condensed matter phenomena. As indicated before the DFT is a fortunate development for the band structure. Though we have stated before that the Kohn-Sham eigenvalues,ε_i are fictitious and are mere mathematical quantities (Lagranges parameters), they are *conventionally* used as the physical energies of the single particle excitations in electronic band theory[11]. A compelling reason to believe the ε_i as band energy lies in the fact that a number of experimental results,e.g. electronic transport, optical and photoemission spectroscopies , are successfully interpreted in terms of the Kohn-Sham ε_i's.

One can calculate the physical energies of the excitations in the self-energy formalism (see for example, Louie in ref. 11). This method is quite involved and it would require a great amount of computational efforts to interpret the experimental results. Since the Kohn-Sham LDA method has already succeded in explaining many experimental data, the Kohn-Sham ε_i's need interpretation. For this we need the help of the variational principle again.

Let f_i be the occupation number of the state i. In the DFT presented till now f_i is 0 for the unoccupied orbital and is 1 for the occupied orbital for the pure states. If we use ensemble states ,where the density $n(\mathbf{r})$ is given by $n(\mathbf{r}) = \Sigma f_i |\psi_i(\mathbf{r})|^2$ with the constraint $\Sigma f_i = N$ but $0 \le f_i \le 1$. ψ_i's are normalised eigen functions. This situation at nonzero temperature is very easy to visualize; here f_i is the Fermi function.

Now we calculate the total energy E_T, which is the internal energy for a thermodynamic system. In the zero temperature limit where the entropy contribution is neglected,the variation of E_T with f_i gives

$$\partial E_T / \partial f_i = \varepsilon_i \ .$$

This result was first derived by Slater and later by Janak. For more details see Ref. 9 and 18.

Calculations of the energy bands and Fermi surfaces (FS) of metals and alloys with the LDA have generally shown good agreements with the experiments, in spite of the fact that the exact Kohn-Sham FS is inequivalent with the physical FS (see Mearns, Phys. Rev. B **38**, 5906 (1988)). The electronic structure of the heavy fermion metals and high T_c superconductors have been studied within the LDA with the result that the LDA can not successfully account for the electronic behaviour in a strong interaction regime. Strong interaction in the DFT is still an open problem.

3. Approximations

In the previous section we have discussed mainly on the formal aspects of the theory. We have not said anything about the functional F[n], except that proving its existence. In order to study a many-body system by using this formulation, a knowledge of the functional is necessary. From a philosophical point of view, except for some idealized models, a realistic many-body problem may not be exactly solvable. Since a single functional, F[n] contains very complicated many-body effects and it is not exactly known *a priori* , one needs to resort to approximations. In this section we point out all commonly used approximations with their validities. Details can be found from the original papers cited.

3.1 Simple Approximations for G[n]

If we use the Euler-Lagrange eq. (20), which is the direct density equation, we need to know the kinetic and XC energy functionls. In the beginning of Sec.2.3 we have said, that these two together are put in G[n]. We note here that the contributions to G[n] from various sources involve integrations of the density functions from differtent space points. This aspect makes the functional generally nonlocal. In the realm of local density functional theory G[n] is expressed in terms of density only. This simplifying feature need to be understood carefully. For the treatment beyond the LDA refer to von Barth (elsewhere in this volume).
We discuss here simple approximations appropriate for the weakly inhomogeneous electron gas[1].

A. The Electron Gas of Almost Constant Density

Let the electron density $n(\mathbf{r})$ have a deviation from the constant density n_0, in presence of a very weak external potential

$$n(\mathbf{r}) = n_o + n(\mathbf{r}) \ , \tag{36}$$

such that $n(\mathbf{r})/n_0 << 1$ and

$$\int d\mathbf{r}\, n(\mathbf{r}) = 0 \ . \tag{37}$$

Since $n(\mathbf{r})$ is a small parameter a functional series expansion of G[n] is made.

$$G[n] = G[n_o] + \int d\mathbf{r}\, K_1(\mathbf{r})\, n(\mathbf{r}) + \frac{1}{2}\int d\mathbf{r}\, d\mathbf{r}'\, K_2(\mathbf{r},\mathbf{r}')\, n(\mathbf{r})\, n(\mathbf{r}') + \dots \ , \tag{38}$$

where the kernels of Taylor expansion are

$$K_1(\mathbf{r}) = [\delta G/\delta n(\mathbf{r})]_{n=n_o}, \quad K_2(\mathbf{r},\mathbf{r}') = [\delta^2 G/\delta n(\mathbf{r})\, \delta n(\mathbf{r}')]_{n=n_0} \quad etc.$$

The unperturbed system is translationally and rotationally invariant. K_1 is independent of $\mathbf{r}$ and $K_2(\mathbf{r},\mathbf{r}') = K_2(\mathbf{r}-\mathbf{r}')$. The second term on the right hand side of (38) vanishes because of (37). Thus, eq.(38) becomes

$$G[n] = G[n_o] + \frac{1}{2}\int d\mathbf{r}\, d\mathbf{r}'\, K_2(\mathbf{r}-\mathbf{r}')\, n(\mathbf{r})\, n(\mathbf{r}') + \dots \ . \tag{39}$$

From a calculation of energy shift in a second order perturbation theory, it is easy to show[1] that $K(\mathbf{q})$, the Fourier transform of $K_2(\mathbf{r}-\mathbf{r}')$, is related to the static irreducible polarization function, $\pi(\mathbf{q})$ and the dielectric function, $\epsilon(\mathbf{q})$.

$$K(\mathbf{q}) = 1/\pi(\mathbf{q}) \ , \tag{40}$$

and

$$K(\mathbf{q}) = 2\pi q^{-2} [\epsilon(\mathbf{q}) - 1]^{-1} . \tag{41}$$

In the absence of many-body effect, the kernel $K_o(\mathbf{q}) = 1/\pi_o(\mathbf{q})$,where $\pi_o(\mathbf{q})$ is the well-known Lindhard function. The difference $[K(\mathbf{q}) - K_o(\mathbf{q})] \equiv K_{xc}(\mathbf{q})$ accounts for the exchange-correlation effects and it contributes to the energy in the second order of the density fluctuations. This treatment is amply covered under the linear response method of the electron gas theory. From the knowledge of $\epsilon(\mathbf{q})$ of the homogeneous systems, the energy functional (38) is calculated.

B. Gradient Expansion for a System of Slowly Varying Density

If the density is slowly varying on the scale of the local Fermi momentum $[k_F(\mathbf{r})]^{-1}$ (local Fermi momentum is defined as $k_F(\mathbf{r}) = [3\pi^2 n(\mathbf{r})]^{1/3}$), which means that the density possesses only small wavenumber components the functional, G[n] can be expanded in terms of gradient of density[1]. The coefficients of this expansion (up to fourth order) can be identified with the small $\mathbf{q}$ expansion of the inverse static dielectric function $\epsilon^{-1}(\mathbf{q},n)$ and other nonlinear response functions.

Since the functional G[n] can be separated into $T_s[n]$ and $E_{xc}[n]$, individual gradient expansion of $T_s[n]$ and $E_{xc}[n]$ have been considered. For details see refs. 13,37-39 and references therein. Generally there are doubts about the convergence of the G[n], since the density variations are not so slow in the physical systems. In spite of it, lower order gradient expansion seem to give surprizingly good results for the metal surfaces and atomic calculations[39, 16] (more on kinetic energy functional see ref. 40).

3.2 Approximations to $E_{xc}[n]$

The advantage of proceeding via the exact single particle Kohn-Sham equations is that it requires only to approximate the XC energy functional. In the weakly inhomogeneous case, where the deviation of the density is small from its homogeneous value, Kohn and Sham[41] proposed that the $E_{xc}[n]$ can be written as

$$E_{xc}[n] \approx \int d\mathbf{r}\, n(\mathbf{r})\, \varepsilon_{xc}[n(\mathbf{r})] , \tag{42}$$

where $\varepsilon_{xc}[n(\mathbf{r})]$ is the XC energy per particle of a homogeneous system of density n. This approximation implies that an inhomogeneous system is replaced by a piece-wise homogeneous system. As mentioned before a reasonable knowledge of $\varepsilon_{xc}[n]$ is available in literature[27] for the homogeneous gas for possible use in the eq.(42). This approximation is the so-called local density approximation (LDA), which has been proved to demonstrate outstanding successes of the DFT. We shall return to the

LDA later.

We have given an heuristic derivation of the XC energy in Sec.2.2. An exact expression[42-45] is given here.

$$E_{xc}[n] = \frac{1}{2}\iint dr\, dr'\, v(r-r') \int_0^1 d\lambda\, n(r)\, [g_\lambda(r,r') - 1]\, n(r') \ . \qquad (43)$$

v(**r**) is the Coulomb potential. λ is a measure of the strength of interaction varying from zero to its full strength 1 (in atomic unit). The pair-correlation function depends on λ. The Hartree result can be trivially derived from λ= 0. A quantity known as 'exchange-correlation hole' is defined by

$$n_{xc}(r,r') = n(r') \int_0^1 d\lambda\, [g_\lambda(r,r') - 1] \ . \qquad (44)$$

Now (43) is rewritten as

$$E_{xc}[n] = \frac{1}{2}\iint dr\, dr'\, n(r)\, v(r-r')\, n_{xc}(r,r') \ . \qquad (45)$$

It is easy to see in analogy with the Hartree energy, that the $E_{xc}[n]$ arises from the Coulomb interaction of the electrons, where each electron at **r** is surrounded by a charge density $n_{xc}(\mathbf{r},\mathbf{r}')$. It looks as if there is a distributed hole charge (which is the depletion of the charge out of the XC processes in the vicinity of the electron) carried by the electron in question. The total hole charge correponds to one unit, which is expressed as a sum rule

$$\int dr'\, n_{xc}(r,r') = -1 \qquad (46)$$

for each **r**. The shape of the XC hole depends on the inhomogeneity of the density. If it is a homogeneous system, the shape is spherical. A number of approximations have been devised based on the treatment of $n_{xc}(\mathbf{r},\mathbf{r}')$ by Gunnarsson and co-workers[44]. The pair-correlation functions $g(\mathbf{r},\mathbf{r}')$ can be modelled for various situations. Here we discuss the most commonly used approximation known as the 'local density approximation'.

3.3 Local Density Approximation

To treat G[n] in the simplest and practical approach one makes the following assumptions. The electron density varies slowly in the real space. The space occupied by the electrons is divided into a set of discrete cells of volume Ω_i. The electron density in Ω_i is constant with an average value n_i, so that G[n] can be expressed as a Riemann sum, $G(\{n_i\}) \approx \Sigma\ \Omega_i\, g^h(n_i)$. $g^h(n_i)$ is the energy density for a strictly *homogeneous* distribution of electrons in the ith cell, which possesses charge neutrality within the cell. The quantity $g^h(n_i)$ is written as $g^h(n_i) = n_i\,\varepsilon(n_i)$, where $\varepsilon(n_i)$ is the relevant energy per electron.

In this *locally homogeneous approximation* G[n] in the continuous limit is expressed as

$$G[n] \approx \int dr\, g^h(n(r)) = G^{LDA}[n].$$

We note here that $g^h(n(r))$ is an ordinary function of the density variable. Though the various energy contributions to G[n] are intrinsically nonlocal, we need to understand how good are the local approximations to nonlocal problems.

We now go back to the exact eqns. (43) and (44) to investigate the exchange-correlation energy in the LDA limit. In the weakly inhomogeneous case

$$g(\mathbf{r},\mathbf{r}') \rightarrow g^h(\mathbf{r}-\mathbf{r}'; n(\mathbf{r}))$$

and

$$n_{xc}(\mathbf{r},\mathbf{r}') \rightarrow n^h_{xc}(\mathbf{r}-\mathbf{r}'; n(\mathbf{r}))$$

$$= n(\mathbf{r})\,[g^h(\mathbf{r}-\mathbf{r}'; n(\mathbf{r})) - 1]\ .$$

By making a spherical average[43] of the $n_{xc}(\mathbf{r},\mathbf{r}')$ eq.(44), about the point **r**, we have

$$n_{xc}(\mathbf{r},R) = (1/4\pi)\int n_{xc}(\mathbf{r},\mathbf{r}+\mathbf{R})\, d\Omega_{\mathbf{R}}\ . \tag{47}$$

Therefore,

$$E_{xc}[n] = (1/2)\int d\mathbf{r}\; n(\mathbf{r}) \int v(R)\, n_{xc}(\mathbf{r},R)\, d\mathbf{R} \tag{48}$$

$$\equiv \int d\mathbf{r}\; n(\mathbf{r})\, \varepsilon_{xc}[n(\mathbf{r})]\ . \tag{49}$$

Eq.(49) is the derivation of the result of Kohn and Sham[41], their eq.(42). Thus, in the weakly inhomogeneous case the local density approximation is justified. But we need to remind that though the XC processes are nonlocal, they are treated here in the local approximation.

Now we point out the exchange-only approximation of $E_{xc}[n]$ due to the parallel-spins of the electrons (i.e. at the Hatree-Fock level). We can calculate the pair-correlation function g(r,n) of a homogeneous gas of density, n exactly from the Slater determinant of the plane waves. Using this result[46]

$$g(r,n) = 1 - \frac{9\pi}{4x^3} J^2_{3/2}(x) , \tag{50}$$

in (45) leads to

$$E_x[n] = \alpha \int dr\, n^{4/3}(r) , \tag{51}$$

where $J_{3/2}$ is the Bessel function of order 3/2, $x = k_F r$ and $\alpha = -(3/4)\,(3/\pi)^{1/3}$. This is the well-known result for the exchange energy of the homogeneous electron gas due to Kohn and Sham[41]. Slater[46] proposed a method, known as X-α method, in which α is varied to take into account some correlations in a phenomenological way.

As an example we point out a comparison of the exact exchange hole with the exchange hole in LDA for Ne atom (see Gunnarsson, Johnson and Lundqvist, Ref.44). Though the LDA result is poor compared to the exact result, the spherical average is unexpectedly good.

A number of expressions for the XC energy and potential are available to be used in the LDA (both spin polarized and unpolarized) calculations. The most commonly used forms are due to Kohn-Sham (exchange-only)[41], Kohn-Sham and Wigner[2], von Barth-Hedin[29], Gunnarson-Lundqvist[43], Vosko et al.[47] and Perdew and Wang[48]. The more recent correlation energy functionals are constructed from fitting the Monte Carlo results in suitable form amenable for easy computations.

4. Conclusions

We have pointed out before that the LDA is restricted to be applied to systems with slowly varying density. It is surprising to see that when applied to a variety of inhomogeneous systems, the ground state properties are rather satisfactorily described. The successes of the theory beyond its limitations must be understood in the right context. Apparently there are number of cancellations, which need to be ascertained.

Going beyond the LDA is a rich area of research. Various aspects of it (e.g. nonlocal functionals in real space and reciprocal space, self-interaction corrected functional and self-energy formalism using G-W approximation etc.) can be found from (ref. 24) and are discussed by von Barth in his lectures.

In this paper we have presented some basic aspects of the density functional theory, which is of great importance in the study of ab-initio electron structure and dynamics. The DFT helps to construct an effective potential which contains the many-body effects in one-body form. The one-body Schrödinger or Dirac equation with the effective potential is solved with a bag of ingeneous techniques (see for example; papers by Andersen and co-workers in this volume).

First principles molecular dynamical simulations in conjunction with the LDA-DFT has opened new avenues to study a variety of molecules, clusters, liquids and other ordered and disordered systems. Now it is time to understand carefully what do these results tell us about the systems and how good are the approximations for quantitative analysis.

REFERENCES

1. Hohenberg, P.C. and Kohn,W., Phys. Rev. **136**, B864 (1964).
2. Lundqvist, S. and March, N.H. (Ed) *Theory of Inhomogeneous Electron Gas*, Plenum Press, New York (1983).
3. Keller, J. and Gazquez, J. (Ed) *Density Functional Theory*, Lecture Notes in Physics,Vol.187, Springer-Verlag, Berlin (1983).
4. Dreizler, D. M. and da-Provincia, X. (Ed) *Density Functional Methods in Physics*, Plenum Press, New York (1984).
5. Dahl, J.P. and Avery, J. (Ed) *Local Density Approximation in Quantum Chemistry and Solid State Physics*, Plenum Press, New York (1984).
6. March, N.H. and Deb, B.M.(Ed) *The Single Particle Density in Physics and Chemistry* (Academic, New York) 1987.
7. Erdahl, R. M. and Smith,V. H. (Ed) *Density Matrices and Density Functionals* (Reidel,Dordrecht) 1987.
8. Kryachko, E.S. and Ludena, E.V. (Ed) *Density Functional Theory in Quantum Chemistry* (Reidel,Dordrecht) 1990.
9. Rajagopal,A.K., Advances in Chemical Physics, Ed.I.Prigogine and S.A.Rice ,Vol 41, 59 (Wiley, New York,1980).
10. Langreth,D.C. and Suhl, H. (Ed) *Many-Body Phenomena of Surfaces*, Academic, New York,1983.
11. Youssouff, M. (Ed) *Electronic Band Structure and Its Applications*, Lecture Notes in Physics,Vol.283, Springer-Verlag, Berlin (1986).
12. Morruzzi,V. L., Janak, J. F. and Williams, A. R., *Calculated Electronic Properties of Metals*, Pergamon, New York,1978.
13. von Barth, U., in *Electronic Structure of Complex Materials*, (Ed) P.

Phariseau and W.Temmermann, Plenum, New York (1983) p. 67.
14. Pickett, W.E., Comm. Solid State Phys. **12**, 1, 57 (1985).
15. Dreizler, R. M. and Gross, E. K. U. *Density Functional Theory : An Approach to the Quantum Many-Body Problem*, Springer Verlag,Berlin (1990).
16. Parr, R. G. and Yang, W. *Density Functional Theory of Atoms and Molecules*, Oxford University Press, New York (1989).
17. Trickey, S. B. (Ed) Advances in Quantum Chemistry, Vol **21**, Academic Press, San Diego (1990).
18. Callaway, J. and March, N.H., in Adv. in Solid State Physics, Vol. 38 Ed. H. Ehrenreich, F. Seitz and D. Turnbull, Academic, New York (1983).
19. Phariseau, P. and Temmerman,W.M., (Ed) *The Electronic Structure of Complex Systems*, Plenum, New York (1982).
20. Devreese, J.T. and Van Camp, P. (Ed) *Electronic Structure, Dynamics and Quantum Structural Properties of Condensed Matter*, Plenum Press, New York (1984)
21. Bassani, F., Fumi, F. and Tosi, M.P. (Ed) *Highlights of Condensed Matter Theory*, North Holland, Amsterdam (1985).
22. Geldart, D. J. W. and Rasolt, M. in *Strongly Correlated Electron Systems* (Ed) M. P. Das and D. Neilson, Nova Science, New York (1992).
23. Katsnelson, A. A., Stepanyuk, V. S. and Szasz A. I. and Farberovich, O. V., *Computational Methods in Condensed Matter: Electronic Structure*, American Institute of Physics, New York (1992).
24. Das, M. P. in *Condensed Matter Physics*, (Ed) J. Mahanty and M. P. Das , World Scientific, Singapore (1998).
25. Levy, M., Phys.Rev. A **26**, 1200 (1982)
26. Lieb, E., in *Physics as Natural Philosophy: Essays in Honor of Laszlo Tisza on his 75th Birthday*, (Ed) A. Shimony and H. Feshbach, Cambridge, Mass. (1982) p.111, see also ref.4, p.31.
27. Singwi, K.S. and Tosi, M.P., in Adv. in Solid State Physics, Vol. 36 Ed. H. Ehrenreich, F. Seitz and D. Turnbull, Academic, New York (1981) p.177; Devreese, J.T. and Brosens, F. (Ed) *Electron Correlations in Solids, Molecules and Atoms*, Plenum, New York (1983).
28. Mermin, N.D., Phys. Rev. A **137**, 1441 (1965).
29. von Barth, U. and Hedin, L., J. Phys. C **5**, 1629 (1972); Stoddart, J. and March, N.H., Ann.Phys.(N.Y.) **64**, 174 (1971); Pant, M.M. and Rajagopal, A.K., Solid State Comm. **10**, 1157 (1972)
30. Rajagopal, A.K. and Callaway, J. Phys. Rev. B **7**, 1912 (1973)
31. Kalia, R. and Vashishta, P., Phys. Rev. B **17**, 2655 (1978)

32. Kohn,W. *ITP* preprint 83-118, see also ref.35

33. Gilbert, T.L., Phys. Rev. B **12**, 214 (1975)

34. Rajagopal, A.K., J. Phys. C **11**, L943 (1978); MacDonald, A.H. and Vosko, S.H., J. Phys. C **12**, 2977 (1979)

35. Theophilou, A.K., J. Phys. C **12**, 5419 (1979), see also in Ref.6; Kohn,W., Phys. Rev. B **34**, 737 (1986); English, H. and English, R., Physica **121A**, 253 (1983)

36. Runge, E. and Gross, E.K.U., Phys. Rev. Lett. **52**, 997 (1984); Gross, E.K.U. and Kohn,W., Phys. Rev. Lett. **55**, 2850 (1985); Ng,T.K. and Singwi, K.S., Phys. Rev. Lett. **59**, 2627 (1988)

37. DePristo, A.E. and Kress, J.D., Phys. Rev. A **35**, 438 (1987)

38. Geldart, D.J.W. and Rasolt, M., in ref. 6 (see their Sec.6.3)

39. Langreth, D.C. and Perdew, J.P., Phys. Rev. B **15**, 2884 (1977), B **21**, 5469 (1980), B **26**, 2810 (1982), Langreth, D.C. and Mehl, M.J., Phys. Rev. B **28**, 1809 (1984).

40. Wang, L. W. and Teter, M. P. , Phys. Rev. B**45** ,13196 (1992)

41. Kohn,W. and Sham, L.J., Phys. Rev. A**140**, 1133 (1965)

42. Harris,J. and Jones, R.O., J. Phys. F **4**,1170 (1974).

43. Gunnarsson, O. and Lundqvist, B.I., Phys. Rev. B **13**, 4274 (1976)

44. Gunnarsson, O., Lundqvist, B.I. and Wilkins, J.W., Phys. Rev. B **10**, 1319 (1974); Gunnarsson, O., Jonson, M. and Lundqvist, B.I., Phys. Rev. B **20**, 3136 (1979); Gunnarsson, O. and Jones, R.O., Phys. Scripta, **21**, 394 (1980)

45. Langreth, D.C. and Perdew, J.P., Phys. Rev. B **15**, 2884,(1977)

46. Slater, J.C. *Quantum Theory of Atomic Structure* , McGraw Hill, New York (1960), *Self-consistent Field for Molecules and Solids*, McGraw Hill, New York (1974).

47. Vosko,S.H.,Wilk,L. and Nusair, M. , Can. J. Phys. **58**,1200 (1980)

48. Perdew, J. P. and Wang, Y., Phys. Rev. B **45**,13244 (1992)

DIFFERENT APPROXIMATIONS WITHIN DENSITY FUNCTIONAL THEORY, THEIR ADVANTAGES AND LIMITATIONS.

U. von Barth

Lund University, S-22362 Lund, Sweden

ABSTRACT

In these short lecture notes we will try to give a short resume of the successes and failures of density-functional theory as such as well as of different approximation schemes within the density-functional framework. Emphasis will be given to fundamental principles rather than to details or completeness.

1. Introduction.

Over the past fifteen years Density-Functional (DF) theory has emerged as the leading tool for the calculation of ground-state properties of complex systems. Its simplicity as compared to other many-body techniques has made it possible to obtain quantitative results for geometries and binding energies of systems with a large number of atoms and a low symmetry such as, *e.g.*, molecules adsorbed on metal surfaces. Despite their simplicity, the accuracy of existing approximations within DF theory is surprisingly good and some times good enough for a successful prediction of experimental facts. In many other cases, however, a better accuracy is required and in later years a large effort has been directed towards designing improved approximations. This task has proven to be exceedingly difficult but some progress has been made in recent years.

The present short article is a compilation of lecture notes from lectures given at a workshop in Trieste in August of 1992. They were intended as a quick introduction into the field of DF theory and a broad overview of its successes and failures. I want to emphasize that I have made no attempt towards completeness and the choice of topics and points of discussion are not intended to do justice to the work of other people. These choices rather reflect my own personal views on which problems that presently are interesting and worthwhile pursuing. Other peoples work is referred to when it has a bearing on these problems. In recent years, a large number of review articles on DF theory and applications thereof have been published [1-13] and I refer the reader to these article for comprehensive treatment of the subject.

I will start by introducing some basic formalism and then go on to present several numerical results for different ground state properties of different systems as obtained from the Local-Density Approximation (LDA). These results are intended to demonstrate the kind of accuracy that can be achieved in different situations and they will point towards both successes and failures. I will then discuss several attempts to go beyond the LDA and show results that indicate the quality of these recent developments. I will further make some personal remarks on possible future developments and, finally, I will discuss excited states, in particular with regard to one-electron excitation energies, and give my views on the so-called band-gap problem.

2. Basic formalism.

Within Density-Functional Theory (DFT) of many-electron systems the basic idea is to consider all quantities in general and the total ground-state energy in particular as functionals of only one variable - the charge density $n(\mathbf{r})$. This latter functional can for each external potential $w(\mathbf{r})$ be shown to have a minimum at the true ground-state density and the total energy can thus be obtained from this variational principle. The possibility of this approach was first proven by Hohenberg and Kohn. [14] Kohn and Sham [15] then showed how the variational problem could be solved by means of an equivalent one-electron scheme, a crucial step which really accounts for the utility and in part for the accuracy of the method. In short, Hohenberg, Kohn, and Sham demonstrated how to transform the full many-electron problem into an equivalent one-electron problem. In my mind, this idea represents one of the greater achievements of the latter part of this century.

In the original papers, the discussion was limited to cases with non-degenerate ground states, a situation rarely occurring in nature. Levy later [16] showed how to circumvent this limitation and his work also contains an explicit construction of the rather abstract functionals introduced by the originators. In addition, I believe a beginner in the field would find it easier to grasp the basic ideas from the work by Levy and I consider this work as one of the cornerstones of the DFT.

Within the DFT the total energy of the many-electron system is written:

$$E[n] = T_o[n] + \tfrac{1}{2}\int n\, v\, n + \int n\, w + E_{xc}[n] \tag{1}$$

We have here and often later suppressed integration variables (d^3r) as well as spatial dependencies $n(\mathbf{r})$, $w(\mathbf{r})$. Here,

$T_o[n]$ is the kinetic energy of a non-interacting system of N electrons having the number density $n(\mathbf{r})$.

$w(\mathbf{r})$ is the externally applied potential, *e.g.*, the attraction from the nuclei.

$v = v(\mathbf{r},\mathbf{r}') = |\mathbf{r}-\mathbf{r}'|^{-1}$ is the Coulomb interaction.

$E_{xc}[n]$ is usually defined by Eq. (1). It is called the exchange-correlation (xc) energy and thus adds everything left out of the first three terms in order to yield the well defined total energy $E[n]$.

If we had replaced $T_o[n]$ with $T[n]$ in Eq. (1), *i.e.*, with the kinetic energy of a fully interacting system with density n, E_{xc} would have been the xc-part of the interaction energy $U_{xc}[n]$ which can be written

$$U_{xc}[n] = \tfrac{1}{2}\int n(\mathbf{r})n(\mathbf{r}')\{g(\mathbf{r},\mathbf{r}')-1\}\, v(\mathbf{r},\mathbf{r}')\, d^3r\, d^3r' \tag{2}$$

where $g(\mathbf{r},\mathbf{r}')$ is the pair-correlation function of the system. The difference between T

and T_o can, however, be accounted for by means of a Hellman-Feynman-like trick consisting of an integration over the strength of the Coulomb potential (e^2). During the integration, the density is kept constant by varying the external potential w_λ. [17] Formally

$$E_\lambda = <\Psi_\lambda | \hat{T} + \hat{W}_\lambda + \lambda \hat{U} | \Psi_\lambda > \tag{3}$$

$$\frac{\partial E_\lambda}{\partial \lambda} = <\Psi_\lambda | \frac{\partial \hat{W}_\lambda}{\partial \lambda} + \hat{U} | \Psi_\lambda > \tag{4}$$

Note that $n(\mathbf{r}) = <\Psi_\lambda | \hat{n} | \Psi_\lambda >$ does not depend on lambda!

$$\frac{\partial E_\lambda}{\partial \lambda} = \int n \frac{\partial w_\lambda}{\partial \lambda} + \tfrac{1}{2} \int n\, n'\, g_\lambda(\mathbf{r}, \mathbf{r}')\, v \tag{5}$$

From Eq. (1)

$$\frac{\partial E_\lambda}{\partial \lambda} = \tfrac{1}{2} \int n\, v\, n' + \int n \frac{\partial w_\lambda}{\partial \lambda} + \frac{\partial E_{xc}(\lambda)}{\partial \lambda} \tag{6}$$

Thus

$$\frac{\partial E_{xc}(\lambda)}{\partial \lambda} = \tfrac{1}{2} \int n\, n' \{g_\lambda(\mathbf{r}, \mathbf{r}') - 1\}\, v \tag{7}$$

But when $\lambda = 0$ there is no xc-energy ($E_{xc}(0) = 0$) so that

$$E_{xc}[n] = \tfrac{1}{2} \int n(\mathbf{r})\, n(\mathbf{r}') \{\tilde{g}(\mathbf{r}, \mathbf{r}') - 1\}\, v(\mathbf{r} - \mathbf{r}')\, d^3r\, d^3r' \tag{8}$$

where $\tilde{g} = \int_0^{e^2} g_\lambda d\lambda$. From the exact expression Eq. (8) we also obtain an exact expression for the xc-potential, $v_{xc}(\mathbf{r})$:

$$v_{xc}(\mathbf{r}) = \frac{\delta E_{xc}[n]}{\delta n(\mathbf{r})} \tag{9}$$

$$\begin{aligned} v_{xc}(\mathbf{r}) = & \int n(\mathbf{r}') \{\tilde{g}(\mathbf{r}, \mathbf{r}') - 1\} v(\mathbf{r} - \mathbf{r}')\, d^3r' + \\ & \tfrac{1}{2} \int n(\mathbf{r}_1)\, n(\mathbf{r}_2) \frac{\delta \tilde{g}(\mathbf{r}_1, \mathbf{r}_2)}{\delta n(\mathbf{r})} v(\mathbf{r}_1 - \mathbf{r}_2)\, d^3r_1 d^3r_2 \end{aligned} \tag{10}$$

where $\delta \tilde{g} / \delta n$ is a three-particle correlation function.

The reason why we here have spent some time on the expression Eq. (8) is that almost all practical approximations for E_{xc} are derived from it and arguments in favor of, or against certain approximations, are frequently based on this expression. The widely used and powerful local-density approximation (LDA) is, *e.g.*, obtained from Eq. (8) by replacing the true pair correlation function by that of the homogeneous but interacting electron gas at the local density $n(\mathbf{r})$ and by also replacing $n(\mathbf{r}')$ by $n(\mathbf{r})$. Thus

$$E_{xc}^{LD}[n] = \frac{1}{2}\int n(\mathbf{r})n(\mathbf{r})\,\{\tilde{g}_h(\mathbf{r}-\mathbf{r}';n(\mathbf{r}))-1\}\,v(\mathbf{r}-\mathbf{r}')\,d^3r d^3r' \tag{11}$$

As in the case of the gas, the $\mathbf{r}'$-integral can now be performed leading to the more familiar expression

$$E_{xc}^{LD}[n] = \int n(\mathbf{r})\,\varepsilon_{xc}(n(\mathbf{r}))\,d^3r \tag{12}$$

where $\varepsilon_{xc}(n)$ is the exchange-correlation energy per particle of the gas at density n.

As we shall see shortly, the LDA is far more accurate than there are *a priori* reasons to expect. In the first attempts towards explaining this phenomenon, Gunnarsson and Lundqvist[18] also started from the basic exact expression, Eq. (8), and defined the quantity

$$n_{xc}(\mathbf{r},\mathbf{r}') = n(\mathbf{r}')\,\{\tilde{g}(\mathbf{r},\mathbf{r}')-1\} \tag{13}$$

referred to as the exchange-correlation (xc) hole. Physically, this hole represents a spatially varying depletion of charge around the electron at $\mathbf{r}$. The depletion is a consequence of the Pauli exclusion principle forbidding two electrons of the same spin to occupy the same region in space, and of the Coulomb repulsion forcing electrons of both spins to avoid each other. These effects are reflected in the shape of the pair-correlation function $\tilde{g}$ in such a way that the depletion is large close to the electron $\mathbf{r}' \approx \mathbf{r}$ and goes to zero further away, $\tilde{g} \to 1$. We can say that every electron "digs a hole" in the average density surrounding it. It is clearly seen from Eq. (8) that the xc-hole plays an important role in obtaining the xc-energy. In fact, this energy is seen to be the sum of the Coulomb energies of every electron interacting with its xc-hole. Due to the spherical symmetry of the Coulomb interaction a complete knowledge of the xc-hole is not required for obtaining E_{xc}, only the spherical average of the xc-hole contributes to E_{xc}. Defining the spherical average of the hole by

$$\bar{n}_{xc}(\mathbf{r},R) = \int n_{xc}(\mathbf{r},\mathbf{r}+R\,\hat{\mathbf{r}})\,\frac{d\hat{\mathbf{r}}}{4\pi} \tag{14}$$

we obtain[18]

$$E_{xc}[n] = \frac{1}{2}\int n(\mathbf{r})\,\bar{n}_{xc}(\mathbf{r},R)\,v(R)\,d^3R\,d^3r. \tag{15}$$

Notice, that the spherical average of the hole only depends on the distance (R) from the electron considered.

The success of the LDA is now explained[18] by the fact that the exact xc-hole has some important properties in common with that of the LDA defined by

$$\bar{n}_{xc}^{LD}(\mathbf{r},R) = n(\mathbf{r})\,\{\tilde{g}_h(R;n(\mathbf{r}))-1\} \tag{16}$$

They both obey the sum rule

$$\int \bar{n}_{xc}(\mathbf{r},R)\,d^3R = -1 \tag{17}$$

expressing the fact that every electron pushes away one unit of negative charge from its immediate neighborhood. Furthermore, both holes obey the relations

$$\bar{n}_{xc}(\mathbf{r}, R) \leq 0 \tag{18}$$

and

$$\bar{n}_{xc}(\mathbf{r}, 0) \geq -n(\mathbf{r}) \tag{19}$$

(Eq. (18) is strictly valid for the exchange hole but only approximately obeyed by the xc-hole.) In the case of the LD hole, these relations are a consequence of the fact that this hole is modeled by that of the homogeneous gas. Finally, as shown by Gunnarsson *et al.*[19], whereas the LD hole sometimes can be rather different from the true hole the spherical average of the LD hole closely resembles the corresponding exact quantity even in strongly inhomogeneous cases like, *e.g.*, the interior of atoms. These observations, although important, do not, however, provide the full story as can be understood from the fact that other approximations within DF theory[19] have similar properties but yield less accurate answers as compared to the LDA.

We end this section on the basic formalism by another twist of the subject which has proven useful in connection with gradient expansions. The discussion above focuses on a description of the spherical average of the xc-hole in real space. Let us now Fourier transform this hole:

$$\bar{n}_{xc}(\mathbf{r}, k) = \int \bar{n}_{xc}(\mathbf{r}, R)\, e^{-i\mathbf{k}\cdot\mathbf{R}}\, d^3R. \tag{20}$$

Notice that only a scalar k is needed since the spherically averaged hole only depends on the scalar radius R. Since the Fourier transform of the Coulomb potential is $4\pi/k^2$ we obtain

$$E_{xc}[n] = \frac{1}{\pi} \int n(\mathbf{r})\, \bar{n}_{xc}(\mathbf{r}, k)\, dk\, d^3r. \tag{21}$$

for the xc-energy. In this way, E_{xc} is written as a sum of contributions from different wave vectors, a decomposition which has been used extensively by Langreth and Perdew.[20]

3. The performance of the LDA.

3.1. Systems suitable to the LDA.

The extreme usefulness of the LDA has been common knowledge for many years. We will here give a short resume of results obtained from the LDA for different physical properties and state some general conclusions concerning the quality of these results.

We will, however, first remind the reader that DF theory, with a few exceptions, is strictly applicable only to ground-state properties and, secondly, we will say a few words about to what systems the theory should be applied.

As a test of the quality of new approximations to E_{xc}, DF theory is often applied to atoms and small molecules. With regard to this practice we offer the following remarks:

i) These systems are, for their physical properties, best treated by other many-body techniques such as Many-Body-Perturbation Theory (MBPT) or Configuration-Interaction expansions (CI) which offer a well defined route to successively more accurate answers. ii) The interior of atoms is extremely difficult to handle by DF techniques due to the rapid density variations in these regions. Fortunately, the physics of these regions have little to do with the physics of everyday life (chemical bonding etc.). iii) The almost non-vanishing densities in the exterior of atoms cause gradient corrections to diverge and approximate xc-holes to be misplaced and deformed but contain little energy. More importantly - such regions do exist only in very sparse solids. Consequently, atoms and small molecules represent very severe tests on the quality of different approximations and care must be exercised in order not to have the conclusions obscured by irrelevant difficulties.

The application of CI expansions or MBPT to complicated systems of considerable interest such as solids with many atoms per unit cell, surfaces, atoms or molecules adsorbed on surfaces, and large molecules, is computationally prohibitive. This is the realm of DF theory provided it can be made accurate enough.

3.2. Results.

The following pages demonstrate the successes of the Local-Density Approximation (LDA) as applied to a variety of physical properties of solids. These pages also indicate some of the deficiencies of the LDA. Some of the results have been taken from the literature and in most cases I have given the source. Other results have been produced in the course of my own research.

Table 1. Total energies (in eV) of a few atoms.

Atom	LDA	exp
H	13.3	13.6
H^-	14.4	14.4
Al	6567	6592

Table 2. Ground-state properties of the molecules H_2O, NH_3, and CO_2, as obtained from the LDA and from experiment.
The results are taken from the work by Müller, Jones, and Harris. [21]
We assume that the numerical errors involved in obtaining the LDA results are negligible in comparison to the deviation between theory and experiment.

	H_2O		NH_3		CO_2	
	LDA	exp	LDA	exp	LDA	exp
d	1.84	1.81	1.94	1.91	2.21	2.20
θ	106	105	108	107	180	180
ω_s	3680	3657	3335	3337	1420	1388
ω_b	1590	1595	820	950	730	667
μ	0.732	0.730	0.564	0.583	0	0

d is the equilibrium distance in atomic units.
θ is the equilibrium bond angle in degrees.
ω_s is the stretching frequency in cm^{-1}.
ω_b is the bending frequency in cm^{-1}.
μ is the dipole moment in atomic units.

Table 3. Ionization potentials (in eV) of a few atoms. LPM designates results from the gradient corrected scheme by Langreth, Perdew and Mehl [20,22,23] and Langreth and Hu. [24] In the cases of Mn and Fe we refer to the removal energy of a 4s electron.

Atom	exp	LDA	LPM
He	24.6	24.2	24.8
Li	5.4	5.6	5.6
Be	9.3	8.4	9.0
Na	5.1	5.3	5.1
Ca	6.1	6.3	6.0
Mn	7.4	7.9	7.1
Fe	7.9	8.4	7.8

Table 4. Cohesive energies (in eV) of a few solids as obtained from the LDA and from experiment. The Si result is from Ref. 25. The other results are taken from the book by Moruzzi *et al.*[26] and from Refs. 27-29.

	Na	Mg	Si	Ti	Zr	Ni
exp	1.1	1.5	4.6	4.9	6.3	4.4
LDA	1.1	1.6	5.1	6.1	6.8	5.7

Table 5. Lattice parameters in atomic units for a few solids as obtained from the LDA and from experiment. With one exception, Si, the data are taken from the book by Moruzzi *et al.*[26] The Si result is from Ref. 25.

	Na	Mg	Si	Ti	Zr	Ni
exp	8.0	8.5	10.3	7.8	8.2	6.7
LDA	7.7	8.4	10.2	7.6	8.2	6.6

Table 6. The binding energies (in eV) of the first-row dimers as obtained from the LDA, from the LPM scheme, and from experiment. The results are taken from the work by Becke.[30,31]

	H_2	Li_2	B_2	C_2	N_2	O_2	F_2
exp	4.8	1.1	3.0	6.3	9.9	5.2	1.7
LDA	4.9	1.0	3.9	7.3	11.6	7.6	3.4
LPM	5.0	0.6	3.3	6.1	10.2	6.4	2.4

Table 7. Equilibrium distances in Bohr of the first-row dimers. The LDA results are from the work by Becke.[30]

	H_2	Li_2	B_2	C_2	N_2	O_2	F_2
exp	1.40	5.05	3.00	2.35	2.07	2.28	2.68
LDA	1.45	5.12	3.03	2.35	2.07	2.27	2.61

Table 8. Vibrational frequencies in cm^{-1} of the first- and second-row dimers as obtained from the LDA and from experiment. The results are taken from the work by Becke.[30]

	H_2	Li_2	B_2	C_2	N_2	O_2	F_2
exp	4400	350	1050	1860	2360	1580	890
LDA	4190	330	1030	1880	2380	1620	1060

Table 9. Heat of formation (in eV) of a few compounds as obtained from the LDA and from experiment. The data are taken from the work by Williams, Kübler, and Gelatt.[32,33]

	MgAg	AlZr	SiZr	NiAl	CuZn
exp	0.19	0.44	0.81	0.61	0.12
LDA	0.24	0.44	0.73	0.74	0.14

Table 10. 10a) The saturated magnetic moment (Bohr magnetons), 10b) the hyperfine field (kiloGauss), and 10c) the spin-susceptibility enhancement factor for a few metals. The data are taken from the work by Janak *et al.*[28,34-36]

Magnetic moment			
	Fe	Co	Ni
exp	2.22	1.56	0.61
LDA	2.15	1.56	0.59

Table 10b.

	Hyperfine field		
	Fe	Co	Ni
exp	339	217	75
LDA	260	220	80

Table 10c.

	Susceptibility enhancement		
	Li	Na	K
exp	2.50	1.65	1.70
LDA	2.25	1.71	1.95

3.3. Successes.

The most striking feature discernible in the data shown above, is the remarkable accuracy of the LDA for a variety of physical properties in many different systems. We will here summarize the existing experience from numerous applications of the LDA as follows:

a. Binding energies are often better than 1 eV but in some s-d bonded systems the error can be twice or even three times as large. The error is systematic.
b. Equilibrium distances are generally accurate to within 0.1 Å. The error is systematic.
c. Vibrational frequencies are accurate to within 10-20%. There exist occasional cases with larger errors.
d. Charge densities are better than 2 %.
e. Geometries are accurate.
f. LDA results are nearly always much better than those of the Hartree-Fock (HF) approximation.
g. Most importantly, physical trends are generally correct.

We end this subsection on successes by a few remarks on why the binding distances are so accurate in the LDA when the corresponding energies are not. We believe that there are essentially two reasons for this fact. The obvious answer is, of course, that a binding energy of a molecule involves the subtraction of the energies of the constituent atoms. The LDA energies of the latter are often less accurate as compared to case of the molecules due to degeneracy effects and to non-spherical and more rapidly varying densities. A more interesting explanation can, however, be given in terms of a recent theorem by Levy and Perdew.[37] They show, that in the special but God-given case of the Coulomb interaction, the approximate variational energy surface (energy as a function of geometry) will parallel the true surface up to second order in the deviation between the exact and the approximate scheme. This theorem provides a hint towards the understanding of the empirical observation, that the LDA errors for the binding energies of many molecules are rather insensitive to the nuclear separation.

3.4. Deficiencies.

The results shown above, however, also indicate some cumbersome deficiencies. Most notable is the systematic overbinding predicted by the LDA, particularly for the s-d bonded systems. The overbinding is also, although to a much lesser extent, reflected in a small but relatively systematic underestimate of the bonding distances. We end this section with a small list of systems for which the LDA predicts the wrong ground states.

a. The transition-metal oxides FeO and CoO are erroneously predicted to be be metallic (MnO and NiO comes out as an antiferromagnetic insulator in accordance with experiment).[38]
b. Solid Fe is predicted to be an fcc paramagnet[39] but is a bcc ferromagnet at low temperatures.
c. In many semiconductors, the LDA gives the metal-insulator transition at much too large volumes.[40]
d. The LDA predicts the wrong dissociation limits for a large number of molecules.
e. The LDA predicts incorrect ground states for many atoms.
f. The LDA gives unstable negative atomic ions in many cases when these are stable.
g. Et cetera, et cetera ..

4. Sources of error in the LDA.

4.1. s-p transfer energies.

Gunnarsson and Jones[41] suggest that the root of the problem is the inability of the LDA to properly account for so called s-to-p and s-to-d transfer energies. When a solid or molecule is formed even from rather "round" atoms the electronic density stretches out towards neighboring atoms to form bonds. In one-electron calculations the stretching is accomplished by transferring electrons from states of lower angular momentum to states with higher angular momentum (*e.g.* s-d) in the bonding process. The physical picture is illustrated, *e.g.*, by the case of Si. The Si atom has four valence electrons $3s^2,3p^2$. In forming the solid we would, to a first approximation, promote one of the s electrons to a p orbital and then form the tetrahedrally arranged hybrids. Thus, the atom in the solid has the approximate configuration $3s,3p^3$ and bonding has resulted in a transfer of one electron from an s- to a p-orbital. This costs energy but, in the LDA, we pay too small a price for the transfer resulting, on an absolute scale, in too low an energy for the solid. Consequently, the LDA overestimates the cohesive energy of Si and in this case the actual error is 0.5 eV or 11% (See Table 4). A similar situation exists in the transition metals although the transfer in that case is from s,p to d.[26] According to Gunnarsson and Jones[41] the errors in the s-p or s-d transfer energies are due to the insensitivity of the LDA to the nodal structure of the one-electron wave functions. Consider *e.g.* the fluorine atom which has a ^2P ground state constructed from the configuration $2s_\uparrow 2s_\downarrow 2p_\uparrow^3 2p_\downarrow^2$. An s-p transfer is achieved by promoting one of the 2s electrons to the empty hole in the 2p-shell resulting in a ^2S excited state. From HF theory, the accompanying change in the exchange energy of the L-shell is easily found to be $\frac{2}{3}G_1(s,p) - \frac{9}{25}F_2(p,p) + \frac{1}{2}[F_o(s,s) - F_o(p,p)]$ in terms of the usual Slater[42] integrals. Inserting the actual orbitals of F gives a 6.7 eV increase in the exchange energy. This rather large value is associated with the extra angular node of the 2p- relative to the 2s-orbital. In the LDA the exchange energy is given by

$$E_x^{LD}[n_\uparrow, n_\downarrow] = A_x \sum_\sigma \int [n_\sigma(\mathbf{r})]^{4/3} d^3r \tag{22}$$

where $n_\uparrow$ and $n_\downarrow$ are the spin-up and spin-down densities and where A_x is equal to $-(3/4)(6/\pi)^{1/3}$ in Hartrees. Thus, a substantial change in the exchange energy associated with the s-p transfer requires the square of the s- and p-orbitals to be rather different. This is not the case in the F atom because both orbitals belong to the same principal-quantum number shell (the L-shell). Consequently, the cost of the transfer is much too small in the LDA (~ 0.6 eV). Although the total error is considerably reduced[8] (to ~ 2.6 eV) by correlation effects and self-consistency, it is still large enough to explain a large part of error (1.7 eV from Table 6) in the binding energy of the F_2 molecule.

Clearly, the LDA in which the exchange-correlation energy only depends on the local density cannot account for changes in the nodal structure of the wave functions. One could, however, hope that xc-functionals based on gradients would be capable of picking up such nodal dependencies. Unfortunately, this does not seem to be the case as can be seen in Table 11 showing results for the aforementioned s-p transfer treated by different methods. The Generalized Gradient Approximation (GGA) by Langreth, Perdew, Mehl and Hu (LPM)[20,22-24] in conjunction with spherically averaged densities perform almost as badly as the LDA.

Table 11. Exchange energies (E_x , in eV) in two different configurations of the fluorine atom as obtained from Hartree-Fock (HF), from the LDA, and from the LPM scheme (Ref. 22) evaluated at spherically averaged densities. LPM-NS designates LPM exchange energies evaluated at the correct non-spherical densities. The numbers illustrate the failure of all DF approximations in describing the correct (HF) s-to-p transfer energy.

	$2s_\uparrow 2s_\downarrow 2p_\uparrow^3 2p_\downarrow^2$	$2s_\uparrow 2p_\uparrow^3 2p_\downarrow^3$	ΔE_x
HF	272.5	264.4	8.1
LDA	246.4	244.1	2.2
LPM	267.4	264.6	2.8
LPM-NS	267.6	264.6	3.0

One could guess that this disappointing result might be a consequence of the spherical averaging, a procedure which certainly softens the nodal structure closely connected to the angular dependence. Therefore, it is even more disappointing to see the results marked LPM-NS in Table 11. They are obtained from the LPM scheme by taking full account of the asphericities of the spin-densities. The results are marginally better than those of the spherical approximation and we conclude that a functional constructed from the density and its first spatial derivatives is not able to respond to a change in the nodal structure of the wave functions. Indeed, s-d transfer energies have been calculated in transition-metal atoms by Kutzler and Painter [43] with the same disappointing outcome. They also tested an improved generalized-gradient approximation by Perdew and Wang[44,45] to be discussed in Section 5.3 and there referred to as the PW86 scheme. Although this scheme is somewhat better than both the LDA and the LPM scheme, the improvement is marginal. (In this case the LDA is actually better than the LPM scheme.)

The s-d transfer errors in atoms is one possible mechanism responsible for the cohesive energy errors of extended systems. Correlations, however, strongly reduce the effect already in atoms and screening as well as smoother densities in the solids will reduce the effect even further. As we saw above, two proposed generalized-gradient approximations which are known to be superior to the LDA with regard to binding energies of molecules and solids (see Section 5) fail to improve the s-d transfer energies. From this we conclude that the s-p or s-d transfer mechanism is probably not the most important mechanism behind the failure of the LDA with regard to cohesive energies.

4.2. The near-degeneracy problem.

CI and MBPT dominate the world of Quantum Chemistry (QC), *i.e.* the physics of small to large molecules. These methods are reliable and can often provide the 1 kcal/mol (1 eV = 23 kcal/mol) accuracy of interest to the physics of chemical reactions.

They are, however, enormously time-consuming and there would be a breakthrough in QC if DF methods could be brought to chemical accuracy. It is therefore rather disappointing to see the LD results for the binding energies of the first row dimers in Table 6. The much larger binding-energy errors in these finite systems can be understood in terms of their more rapidly varying densities. Clearly, the explanations based on the s-p transfer mechanism becomes more relevant but also other effects come into play. One such effect can be referred to as the near-degeneracy problem in the DF theory. Consider the H_2^+-molecule at large internuclear separation R. The correct wave function is close to

$$\psi(\mathbf{r}) = \sqrt{\tfrac{1}{2}}\,\{\phi(\mathbf{r}) + \phi(\mathbf{r}-\mathbf{R})\} = \sqrt{\tfrac{1}{2}}\,\{\phi_a + \phi_b\} \tag{23}$$

where $\phi(\mathbf{r})$ is the wave function of hydrogen: $\phi = e^{-r}/\sqrt{\pi}$. The expectation value of the full Hamiltonian

$$H = -\tfrac{1}{2}\nabla^2 - \frac{1}{r_a} - \frac{1}{r_b} + \frac{1}{R} \tag{24}$$

with respect to this wave function is, to exponential accuracy, equal to $<\phi|H|\phi> = -1$ Ry. Thus, in real life, the system cares very little about whether we break the symmetry of the molecule and put the electron on one of the sites or we keep the symmetry and put half an electron on each site - the corresponding energies are very much the same. In all approximations to DF theory which are constructed by using only the local density and its gradients, however, the situation is quite different. In addition to the energy terms considered above, the effective one-electron Hamiltonian of DF theory contains the terms

$$\tfrac{1}{2}\int \psi^2 v\, \psi^2 + E_{xc}[\psi^2] \tag{25}$$

If we have found a functional which is very accurate for the atom and which achieves a high degree of cancellation of the "self-interaction" this becomes

$$\tfrac{1}{2}\int \phi^2 v\, \phi^2 + E_{xc}[\phi^2] \;=\; U - J \;\approx 0 \tag{26}$$

in the hydrogen-plus-proton case. Thus, for this case, the LDA provides a good answer. In the symmetric case, we instead obtain

$$2\tfrac{1}{2}\tfrac{1}{4}\int \phi^2 v\, \phi^2 + 2\tfrac{1}{2}\tfrac{1}{4}\int \phi_a^2 v\, \phi_b^2 + E_{xc}[\tfrac{1}{2}(\phi_a^2 + \phi_b^2 + 2\phi_a\phi_b)] \approx$$

$$\approx \tfrac{1}{2}U + \frac{1}{4R} + 2E_{xc}[\tfrac{1}{2}\phi^2] \tag{27}$$

i.e., a quantity with a slow variation with R. This result is fundamentally incorrect, thus demonstrating that the correct functional must depend on the density in a more complicated way than just through its local value and its gradients. If we, *e.g.*, use the LDX approximation (Local-Density-eXchange-only) given by Eq. (22), we obtain an LD-error of

$$\Delta E^{LD} = \tfrac{1}{2}U + \frac{1}{4R} - 2^{-1/3}J \approx 0.25/R - 0.3U, \tag{28}$$

and, since U is ~ 8 eV for hydrogen, we obtain ~ 2.4 eV overbinding in the LDA at large separation. Even though the other spurious term (1/4R) partly reduces the error at smaller R, an effect might still remain at the equilibrium distance. (As a matter of fact, the LD error in the binding energy of H_2^+ is only 0.15 eV at equilibrium.) [46] Notice, however, that an essential ingredient of the mechanism discussed above is the existence of an unpaired electron. Consequently, the effect does not affect the binding energy of the dimers but it does affect their ionization potentials and also the binding energies of the dimer cations as demonstrated by Merkle *et al.* [47]

The accurate results achieved by the LDA for the "billiard-ball" systems Be_2 (error ~ 0.46 eV) [31] and Mg_2 (error ~ 0.12 eV) [48], and the difficulty for the LDA to provide good answers in cases with near degeneracies is opposite to the situation encountered with CI methods. The so called dynamical correlations responsible for, *e.g.*, binding in Be_2 require a very large number of configurations for their accurate description whereas the near-degeneracy cases are often taken care of by a few configurations. This immediately suggests a hybrid method[49] in which a local-density correction is added to a small CI calculation.

The discussion above illuminates another peculiar feature of the LDA and approximations based on density gradients. Even though the LDA does not work well for a system with some particular symmetry, one can often find a "neighboring" system with nearly the same energy (*e.g.* a symmetry broken solution) for which the LDA provides an accurate answer. For example, the LDA gives a reasonable description of the H_2^+ molecule at large nuclear separation if we break the symmetry and put one electron on one of the protons. A more interesting example is furnished by the Cr_2 molecule who's bond length is accurately given by the LDA provided the molecule is assumed to be antiferromagnetic, *i.e.*, symmetry broken. [50] The binding energy is reasonable but too large as usual. In this case, one of the largest and most sophisticated CI calculations is inferior to the LDA. [51]

5. Improvements.

As we saw in the previous paragraph, the LDA is often quite adequate while, in other cases, a higher accuracy is desired. An error in a binding energy of the order of 20% or 1 eV is, *e.g.*, not acceptable in the study of chemical reactions. In this field we would like binding-energy errors to be of the order 0.1 eV or less. The simplicity of DF methods as compared to traditional many-body techniques has, however, spawned considerable efforts to improve on the LDA. It would, *e.g.*, mean a breakthrough in quantum chemistry if DF methods could be brought to the same level of accuracy as large-scale configuration-interaction (CI) calculations for larger molecules. One could then confidently go on to treat very large molecules as well as extended systems which presently are beyond our computational capabilities.

Attempts to go beyond the LDA are based either on an improved description of the exchange-correlation hole (see Eqs. (13) and (14)) in real space or on a description of exchange-correlation energies in reciprocal space usually leading to so-called generalized gradient corrections. Some times a mixture of the two approaches have been considered. [44,45] In recent years, the largest effort has gone into the reciprocal-space approach which so far has been the most successful *ab initio* DF method. As we shall see later, *ab initio* density functionals are still far from the 0.1-eV goal mentioned above but the results of new semi-phenomenological functionals are not too far away.

5.1. Gradients.

By construction, the LDA is correct for the homogeneous electron gas and accurate for densities that vary little over distances smaller than an inverse Fermi wave vector. A natural step beyond the LDA would therefore be to allow the energy functional to depend not only on the local value of the density but also on its different gradients. We will now use linear response theory to see how such gradient corrections can be obtained. [14] As discussed in Section 2 one obtains the density $n(\mathbf{r})$ i the DFT by minimizing the total energy $E[n]$ with respect to the densities constrained to contain a fixed number of electrons. The corresponding Euler equation reads (see Eq. (1))

$$\frac{\delta T_o}{\delta n} + \int v\, n + w + v_{xc} = \mu \tag{29}$$

where μ is a Lagrangian parameter that ensures particle conservation. If we change the external potential from w to $w + \delta w$ the change δn in the ground-state density is by definition given by

$$\delta n = \chi\, \delta w \tag{30}$$

where $\chi(\mathbf{r}, \mathbf{r}')$ is the static linear density response function of the system. The small change in external potential w leads to a new Euler equation from which the original equation (Eq. (29)) can be subtracted leading to

$$\frac{\delta^2 T_o}{\delta n\,\delta n'}\,\delta n' + v\,\delta n' + \delta w + K_{xc}\,\delta n' = 0 \tag{31}$$

where we have defined the kernel K_{xc} through

$$K_{xc}(\mathbf{r},\mathbf{r}') = \frac{\delta v_{xc}(\mathbf{r})}{\delta n(\mathbf{r}')} = \frac{\delta^2 E_{xc}[n]}{\delta n(\mathbf{r})\delta n(\mathbf{r}')} \tag{32}$$

In order to find the quantity $\delta^2 T_o/\delta n \delta n'$, we apply exact DF theory also to a non-interacting system. (Notice that the basic theorems of Hohenberg, Kohn and Sham[14,15] make no reference to the strength of the Coulomb interaction and consequently these theorems are equally valid for non-interacting electrons.) The equations corresponding to Eq.(30) and Eq.(31) are

$$\delta n = \chi_o\,\delta w \tag{33}$$

$$\frac{\delta^2 T_o}{\delta n\,\delta n'}\,\delta n' + \delta w = 0 \tag{34}$$

From the the last two equations we conclude that

$$\frac{\delta^2 T_o}{\delta n(\mathbf{r})\,\delta n(\mathbf{r}')} = -\,\chi_o^{-1}(\mathbf{r},\mathbf{r}') \tag{35}$$

and combining this with Eq.(30) and Eq.(31) we obtain the integral equation

$$\chi = \chi_o + \chi_o\,(v + K_{xc})\,\chi \tag{36}$$

Consequently, the static density response function of the interacting system obtains from that of the non-interacting system provided we know the exchange-correlation kernel K_{xc}. Conversely, a knowledge of the density response function of the interaction system allows us to infer the second derivative of the exchange-correlation energy with respect to the density. It is the latter property which allows us to construct a gradient expansion for E_{xc}. Suppose we perturb the homogeneous electron gas by a weak external potential and let us compute the change in E_{xc} to second order in the resulting change δn in the density. We find

$$E_{xc}[n] = E[n_o] + \int \frac{\delta E_{xc}}{\delta n}\,\delta n + \tfrac{1}{2}\int \delta n\,\frac{\delta^2 E_{xc}}{\delta n \delta n'}\,\delta n' + \ldots.. \tag{37}$$

$$E_{xc}[n] = \int n_o \varepsilon_{xc}(n_o) + \int \mu_{xc}(n_o)\,\delta n + \tfrac{1}{2}\int \delta n\, K_{xc}\,\delta n' \tag{38}$$

where the subscript o indicates the homogeneous limit for which the LDA is exact by construction. The corresponding result within the LDA reads

$$E_{xc}^{LD}[n] = \int n_o \varepsilon_{xc}(n_o) + \int \frac{\partial(n\varepsilon_{xc})}{\partial n_o}\,\delta n + \tfrac{1}{2}\int \frac{\partial^2(n\varepsilon_{xc})}{\partial n_o^2}\,(\delta n)^2 + \cdots \tag{39}$$

Now, subtracting the LDA result from the exact one and noting that $\mu_{xc} = \partial(n\varepsilon_{xc})/\partial n$ gives the following correction to the LDA

$$E_{xc}[n] = E_{xc}^{LD}[n] + \tfrac{1}{2}\int \delta n\,\{K_{xc} - \mu'_{xc}\cdot\delta(\mathbf{r}-\mathbf{r}')\}\,\delta n' \tag{40}$$

In Fourier space this becomes

$$E_{xc}[n] = E_{xc}^{LD}[n] + \tfrac{1}{2}\int \frac{d^3q}{(2\pi)^3}\,\{K_{xc}(\mathbf{q}) - K_{xc}(0)\}\,|\delta n_{\mathbf{q}}|^2 \tag{41}$$

where we have used the exact relation $K_{xc}(0) = \partial\mu_{xc}/\partial n$ which follows from the compressibility sum rule of the electron gas.[52] If we now make the crucial assumption that the perturbed density has appreciable Fourier components only at small q we can approximate K_{xc} by its small-q Taylor series

$$K_{xc}(q) = K_{xc}(0) + 2\,B_{xc}(n_o)\,q^2 + \cdots \tag{42}$$

and going back to real space we obtain

$$E_{xc}[n] = E_{xc}^{LD}[n] + \int B_{xc}(n_o)\,|\nabla n|^2 d^3r \tag{43}$$

demonstrating how the lowest order gradient correction is related to the density response function of the homogeneous electron gas. Notice also that the density argument in the coefficient B_{xc} is easily replaced by the local density $n(\mathbf{r})$ in a strongly inhomogeneous system where we would have difficulties in defining n_o.

In cases when the density varies more rapidly and does have appreciable amplitude at larger **q**-vectors, the Taylor expansion in Eq. (42) cannot be used. The derivation above suggests, however, that it is not at all necessary to use a gradient approximation. Using the fact that

$$\int \{K_{xc}(\mathbf{r}-\mathbf{r}'; n_o) - \mu'_{xc}\cdot\delta(\mathbf{r}-\mathbf{r}')\}\,d^3r' = 0 \tag{44}$$

Eq. (42) can readily be rearranged to read[15]

$$E_{xc}[n] = E_{xc}^{LD}[n] - \frac{1}{4}\int K_{xc}(\mathbf{r}-\mathbf{r}'; n_o)\,\{n(\mathbf{r}) - n(\mathbf{r}')\}^2\,d^3r\,d^3r' \tag{45}$$

This correction to the LDA is obviously valid for arbitrarily rapid but small density variations. It represents an infinite summation of the gradient terms. When, however, we want to apply this correction to a strongly inhomogeneous system like, *e.g.*, an atom we are forced to decide what to use as an average density n_o in the kernel K_{xc}. In a solid we could, *e.g.*, use the average density of the valence electrons. In an atom it becomes more natural to think of a local density but, since the correction involves two points in space ($\mathbf{r}$ and $\mathbf{r}'$), there are a number of possible choices, *e.g.*, $n((\mathbf{r}+\mathbf{r}')/2)$ or $(n(\mathbf{r}) + n(\mathbf{r}'))/2$. Gunnarsson *et al.*[19] have shown that the first choice gives an infinite correction for an atom whereas the second choice gives a reasonable result. These conflicting results of two similar and seemingly sound approximations could be viewed as a complete breakdown of gradient expansions. The origin of the sensitivity to the choice of average density is the use of linear response theory underlying the entire discussion above. Thus, our conflicting results is merely a reflection of the fact that an atom or, for that matter, a solid is *not* a linear perturbation of the homogeneous electron gas. As we shall see shortly, things are not as bad as they seem. Gradient expansions are not convergent in the sense that successively more accurate results can be obtained by adding more terms. They represent asymptotic expansions, meaning that an approximation with a few gradient terms added will be more accurate than the LDA in cases with relatively slow density variations. When the density varies more rapidly, we might be better off without gradient terms. As we shall see, the criterion for the validity of the LDA is not as strict as the corresponding criterion for the gradient expansion. Recently, people have tried to circumvent this problem by applying different cut-off procedures designed to retain the advantages of gradient corrections for moderate density variations without displaying the breakdown characteristic of the straight-forward gradient expansion in cases with rapid density variations.

Due to the difficulties mentioned above there are very few applications of a straight-forward gradient correction to the LDA in realistic physical systems. The coefficient B_{xc} in front of the lowest order correction (Eqs. (42) and (43)) was first calculated by Sham[53] in the exchange-only approximation[22,54] but Sham used a statically screened Coulomb interaction and let the screening length become infinitely large at the end of the calculation. Kleinman and Lee [55] later discovered that the exchange-only approximation is highly singular in the sense that working directly with an unscreened Coulomb interaction gives a coefficient which is 10/7 times larger than the value obtained by Sham. The Kleinman value certainly gives better atomic exchange energies but this value is still much smaller than the empirical values used by Herman *et al.* [56] in order to reproduce the Hartree-Fock results for the total energies of several atom. Of course, in the calculations by Herman *et al.* there is no room for higher order gradient corrections which can be large.

The singularities present in the exchange-only approximation is generally believed to disappear when the contribution from the correlation energy is added. In fact the coefficient B_{xc} has been calculated within the RPA by Ma and Brueckner[57] in the high-density limit and by Geldart and Rasolt[58,59] at metallic densities.

5.2. The LPM scheme.

The title of this subsection refers to the generalized gradient correction introduced by Langreth, Perdew, Mehl, and Hu [20,22-24] (LPM). It would carry too far to present the rather complicated theory underlying the LPM correction to the LDA. We will here be content with presenting some of their key ideas and some results for different physical systems. Langreth and Perdew[20] also start from the basic equation for the exchange-correlation energy of the inhomogeneous system (Eq.(8)). In order to find the interaction-averaged pair-correlation function $\tilde{g}$ of the inhomogeneous system they first use the RPA and compute the frequency dependent density-density response function to second order in a perturbing external potential w. Within the RPA, the required integration over the strength of the Coulomb interaction can be carried out analytically and from the zero-temperature version of the fluctuation-dissipation theorem[52] they then obtain E_{xc} for the slightly inhomogeneous system in the form of an integral in reciprocal (**k**-) space. In this way they can study the contributions from different wave vectors both to the LDA (E_{xc}^{LD}) and to the xc-energy obtained by adding the lowest order gradient correction. They find that the gradient correction works well for large wave vectors but strongly overestimates the contributions to the xc-energy at small **k**. The crossover occurs around a **q**-vector approximately equal to $q = |\nabla n(\mathbf{r})|/(6\,n(\mathbf{r}))$. In principle, their full result for E_{xc} could be used in actual calculations although the computational effort would be considerable. Therefore, in order to obtain an easily applicable scheme based on density gradients, Langreth, and Mehl[22] suggested an approximation to the full results by Langreth and Perdew. They proposed to keep the full gradient correction down to a cut-off q and to neglect the gradient correction altogether below this cut-off. In real space this results in the following approximation for E_{xc}:

$$E_{xc}[n] = E_{xc}^{LD}[n] + a\int [n(\mathbf{r})]^{-4/3}\,[\nabla n(\mathbf{r})]^2\,\{e^{-F} - \frac{7}{18}\}\,d^3r \qquad (46)$$

where

$$F = b\cdot|\nabla n(\mathbf{r})|\cdot[n(\mathbf{r})]^{-7/6}\ ;\ a = \frac{\pi}{8}(3\pi^2)^{-4/3}\ ;\ b = (9\pi)^{1/6} f \qquad (47)$$

in atomic units, *i.e.*, Hartrees (= 27.21 eV).

The quantity f is a "fudge" factor of order 0.15 - 0.17 and results are relatively insensitive to the choice of f in this range. Notice that surface energies indeed are sensitive to the choice of f, indicating, not a breakdown of the above theory, but rather of the simple *cut-off procedure leading to the practical formula*, Eq. (46).

It is important to stress one of the basic assumptions underlying the simplified version represented by Eq. (46) of the wave vector decomposition theory of LPM. It is assumed that the physical system under study is dominated by a single length scale approximately given by $1/q$. This is probably correct in an atom where that length would be, *e.g.*, the radius of an atomic orbital. As a result, the LPM scheme yields quite accurate total energies for atoms, the error being of the order of a few parts in a thousand. As it turns out, the major source of error comes from the treatment of exchange. For this part of the energy the LPM approximation gives an error of ~ 3% which should be compared to an error of the order of ~ 10% in the LDA. In the case of atomic correlation energies only, the LPM errors are ~ 10% compared to the infamous >100% overestimate resulting from the LDA.

In molecules, on the other hand, there is clearly at least one additional important length scale namely that of the molecular bond. As a result, the performance of the LPM scheme is not nearly as good in the case of binding energies of molecules as illustrated in Table 6. Also, in this case, there does not seem to be any advantage in treating exchange exactly and only leave the correlation part of the binding energies to the LPM scheme (see Table 13).

As pointed out in Section 3, approximations based on a theory for systems with relatively small and slow density variations like, *e.g.*, the LDA and different gradient corrections to it, can perhaps not be expected to work well for strongly inhomogeneous systems like atoms and small molecules. Thus, tests on the ground-state properties of solids are highly desirable. Unfortunately, such tests are rare in the literature. The first such test was carried out by von Barth and Car[25] in bulk silicon and they obtained a cohesive energy of 4.89 eV compared to 5.19 eV in the LDA and 4.63 eV from experiment. The lattice parameter of the LPM calculation agrees with experiment and is 0.5% larger than that of the LDA. Within the numerical accuracy, the bulk moduli of the LDA and the LPM calculations both agree with experiment. Later von Barth and Pedroza[60] tried the LPM approximation in bulk beryllium and obtained a cohesive energy of 3.20 eV compared to 3.65 eV in the LDA and 3.32 eV from experiment. More recently Bagno *et al.*[61] have applied the LPM scheme to the structural properties of several elemental solids including transition metals in which the general tendency of the LDA to overbind is particularly pronounced. Their results are similar to those reported above. The overbinding is reduced and the lattice parameters are slightly improved by the LPM scheme. In particular, this scheme gives the correct ground state of iron, *i.e.*, a ferromagnetic bcc structure whereas the LDA erroneously predicts a paramagnetic fcc structure.

We can summarize our experience of the performance of the LPM scheme in solids and molecules by saying that this scheme is superior to the LDA and that errors of the latter are typically reduced by a factor of two by the LPM approximation. Thus, the new scheme is an improvement but does certainly not represent a breakthrough within DF theory.

Before ending this section on the LPM scheme, we must mention a few technical details of relevance to the practical application of the scheme. In most cases, especially in finite systems, it is necessary to have a spin-polarized version of the theory. Such a generalization has been worked out by Hu and Langreth.[24] For self-consistent calculations the potential corresponding to Eq. (46), *i.e.*, the functional derivative of E_{xc} with respect to the density, must be derived and an explicit formula involving second spatial derivatives of the density can be found in Ref. 23.

The LPM theory for the slightly inhomogeneous electron gas is based on the RPA. According to Langreth *et al.*[22,62] there is a large cancellation between local and non-local contributions to the xc-energy also beyond the RPA and in order not to lose this advantage by using a theory beyond the RPA for just the LDA, the RPA version of the latter should be used in conjunction with the LPM gradient correction.

As a byproduct of their wave vector analysis Langreth *et al.* found two very useful criteria determining the validity of the LDA and of the straight-forward gradient correction. They found that the LDA is valid when

$$\frac{|\nabla n(\mathbf{r})|}{k_F(\mathbf{r})\, n(\mathbf{r})} \ll 6 \tag{48}$$

whereas the gradient expansion is valid when

$$\frac{|\nabla n(\mathbf{r})|}{k_{TF}(\mathbf{r})\, n(\mathbf{r})} \ll 1. \tag{49}$$

Here $k_F(\mathbf{r})$ is the local Fermi momentum ($k_F^3 = 3\pi^2 n$) and $k_{TF}(\mathbf{r})$ is the inverse of the local screening length, *i.e.*, the Thomas-Fermi wave vector given by $k_{TF} = 2\sqrt{k_F/\pi}$. In real systems the last criterion is usually an order of magnitude more severe than the first which explains why the LDA can provide reasonable answers while the inclusion of gradient terms will make things worse.

5.3. The Perdew-Wang scheme.

In the discussion of the LPM scheme we noted that the largest errors in that scheme originates in their treatment of exchange. For this part of the energy Langreth *et al.* essentially use a straight-forward gradient correction. Being concerned about the shortcomings of the previous treatment Perdew [63,64] introduced a new idea comprising a combination of real- and reciprocal space arguments. In Section 2 we attributed the success of the LDA to various exact properties of its exchange-correlation (xc) hole like the sum rule (Eq.(17)) or the negativity of the exchange hole (Eq.(18)). In the wave vector analysis of Langreth *et al.* we can say that we study and approximate the xc-hole in reciprocal space and it becomes less evident how to implement, *e.g.*, the sum rule. A particular approximation for the xc-energy in reciprocal space can, however, often be transformed into an approximation for the xc-hole in real space based on the density and its gradients. Perdew showed, *e.g.*, that the second order gradient expansion for the xc-energy does not correspond to an xc-hole that obeys the sum rule. This is yet another way in which we can understand why the LDA in some cases can yield an accurate result when the addition of the lowest order gradient term can destroy that accuracy.

From the work by Gross and Dreizler [65] on gradient expansions it is not difficult to obtain one for the exchange hole. A second order expansion for the hole does not obey either the sum rule (Eq.(17) or the negativity condition (Eq. (18)). Therefore, Perdew [63] multiplied the result by Gross and Dreizler by an appropriate cut-off function to ensure the satisfaction of both the sum rule and the negativity condition. The resulting approximation for the exchange hole looked somewhat complicated and difficult to use but in a later paper by Perdew and Wang [44] the result was simplified by partial integration and parametrized for easy use in DF calculations. Their functional for the exchange energy is (cf. Eq.(22)):

$$E_x[n] = A_x \int [n(\mathbf{r})]^{4/3} F_x(s)\, d^3r \tag{50}$$

where A_x is a constant determined by the electron-gas limit and given by

$$A_x = -\frac{3}{4\pi}(3\pi^2)^{1/3}. \tag{51}$$

$F_x(s)$ is a function of the dimensionless density gradient

$$s = \frac{|\nabla n(\mathbf{r})|}{2\,k_F(\mathbf{r})\,n(\mathbf{r})} \tag{52}$$

The function F_x is originally obtained numerically in order for the x-hole to obey the sum rule and the negativity condition and later fitted to an analytical formula:

$$F_x(s) = (1 + a\,s^2/m + b\,s^4 + c\,s^6)^m \tag{53}$$

with the constants

$$a=\frac{7}{81}\ ;\ b=14\ ;\ c=\frac{1}{5}\ ;\ m=\frac{1}{15} \tag{54}$$

Energies are given in atomic units, *i.e.* Hartrees (1 Hartree = 27.21 eV). Notice that $F_x(s)=1$ corresponds to the LDA for exchange only. Notice also that many systems are spin-polarized, particularly those of finite extent and that, therefore, the spin-polarized versions of the functionals must be used. In the case of exchange only this is easily obtained as

$$E_x[n_\uparrow, n_\downarrow] = \tfrac{1}{2}\{E_x[2n_\uparrow]+E_x[2n_\downarrow]\} \tag{55}$$

As a side issue, we finally notice that the generalized gradient approximations by Becke[31,66,67] can also be cast in the form of Eq. (50). This can be seen in the next subsection.

The particular parametrization given by Eq. (53) has the property that a somewhat incorrect second-order gradient term is obtained in the slowly varying limit (small s). In this context, "incorrect" refers to the gradient coefficient originally obtained by Sham[53] by assuming a screened Coulomb interaction and letting the screening length tend to infinity at the end of the calculation. Due to the highly singular nature of the exchange-only approximation, a different coefficient is obtained if the screening length is allowed to be infinitely large from the start of the calculation.[55,68] Since the small q-limit of the exchange-only kernel $K_x(\mathbf{q})$ (see Eq. (32)) is better described by the latter coefficient over a larger range of q-values, the function $F_x(s)$ (Eq. 53) should be modified to give this slowly varying limit. Seting $a=10/81$ in Eq. (53) will achieve the desired result. In later work,[69] Perdew *et al.* have made additional modifications of the function $F_x(s)$ and essentially made it much more similar to the Becke[67] exchange approximation. To this date however, actual tests in solids have primarily been performed using the approximation given by Eqs. (50,53). The slightly modified functional of Ref. 69 has only been tried in solid Li and Na and in atoms and small molecules.[47,69]

Having improved on the LPM description of non-local contributions to exchange energies, Perdew[45] proceeded to introduce some beyond-RPA effects into the LPM scheme for the correlation energies. Apart from taking away a spurious contribution of exchange origin added by Langreth *et al.* in order to separate exchange and correlation in *finite* systems, Perdew's modification essentially amounts to a rescaling of the LPM result in order to retrieve the correct beyond RPA second-order gradient coefficient calculated by Rasolt and Geldart.[58] The explicit formula is very similar to Eq. (46) without the 7/18:th and we refer the reader to Ref. 45 for details. Having a beyond RPA result for the non-local correlation energy allowed Perdew to add the state-of-the-art version of the LDA for the correlation energy, *e.g.*, a parametrized version of the Monte-Carlo results for electron-gas correlation energies by Ceperley and Alder.[70]

During the past couple of years, the Perdew-Wang (PW86) correction to the LDA described above has been employed in several electronic structure calculations of solids. Kong *et al.*[71] calculated the ground-state properties of solid Al, Si, and C and concluded that the PW86 functional gives a substantial reduction of the overbinding characteristic of the LDA. The lattice parameters and the bulk moduli are almost the same as those of the LDA but there was a small increase in the lattice parameters resulting in closer agreement with experiment. Notice, however, that these quantities are very close to experiment also in the LDA. Bagno *et al.* [61] tried both the PW86 functional and the LPM functional on solid K, Ca, V, Fe, and Cu. Their conclusions were similar to those of Kong *et al.*. The LDA errors in the binding energies are reduced by typically a factor of two and the lattice parameters and the bulk moduli are also improved by the two gradient corrected schemes. In particular these schemes predict the correct ground state of iron, *i.e.*, ferromagnetic bcc. On the other hand one finds that one of the well known failures of the LDA, *i.e.*, the prediction of metallic ground states for the transition metal oxides FeO and CoO, are not corrected by the PW86 functional. [72] Comparing the PW86 and the LPM functionals, Bagno *et al.* found that the first had a slight edge on the latter although they noted a tendency of the PW86 functional to overcorrect the LDA errors especially for the more weakly bound systems. The same tendency has been reported by Garcia *et al.* [73] in a study of the cohesive properties of solid Al, Si, Ge, GaAs, Nb, and Pd. In particular these researchers report that the PW86 functional in many cases gives lattices which are too soft. Similar conclusions have been reached by others. [74,75]

As far as molecules are concerned, the experience with the PW86 functional is rather limited. [76] It seems clear, however, that this functional performs slightly better than the LPM functional also in the molecular case and that it thus represents a substantial improvement on the LDA. As will be discussed in the next section, this is a somewhat spurious result most likely of unphysical origin.

Most recently Perdew *et al.* [69] have made additional improvements of their total xc-functional. They have incorporated the sum rule for the correlation hole, *i.e.*, the fact that this hole must contain zero charge which follows from the sum rule for the total hole (Eq. (17)) and the unit charge contained in the Fermi (or exchange) hole. They have also modified their approximation to the exchange energy through a different choice of the function $F_x(s)$ (Eq. (53)) as discussed above. The latest $F_x(s)$ behaves as $1 + 10/81\, s^2$ at small s corresponding to the Kleinman [55] value for the straight-forward second-order gradient correction for the exchange energy of the homogeneous electron gas. Probably inspired by the success of one of Becke's [67] gradient corrections for exchange (see next section), the function $F_x(s)$ closely follows the Becke approximation at intermediate s. In contrast to the latter approximation, $F_x(s)$ is made to vanish as $const/s^2$ at large s in order to obey certain inequalities for exact exchange. The sum rule on the x-hole is not enforced but is obeyed rather closely by the resulting functional referred to as PW91 in what follows. The PW91 functional was tested by Perdew and his collaborators on solid Li and Na. In both cases the PW91 functional corrects the unusually large LDA errors in the lattice parameters of these systems (0.11 Å in Li and 0.17 Å in Na). Since the PW91 functional is rather new, we are not aware of other tests on its performance in solids but, due to its similarity to the PW86 functional, we would expect the PW91 to give results of comparable accuracy or slightly better. Also in molecules the PW91 functional is comparable to the PW86 functional as reported by Merkle, Savin, and Preuss. [47]

We will now attempt to summarize our experience from the gradient corrected density functional methods. From a theoretical point of view, these schemes are a result of a careful analysis of exchange and correlation effects in a weakly perturbed electron gas (basically linear response theory). Through the use of sum rules and other consistency requirements, effects of higher order responses are brought into the approximations. The results are encouraging and usually much superior to the LDA. Errors in binding energies are reduced by a factor of two to three or even more. In the case of properties which are very well described already by the LDA the picture is less consistent and gradient corrections can lead to worse results although the difference is marginal. In going from the LPM gradient corrected functional to the PW86 and the PW91 functionals there is an increase in theoretical sophistication. Tests on different systems and for different physical properties indicate a minor and inconsistent improvement going from the LPM to the PW91 scheme. The gradient corrected schemes are marginally more difficult to apply as compared to the LDA and in view of aforementioned properties we recommend that the PW91 functional be used in all DF calculations. This functional represents a substantial improvement on the LDA but we do not consider the accuracy of the PW91 functional to be of the kind that would allow us to talk about a breakthrough in DF theory. In particular, the application to small molecules demonstrate errors much larger than can be tolerated in thermochemical applications.

In Section 4.1, several of the failures of the LDA was blamed on the the inability of the LDA to account for the s-p or the s-d transfer energies. Tests on the gradient corrected functionals[8,43] show that they suffer from the same deficiency but still yield much better binding energies. This somewhat unexpected result is the basis for our comment in Section 4.1 that the s-d transfer problem may not even be the major source of error in the LDA.

In small molecules it is relatively easy to compute the exact exchange energy thus leaving only the much smaller correlation energy to be treated by DF methods. Such a separation leads to disastrous results as can be seen in Table 13. The usual argument for treating exchange and correlation together is based on the large cancellation occurring between exchange and correlation energies due to the much smaller extent of the xc-hole as compared to the exchange and correlation holes taken separately. This argument is certainly valid in extended systems but its validity is doubtful for small systems where already the size of the system limits the extent of both holes. The bad results shown in Table 13 and resulting from a separate treatment of exchange and correlation, rather indicate a poor description of the correlation hole at intermediate distances - a region not amenable to gradient corrections based on electron-gas theory. At this stage, the relatively good results obtained for the combined DF treatment of exchange plus correlation must be considered as fortuitous. Indeed, as we shall see in the next section, Becke[77] has managed to produce accurate correlation energies for a large number of molecules using an approximation based on arguments relevant only to exchange energies.

5.4. The Becke approach.

In the previous subsections we have described attempts to design approximations beyond the LDA which are based on a careful analysis of the effects of exchange and correlation in inhomogeneous systems. From early on, Becke has taken a different approach to DF theory. In principle we seek to map out the functional $E_{xc}[n]$ in systems of interest which, in Becke's case, have been atomic and molecular systems. Suppose that we have designed a functional based on, *e.g.*, the density, its different gradients, and different integrals thereof, containing a large number of parameters. Suppose further, that we were able to determine those parameters in such a way that the chosen functional, within a certain accuracy, reproduces the exact E_{xc} for a large number of physical systems with rather different properties with regard to exchange and correlation. Then, we would certainly believe that we had obtained an accurate description of the true $E_{xc}[n]$, at least for densities not deviating in a qualitative way from those already studied. We would feel comfortable in applying the constructed functional to unknown systems and we would have some confidence in the results thus obtained. There are certainly theoreticians who would be prone to scorn the described procedure but we find such attitudes unwise. First of all, the semi-phenomenological procedure will enable us to treat more complicated systems with much greater ease as compared to *ab initio* methods, we could even attack systems which presently are beyond computational feasibility and we could concentrate on other physical aspects like, *e.g.*, molecular reactions. Secondly, we would have learned something about the functional $E_{xc}[n]$. For instance, if our chosen functional form contains only the density and its first spatial derivatives we would know that it must be possible to cast a sophisticated *ab initio* theory in a form containing the same basic ingredients.

When we, after this philosophical digression, return to Becke's work we must stress that a considerable amount of physical insight is needed in order to invent a reasonable functional form for $E_{xc}[n]$. A better functional form leads to fewer adjustable parameters and Becke's functionals do contain few parameters. He started with the observation that when density gradients become large, gradient corrections are irrelevant. Concentrating on the exchange energy, Becke[31] designed a functional which, at small gradients, amounts to a normal first order gradient correction with an adjustable coefficient and which becomes an adjustable constant at large gradients. As a matter of fact all Becke's gradient corrections for exchange can be cast in the form of Eq. (50) and the first one[31] used

$$F_x(s) = 1 + \frac{\beta s^2}{1+\gamma s^2} \tag{56}$$

The parameters β and γ were adjusted so that this particular form gave an accurate fit to the exchange energies of a large number of atoms (β = 0.2351, γ = 0.2431). To this exchange energy must be added some approximation for the correlation energy and in his first semi-empirical approach towards molecular energies Becke chose one due to Stoll *et al.*[78] This approximation is much superior to the LDA for correlation energies of finite systems but is has the wrong slowly varying limit. The resulting atomization energies of the first row dimers can be studied i Table 12. The maximum error is 0.8 eV and the average error for the seven molecules is 0.3 eV. The corresponding figures for the LDA are 2.4 eV and 1.2 eV. Considering the simplicity of the approximation this represents an extraordinary achievement which to this date has not been surpassed by any generalized gradient approximation.

Based on ideas concerning the behavior of exchange energies in strongly inhomogeneous systems (large gradients) Becke[66] later proposed a minor modification of Eq. (56)

amounting to raising the denominator to the 4/5 power and changing β and γ but, to our knowledge, this functional has never been tested in molecules. This is certainly not true about the next gradient correction for the exchange energy by Becke[67] which reads

$$F_x(s) = 1 + \frac{\beta s^2}{1 + (2.25/\pi)\, \beta\, s\, arsinh(\gamma s)} \tag{57}$$

Here, γ is a well defined constant equal to $\gamma = 2\,(6\pi^2)^{1/3}$ (a conversion between Becke's $x = |\nabla \rho| \cdot \rho^{-4/3}$ and Perdew's $s = |\nabla n|/(2k_F n)$) and only β is an adjustable parameter which is determined by a fit to atomic exchange energies. A value of $\beta = 0.2743$ gives an average exchange-energy error of only 0.11% for six noble-gas atoms - a truly remarkable result considering the simplicity of the ansatz (Eq. (57)). The chosen form for F might appear peculiar but it is constructed to give the correct inverse-distance dependence of the exchange-energy density far outside a finite system. In this region, the density decays as $e^{-\lambda \cdot r}$ and it is not difficult to see from the Eqs. (50,51,57) that this leads to a contribution of the form

$$-\tfrac{1}{2} \int_{r>R} \frac{n(\mathbf{r})}{r}\, d^3r \tag{58}$$

to $E_x[n]$ at large R. In fact, the particular exchange-energy functional of Eq. (57) is presently the only available functional with this correct behavior but we are a little doubtful about the value of this property since the energy contributions from the large-R regions are small due to the very small density in these regions. Nevertheless, the new functional by Becke has proven extremely successful. Becke[77] has tested the functional by calculating the atomization energies of the 55 molecules of the so-called Gaussian-1 data base of Pople *et al.*[79,80] For this test, Becke used the normal LDA for the correlation energy - an approximation known to overestimate the correlation energy by a factor two to three. One can only speculate why Becke refrained from adding one of the existing gradient-type correlation corrections to the LDA. In any case, his results were astonishingly good. The average absolute error was only 0.16 eV (3.7 kcal/mol) and the maximum error was 0.48 eV. The corresponding errors of the LDA are 1.6 eV and 3.7 eV, *i.e.*, an order of magnitude larger. Since there is a large overestimate of the correlation energies in the test, Becke's exchange functional must, somehow, make up for the difference. It is an interesting task for future research to understand how this cancellation of errors comes about. The CI-oriented *ab initio* results for the same set of molecules have average errors of the order of 1 kcal/mol and we clearly see that DF methods are approaching the same quality but with much less effort.

In other tests on atoms and molecules, Becke found that the tested functional had an unsatisfactory performance in the case of electron non-conserving processes like, *e.g.*, ionization. Becke[81] then decided to add to his previous functional the latest gradient correction by Perdew *et al.*[69] (PW91C) for the correlation energy. The approximation for the exchange energy remained the same - Eqs. (50) and (57). Becke also enlarged his data base by including many ionization potentials and several proton affinities and redid the test, again with a very impressive result.[81] The average absolute error and the maximum error for the atomization energies of the molecules of the Gaussian-1 data base was found to be 0.25 eV and 0.8 eV respectively. Thus, an "improved" description of correlation effects results in slightly larger binding-energy errors but there is a big gain as far as ionization potentials are concerned. These show an average error of only 0.15 eV and a maximum error of 0.44 eV. The corresponding numbers for the LDA are 0.23 eV and 0.62 eV for 42 ionization potentials.

The results described above are presently the most accurate available from density functionals based on the local density and its first spatial derivatives. These results are an order of magnitude more accurate than those of the LDA which is indeed encouraging. We are close to achieving "chemical" accuracy but there is still some way to go. Occasional errors of the order of 0.8 eV cannot be tolerated. The Becke functional (Eq.(57)) is specially designed for finite systems (atoms and molecules) and it has the wrong slowly varying limit. What this would mean for solids or larger molecules is hard to say. Since only the gradient terms are in error, we would not expect the Becke functional to be any worse than the LDA in extended systems. From a theoretical point of view, however, we would prefer a functional with all the correct limits built in but no such functional can presently overshadow the one by Becke in smaller molecules.

As mentioned several times in these notes, exchange energies are much larger than correlation energies even in extended systems like metals. This immediately suggests that one should treat exchange exactly and, *e.g.*, use a GGA for only the correlation energy. So far, attempts in in this direction have failed as can be seen in Table 13. Recently, Becke[82] suggested a hybrid method in which half of the exchange energy is treated exactly leaving the other half plus the correlation contribution to approximations of the GGA-type. Theoretical support for this scheme can again be obtained from the basic formula for the xc-energy (Eq.(8)) sometimes referred to as the "adiabatic connection" formula. Using the bare LDA for the correlation contributions Becke found that this new scheme had an accuracy similar to his previous functionals for the atomization energies of the Gaussian-1 data base. In one further step Becke[83] included some of the correlation correction of Perdew *et al.* (PW91C) and decided to treat the amount of exact exchange as a fitting parameter. His latest approximation can be summarized as

$$E_{xc}[n] = E_{xc}^{LD}[n] + a_o \cdot (E_x[n] - E_x^{LD}[n]) + a_x \cdot \Delta E_x^{B88} + a_c \cdot \Delta E_c^{PW91} \tag{59}$$

where a_o, a_x, and a_c are semiempirical coefficients to be determined by an appropriate fit to experimental data. The symbol Δ signifies the gradient part of the corresponding approximation, B88 refers to the Becke[67] gradient correction for exchange (Eq.(57)), and PW91 refers to the latest gradient correction for the correlation energy by Perdew *et al.*[69] The functional $E_x[n]$ is the exact exchange-energy functional, *i.e.*, the Hartree-Fock exchange-energy expression here evaluated using the DF orbitals. The functional defined by Eq.(59) was fitted to the Gaussian-1 data base including also 42 ionization energies and 8 proton affinities and the resulting optimum parameters was found to be $a_o = 0.20$, $a_x = 0.72$, and $a_c = 0.81$. These values resulted in an average absolute error of only 0.10 eV (2.4 kcal/mol) for the atomization energies with a maximum error of 0.33 eV. This result borders on chemical accuracy and we are close to being justified in proclaiming a breakthrough in DF methods.

As mentioned previously, a purist might discard the latest method by Becke for being a fitting procedure but we think that there is a lesson to be learned here. The success of the latest Becke scheme seems to suggest that some sort of more long-range nonlocalities, than can be provided by mere gradients, must be considered in the *construction of an accurate xc-functional.*

Before ending this Section on GGA-type approximations we just mention that there are other available xc-functionals than those discussed here. For instance, Becke[84] has designed a gradient correction for correlation energies and so have Lee, Yang and Parr (LYP).[85] Both these functionals have been tested on small molecules by Miehlich *et al.*[86] *Johnsson and coworkers*[87] *have also applied the LYP correlation functional in con-*junction with the Becke gradient correction for exchange in an extensive test on a large number of smaller molecules. The conclusion from these tests is that the performance of the Becke correlation functional, the LYP functional and the previously discussed PW86 functional for correlation energies is rather similar.

Table 12. The errors (in eV) in the binding energies of the first-row dimers as obtained from different density functionals defined below.
Δ is the average absolute error for each functional.

	Li_2	Be_2	B_2	C_2	N_2	O_2	F_2	Δ
LDA	-0.1	0.5	0.8	1.0	1.7	2.4	1.7	1.2
LPM	-0.5	0.3	0.2	-0.2	0.3	1.2	0.7	0.5
PW86	-	-	0.2	-0.1	-	0.7	0.5	0.4
PW91	-0.1	0.3	0.3	-0.1	0.7	1.0	0.7	0.5
B86	-0.1	0.1	-0.5	-0.8	0.2	0.2	0.0	0.3
B I	-0.1	0.1	-0.3	-0.7	0.3	0.3	0.2	0.3
B II	-0.3	-	-	-	0.4	0.8	0.5	0.5
B III	-0.3	-	-	-	-0.1	0.2	-0.1	0.2

LDA	the local-density approximation. Data from Ref. 30.
LPM	the functional by Langreth, Perdew, Mehl, and Hu.[20,22-24] Data from Ref. 31
PW86	the older functional by Perdew and Wang.[44,45] Data from Ref. 76.
PW91	the latest functional by Perdew *et al.*[69] Data from Ref. 47.
B86	an older exchange aproximation by Becke[31] plus correlation from Stoll *et al.*[78] Data from Ref. 31.
B I	*the functional of "Becke: Thermochemistry I".*[77] Data from Ref. 47.
B II	the functional of "Becke: Thermochemistry II".[81] Data from Ref. 81.
B III	the functional of "Becke: Thermochemistry III".[83] Data from Ref. 83.

Table 13. The errors (in eV) in the correlation-energy contributions to the binding energies of the first-row dimers as obtained from different correlation functionals defined below. Δ is the average absolute error for each functional.

	Li_2	Be_2	B_2	C_2	N_2	O_2	F_2	Δ
LDC	-0.1	-0.3	-1.9	-4.1	-2.4	-2.9	-2.5	2.0
LPMC	0.1	0.1	-0.8	-2.7	-1.0	-1.2	-1.4	1.0
PW91C	-0.3	-0.2	-1.5	-3.9	-2.4	-2.1	-2.4	1.8

LDC the local-density approximation for correlation energies. Data from Savin *et al.* in Ref. 9.

LPMC the correlation part of the functional by Langreth *et al.* [20,22-24] Data from Savin *et al.* in Ref. 9.

PW91C the latest correlation functional by Perdew *et al.* [69] Data from Ref. 88.

6. Excitation energies.

6.1. General remarks.

We will end these notes by a few comments on the possibility of obtaining various excitation energies from DF theory. In the original formulation by Hohenberg, Kohn, and Sham, [14,15] DF theory was a theory for ground-state properties of electronic systems. Not much later, however, several generalizations were suggested. Gunnarsson and Lundqvist [18] realized that the entire Hilbert space for the system considered, could be split up into symmetry channels and a DF theory could be defined for each channel. One could then obtain the "ground" state of each symmetry which would enable one to find, *e.g.*, the energies of the those states of lowest energy with a symmetry different from that of the true ground state. Theophilou [89] proposed a DF theory for subspaces from which energies of excited states of *finite* systems could be obtained successively by subtraction of the energies of subspaces containing lower lying states. Guided by an observation by Ziegler *et al.* [90] von Barth [91] defined a DF theory for mixed-symmetry states that allowed for a rather accurate determination of atomic and molecular term values.

All these ideas on excited-state energies are interesting and valuable but rather limited in scope. They are, *e.g.*, all concerned with *finite* systems. Excited states are intimately connected to time dependent phenomena and a true DF theory for excitation energies requires a time dependent generalization of the ground-state theory. Such a *generalization was first given by* Peuckert [92] *and later in a more general form by* Runge and Gross [93] allowing for arbitrary time dependent external potentials. We will not further *dwell upon these developments here* but rather take up a few specific points that we find intriguing.

6.2. Eigenvalues.

In DF theory, [15] one obtains the electron density by solving a set of one-particle equations of the form

$$\{-\tfrac{1}{2}\nabla^2 + w + \int n v + v_{xc}\}\phi_k = e_k \cdot \phi_k \tag{60}$$

giving orbitals ϕ_k the squares of which sum up to the density

$$n(\mathbf{r}) = \sum_k |\phi_k(\mathbf{r})|^2. \tag{61}$$

The DF eigenvalues e_k arise simply as Lagrangian multipliers intended to guarantee the normalization of the wave functions ϕ_k during the minimization process for the total energy $E[n]$. The eigenvalues have no immediate physical meaning with the exception *of the highest occupied one which , for any system, can be shown to equal the ionization* potential. [15,94] Nevertheless, these eigenvalues are used extensively and rather suc*cessfully for interpreting one-electron excitation spectra of solids, e.g. photoemission.* Thus, the DF eigenvalues are often rather accurate approximations to one-electron excitation energies although many counterexamples are well known by now. [2,95]

In order to prove that the exact DF eigenvalues are not, in general, equal to excitation energies we offer the following illustrative counterexample. We consider the fully interacting but homogeneous electron gas. In this case, the sum of the external potential w (Eq. (60)) and the Hartree potential $\int n v$ vanishes and, due to the complete translational symmetry of the system, the exact exchange-correlation potential v_{xc} must be position independent. Consequently, the exact DF eigenvalues e_k must be

$$e_k = \tfrac{1}{2} k^2 + v_{xc} \tag{62}$$

As mentioned above, the highest occupied DF eigenvalue correctly gives the Fermi energy E_F so that

$$E_F = \tfrac{1}{2} k_F^2 + v_{xc} \tag{63}$$

On the other hand, the correct excitation energies E_k - the quasi-particle energies - are obtained from Dyson's equation[96]

$$E_k = \tfrac{1}{2} k^2 + \Sigma(k, E_k) \tag{64}$$

where $\Sigma(k, \omega)$ is the momentum and energy dependent electronic self-energy. Also the quasi-particles at the Fermi surface have the correct Fermi energy:

$$E_F = \tfrac{1}{2} k_F^2 + \Sigma(k_F, E_F) \tag{65}$$

If we study the energies close to the Fermi level in both cases we obtain (expand Σ around $k = k_F$ and $\omega = E_F$)

$$e_k \approx E_F + k_F \cdot (k - k_F) \tag{66}$$

$$E_k \approx E_F + \frac{k_F}{m^*} \cdot (k - k_F) \tag{67}$$

where the effective mass m^* is given by

$$m^* = \left[1 - \frac{\partial \Sigma}{\partial \omega}\right] \cdot \left[1 + \frac{1}{k_F} \frac{\partial \Sigma}{\partial k}\right]^{-1} \tag{68}$$

From the Eqs. (66) and (67), it is clear that the two sets of eigenvalues are different, in principle. They differ to the extent that the effective mass m^* deviates from unity. In the electron gas this difference is ~ 5 %, which explains why the interpretation of photoemission spectra of simple metals in terms of DF eigenvalues has been so successful. The small deviation of m^* from unity in the gas is sometimes explained by hand-waving arguments about the effects of the Coulomb interaction being small due to the presence of the xc-hole that surrounds each electron. It is more correct to say that m^* is almost one due to a subtle cancellation between the energy and the momentum dependence of the self-energy of the gas as evidenced by Eq. (68). In other Fermion systems m^* can be large. For instance, in ^{3}He, m^* is ~ 3.[97] As we move away from the Fermi energy in simpler metals or move to more complicated systems with localized electrons such as transition metals, there is a more obvious deviation between experimental one-electron excitation energies and DF eigenvalues as they are obtained from the LDA.[2] This deviation is again a consequence of the fundamental difference between the two sets of energies rather than of the inaccuracy of the LDA.

6.3. The Fermi surface.

As we have seen, the DF eigenvalues deviate from excitation energies as soon as we leave the Fermi level. It would, however, be interesting to know the situation right at the Fermi level, *i.e.*, what about the Fermi surface? The DF Fermi surface is that surface in **k**-space for which the DF eigenvalues equal the Fermi energy. On the other hand, the measurable Fermi surface obtains when the true one-electron excitation energies, *i.e.* the quasi-particle energies, agree with Fermi energy. Are these two surfaces one and the same? The following argument [7] suggests that they might be. Suppose we introduce a small test charge into an otherwise perfect lattice. The test charge will be screened and the screening charge will show Friedel oscillations far away from the impurity. It is well known that the frequency of these oscillations in a certain direction is given by the diameter of the Fermi surface in that direction. For a sufficiently small test charge, the long-range properties of the screening charge is governed by possible non-analyticities in reciprocal space of the exact static linear density response function. From many-body theory, we know that the derivative of this function blows up at momentum transfers corresponding to the diameter of the sharp Fermi surface. The present case is similar to that of the perhaps more well known Kohn anomalies [98] encountered in phonon dispersion relations. As shown in Section 5.1, we can obtain (see Eq. (36)) the exact linear density response function χ also from DF theory. The response function χ is given in terms of the response function χ_o of a non-interacting system with the same density as the true system and the exchange-correlation kernel K_{xc}. The non-analyticities of the former are given by the DF Fermi surface, *i.e.*, by the exact one-electron eigenvalues of Eq. (60). Thus, provided the xc-kernel K_{xc} is free from non-analyticities, the two Fermi surfaces must be the same.

Unfortunately, this is not the case as shown by D. Mearns who evaluated both Fermi surfaces to second order in both the lattice potential and the Coulomb interaction. [99] The deviation between the Fermi surfaces in the case of weak lattice potentials and a small Coulomb interaction is an adequate counterexample demonstrating the fundamental inequivalence of the two entities. A similar conclusion was reached by Schönhammer and Gunnarsson [100] in a Hubbard model of an interacting system for which they invented a DF-like theory where the occupation numbers of each site played the role of the density in normal DFT. We will return to this interesting generalization [101] of DFT in the Section 6.5.

Consequently, the correct xc-kernel K_{xc} must have two sets of singularities, one at momentum transfers corresponding to the diameter of the DF Fermi surface in order to remove this singularity in χ_o from the full response function χ, and another one corresponding to the physical Fermi surface in order to give χ the correct singular behavior. This very unphysical behavior of the kernel K_{xc} is obviously the price we have to pay in order to be able to squeeze the full many-body problem into a one-particle framework like the DFT.

6.4. The metal-insulator transition.

The non-physical behavior of the xc-kernel K_{xc}, discussed in the previous subsection, is even more conspicuous in the case of the metal-insulator transition. It is well known[52] that the static density response function changes its analytic structure when a system becomes metallic. In the insulating phase, a test charge is only partially screened at large distances whereas the screening is complete in the metallic case. As a consequence, the response function χ (see Section 5.1) of an insulator approaches zero as q^2 at small momentum transfers whereas χ remains finite in the metallic case. As discussed in Section 5, the exact response function χ obtains from the xc-kernel K_{xc} and the non-interacting response function χ_o. The analytical structure of the latter is, in turn, determined by the DF eigenvalues. The existence of a gap in the spectrum of these eigenvalues will cause a q^2 behavior of χ_o at small q whereas χ_o will remain finite if there is no gap. Assuming K_{xc} to be a well behaved analytic function of q, leads to the conclusion that the system will be metallic or not depending on whether there isn't or is a gap in the spectrum of DF eigenvalues. Unfortunately, investigations by Godby and Needs[102] indicate that, when the system is compressed, the DF gap closes long In terms of the response function χ (see Section 5.1), in the before the system goes metallic. In the work by Godby and Needs, the quasi-particle spectrum used to decide between metallic and insulating behavior was computed from the so called GW-approximation[96] whereas the DF eigenvalues were obtained both from the LDA and from an approximation to the DF potential based on the GW-approximation. Even though the GW-approximation and the corresponding DF local potential generally are believed to be accurate representations of respectively the self-energy and the exact DF effective potential, they are still approximations and the obtained result can not be considered as proof that the exact DFT fails to give the correct metallization point. The large discrepancy between the approximate result of quasi-particle theory and that of the DFT suggests, however, that even the exact DF eigenvalues cannot be used to infer the volume at which metallization occurs. The consequences to the xc-kernel K_{xc} are now clear. When an insulating material is compressed the DF gap will first close and cause χ_o to go metallic (χ_o remains finite as q tends to zero). However, since the full response function χ stays insulating (χ goes to zero as q^2) at this point, K_{xc} in Eq.(36) must have an abrupt change in its analytical behavior in order to restore the correct behavior of χ. Upon further compression, the system, and therefore χ, will become metallic without any change in χ_o. Thus, at this point, K_{xc} must have yet another singularity in order to give χ metallic character through Eq.(36). This peculiar result again demonstrates the high prize that sometimes must be paid for enabling us to convert the many-problem into a equivalent one-electron theory.

6.5. The gap problem.

A theoretically as well as experimentally well defined definition of the top of the valence band, E_v, in a semiconductor is the negative of the minimum energy required to move an electron from the system to infinity. Thus

$$E_v = E_o(N) - E_o(N-1) \tag{69}$$

where $E_o(M)$ is the ground-state energy of the solid containing M electrons and where the neutral solid has N electrons. Similarly, the bottom of the conduction band, E_c, is given by the negative of the energy gained by adding one electron to the solid,

$$E_c = E_o(N+1) - E_o(N). \tag{70}$$

The difference between these two energies is, by definition, the band gap E_g,

$$E_g = E_o(N+1) + E_o(N-1) - 2E_o(N) \tag{71}$$

Consequently, the band gap is related to the ground-state energy of systems differing only in their particle number and is, therefore, strictly obtainable from the DFT. How the gap relates to a difference between DF eigenvalues is, however, quite a different matter. Assuming the functional for the ground-state energy $E[n]$ (see Eq.(1)) to be a differentiable function of particle number, it is not difficult to prove [103,104,105] the relation

$$\frac{\partial E}{\partial n_k} = e_k \tag{72}$$

where e_k is an exact DF eigenvalue of Eq.(60). The occupation numbers n_k are here defined as multipliers of the squares of the DF one-electron orbitals in the definition of the density (cf. Eq.(61)). The physical values of the occupation numbers are thus 0 or 1. In an extended system like a semiconductor, the orbitals ϕ_k are spread over the entire system. Physical intuition then suggests that the change in the effective potential of the Kohn-Sham equation (Eq.(60)) upon removal of one such orbital must be of order $1/N$ and thus negligible. Consequently, an eigenvalue will not change when charge is removed from the corresponding orbital and the above definitions (Eqs. (69) and (70)) together with Eq.(72) leads to the results

$$E_c = e_{N+1}(N+\delta) \tag{73}$$

$$E_v = e_N(N-1+\delta) \tag{74}$$

where δ is any number between zero and one, the subscripts numerate the eigenvalues and the numbers in parenthesis indicate the number of particles in the system. Chosing $\delta = 0$ in the first of these equations and $\delta = 1$ in the second gives

$$E_g = e_{N+1}(N) - e_N(N) \tag{75}$$

meaning that the band gap is given by the gap in the exact DF band structure.

Over the years, evidence has piled up to show that LD band structures routinely underestimate band gaps by a substantial amount and that several improvements on the LDA give similar deviations from experiment. [95] In fact, we now know that the gap is not given by the exact DF band structure. Consequently, there must be an error in the chain of arguments given above and will return to this point below.

Over the years, several many-body theoreticians claimed the DF band gap to be incorrect, *e.g.*, Sham and Inkson to mention just a few. No one was, however, able to prove their assertion although they gave several good arguments supporting it. The true band gap can be obtained from a Schrödinger-like one-particle equation in which the electronic self-energy appears as a non-local and energy dependent potential.[96] It was then, *e.g.*, argued that the eigenvalues of such a potential could never be reproduced by the local energy-independent potential of the exact DFT. It is not difficult to see that there is whole class of local potentials capable of producing the correct eigenvalue difference corresponding to the gap but whether any of these also give the correct ground-state density is, of course, a different matter. In 1983 Sham and Schlüter[106] started from basic many-body theory and arrived at an expression for the difference between the true gap and that of DFT. Their expression is somewhat ambiguous and, interpreted in a strict sense, it was later shown to be incorrect by Schönhammer and Gunnarsson.[107] In fact, the expression turned out to give a vanishing correction to the DF band gap. Interpreted in a more benevolent way, the expression essentially amounts to first-order perturbation theory for the difference between the full self-energy and the DF xc-potential. Unfortunately, the authors did not offer any solid evidence demonstrating their expression to remain finite in some systems. The fact that an approximate calculation for a model system gave a finite result[106] can not be considered as a proof.

From a different starting point, Perdew *et al.*[108,109,110] also discovered that there is a discontinuity in the xc-potential of DFT as function of the number of particles in the system. The first step was to consider the finite-temperature generalization of DFT by Mermin.[111] In this theory, fractional charges are introduced in natural and well defined way as opposed to the rather *ad hoc* fashion in which fractional occupation numbers were defined above. (In fact, the use of fractional occupation numbers in the exact DFT is a dubious procedure.) As long as the temperature T remains finite there are, of course, no singularities but at $T \to 0$ the xc-potential $v_{xc}(\mathbf{r}; N+\delta)$ as function of particle number $N+\delta$ can, in principle, be different depending on whether $\delta \to 0$ from above or from below. In particular, Perdew *et al.* discovered that the two Eqs. (73) and (74) still hold but only in the semi-open interval $0 < \delta \leq 1$. They also argued that the constant C defined by

$$C = \lim_{\delta \to 0^+} e_{N+1}(N+\delta) - e_{N+1}(N), \tag{76}$$

in principle, could be finite also in extended systems. Again, however, no solid evidence was offered in support of this view.

To our knowledge, there exist two proofs that the gap is not given by the exact DF gap. The first one appeared in Ref. 7 and will be described below. The second one is due to Schönhammer and Gunnarsson.[107] Their proof is based on exact results for discrete DF theory on a lattice with an on-site electron-electron interaction of the Hubbard type (cf. Section 6.3). They also demonstrated that the size of the correction to the DF gap is very sensitive to the detailed properties of the system. The same conclusion was reached by Almbladh and von Barth.[7]

Some people might feel uncomfortable about extrapolating results from discrete DF theory to the continuum case although there appears to be a complete analogy between the two cases. Anyway, we will here give a very simple proof that the DF gap is not the true gap.[7,112] Consider first a closed-shell atom like, *e.g.*, Be or Ne. As mentioned previously, the eigenvalue of the highest occupied exact DF orbital is equal to the negative of the ionization potential, $e_N(N) = -I$.[94] For a neutral system, the effective potential of the DFT always has a $-1/r$ -tail[94] which ensures that there is an infinity of bound Rydberg levels below the the continuum edge. Consequently, the lowest unoccupied DF eigenvalue $e_{N+1}(N)$ is positioned a finite energy C below the vacuum level. On the other hand, many of these closed-shell atoms have no electron affinities meaning that, for these atoms, the affinity level is at the continuum edge ($A=0$), *i.e.*, a finite amount C above the lowest unoccupied DF eigenvalue, $A=e_{N+1}(N)+C$. Let us then form a solid of such atoms by placing them on a regular lattice with a very large lattice parameter such that the interatomic interactions are negligible. The true band gap of such a solid is obviously the atomic ionization potential, $I=I-A$. The DF eigenvalues will broaden into bands of negligible width and they might be shifted. In the limit of very large separation, however, the density of the solid must be a simple superposition of atomic densities. By definition, the DF orbitals correctly sum up to this density and consequently the effective Kohn-Sham potential (Eq.(60)) must be a superposition of the exact atomic Kohn-Sham potentials. The formation of the solid might be associated with a constant potential shift of no consequence to the difference between eigenvalues. Therefore, the exact DF band gap of the sparse solid is given by the difference $e_{N+1}(N)-e_N(N)$ between the lowest unoccupied and the highest occupied DF eigenvalues of the isolated atom. The fact that this difference is a finite quantity C smaller than the true gap completes our simple proof that, for the chosen sparse solid, the two gaps are different. Of course, in order to disprove something, one counterexample is sufficient.

Before ending this section, we notice that there is no reference to a change in the number of particles in our proof above. Thus, although the generalization of DFT to fractional number of electrons in finite system can sometimes be helpful, no new physics is to be gained by this procedure. In other words, there is really no need to talk about derivative discontinuities in DF theory.

6.6. Acknowledgements.

I would like to thank the organizers of the miniworkshop on "Methods of Electronic Structure Calculations" (August 1992) for a most enjoyable and rewarding week at the ICTP in Trieste. Thanks are also due to Sverre Svendsen for critical reading of the manuscript and to John Perdew for an interesting discussion. It must be understood that neither of these fine men are responsible for any errors or misconceptions that might be revealed by this manuscript. Finally, I am grateful to the Swedish Natural Science Research Council for financial support.

7. References.

1. A. K. Rajagopal, Advances in Chemical Physics, edited by I. Prigogine and S. A. Rice, **41** , 59, Wiley, New York (1980).
2. A. R. Williams and U. von Barth, *Applications of Density-Functional Theory to Atoms, Molecules and Solids* in *Theory of the Inhomogeneous Electron Gas*, edited by S. Lundqvist and N. H. March, Physics of Solids and Liquids Series, Plenum, New York 1983.
3. U. von Barth, *An Overview of Density-Functional Theory* in *Many-Body Phenomena at Surfaces* edited by D. Langreth and H. Suhl, Academic Press, New York 1984, p. 3.
4. U. von Barth, *Density-Functional Theory for Solids* in *The Electronic Structure of Complex Systems* edited by P. Phariseau and W. Temmerman, NATO ASI Series B: Physics, **113**, p. 67, Plenum, New York 1984.
5. J.P. Dahl and J. Avery, editors of *Local-Density Approximations in Quantum Chemistry and Solid State Physics*, Plenum, New York 1984.
6. J. Callaway and N.H. March, Solid State Phys. **38**, 135 edited by H. Ehrenreich, F. Seitz, and D. Turnball, Academic Press, New York 1984.
7. C-O. Almbladh and U. von Barth, *Density-Functional Theory of Excitation Energies* in *Density-Functional Methods in Physics*, edited by R.M. Dreizler and J. de Providencia, NATO ASI Series B: Physics, **123**, p. 209, Plenum, New York 1985.
8. U. von Barth, Chemica Scripta, **26**, 449 (1986).
9. R.Erdahl and V.H. Smith, editors of *Density Matrices and Density Functionals*, Proceedings of the A.J. Coleman Symposium, D. Reidel Publ. Co., Dordrecht 1987.
10. R.O. Jones and O. Gunnarsson, Rev. Mod. Phys. **61**, 689 (1989).
11. R.G. Parr and W. Yang, *Density-Functional Theory of Atoms and Molecules*, Oxford University Press, New York 1989.
12. R.M. Dreizler and E.K.U. Gross, *Density-Functional Theory: An Approach to the Quantum Many-Body Problem*, Springer-Verlag, Berlin 1990.
13. J.K. Labanowski and J.W. Andzelm, editors of *Density-Functional Methods in Chemistry*, Springer-Verlag, New York 1991.
14. P. Hohenberg and W. Kohn, Phys. Rev. **136**, B864 (1964).
15. W. Kohn and L. J. Sham, Phys. Rev. **140**, A1133 (1965).
16. M. Levy, Proc. Natl. Acad. Sci. USA, **76**, 6062 (1979).
17. C.-O. Almbladh, Technical Report, Lund University (1973).
18. O. Gunnarsson and B.I. Lundqvist, Phys. Rev. B **13**, 4274 (1976).
19. O. Gunnarsson, M. Jonsson, and B.I. Lundqvist, Phys. Rev. B **20**, 3136 (1979).

20. D.C. Langreth and J. Perdew, Phys. Rev. B **21**, 5469 (1980).

21. J.E. Müller, R.O. Jones, and J. Harris, J. Chem. Phys. **79**, 1874 (1983).

22. D.C. Langreth and M.J. Mehl, Phys. Rev. B **28**, 1809 (1983).

23. D.C. Langreth and M.J. Mehl, Phys. Rev. B **29**, 2310 (1984).

24. C.D. Hu and D.C. Langreth, Physica Scripta **32**, 391 (1985).

25. U. von Barth and R. Car, *Ground State Properties of Bulk Si as a Test of the Langreth-Perdew-Mehl Scheme for Exchange and Correlation*, unpublished work (1982).

26. V.L. Muruzzi, J.F. Janak, and A.R. Williams, *Calculated Electronic Properties of Metals*, Pergamon, New York (1978).

27. J.F. Janak, V.L. Muruzzi, and A.R. Williams, Phys. Rev. B **12**, 1257 (1975).

28. J.F. Janak and A.R. Williams, Phys. Rev. B **14**, 4199 (1976).

29. V.L. Muruzzi, A.R. Williams, and J.F. Janak, Phys. Rev. B **15**, 2854 (1977).

30. A.D. Becke, Phys. Rev. A **33**, 2786 (1986).

31. A.D. Becke, J. Chem. Phys. **84**, 4524 (1986).

32. A.R. Williams, J. Kübler, and C.D. Gelatt, Phys. Rev. B **19**, 6094 (1979).

33. C.D. Gelatt, A.R. Williams, V.L. Muruzzi, Phys. Rev. B **27**, 2005 (1983).

34. J.F. Janak, Phys. Rev. B **16**, 255 (1977).

35. J.F. Janak, Solid State Commun. **25**, 53 (1978).

36. J.F. Janak, Phys. Rev. B **20**, 2206 (1979).

37. M. Levy and J.P. Perdew, J. Chem. Phys. **84**, 4519 (1986).

38. K. Terakura, T. Oguchi, A.R. Williams, and J. Kübler, Phys. Rev. B **30**, 4734 (1984).

39. C.S. Wang, B.M. Klein, and H. Krakauer, Phys. Rev. Lett. **54**, 1852 (1985).

40. R.W. Godby and R.J. Needs, Phys. Rev. Lett. **62**, 1169 (1989).

41. O. Gunnarsson and R.O. Jones, Phys. Rev. B **31**, 7588 (1985).

42. J.C. Slater, *Quantum Theory of Atomic Structure*, Vol. I, McGraw-Hill, New York (1960).

43. F.W. Kutzler and G.S. Painter, Phys. Rev. B **43**, 6865 (1991).

44. J.P. Perdew and Wang Yue, Phys. Rev. B **33**, 8800 (1986).

45. J.P. Perdew, Phys. Rev. B **33**, 8822 (1986).

46. O. Gunnarsson and P. Johansson, Int. J. Quant. Chem. **X**, 307 (1976).

47. R. Merkle, A. Savin, and H. Preuss, J. Chem. Phys. **97**, 9216 (1992).

48. P. Mlynarski and D.R. Salahub, Phys. Rev. B **43**, 1399 (1991).

49. A. Savin, Int. J. Quant. Chem.: Quant. Chem. Sympos. **22**, 59 (1988).

50. J. Bernholc and N.A.W. Holzwarth, Phys. Rev. Lett. **50**, 1451 (1983).

51. M.M. Goodgame and W.A. Goddard III, Phys. Rev. Lett. **48**, 135 (1982).

52. D. Pines and P. Nozieres, *The Theory of Quantum Liquids*, Benjamin, New York (1966).

53. L.J.Sham, in *Computational Methods in Band Theory*, edited by P.M. Marcus, J.F. Janak, A.R. Williams, Plenum, New York (1971).

54. J.D. Talman and W.F. Shadwick, Phys. Rev. A **14**, 36 (1976).

55. L. Kleinman and S. Lee, Phys. Rev. B **37**, 4634 (1988).

56. F. Herman, J.P. van Dyke, and I.B. Ortenburger, Phys. Rev. Lett. **22**, 807 (1969).

57. S.-K. Ma and K.A. Brueckner, Phys. Rev. **165**, 18 (1968).

58. M. Rasolt and D.J.W. Geldart, Phys. Rev. B **13**, 1477 (1976).

59. M. Rasolt, Phys. Rev. B **16**, 3234 (1977).

60. U. von Barth and A.C. Pedroza, Physica Scripta **32**, 353 (1985).

61. P. Bagno, O. Jepsen, and O. Gunnarsson, Phys. Rev. B **40**, 1997 (1989).

62. C.D. Hu and D.C. Langreth, Phys. Rev. B **33**, 943 (1986).

63. J.P. Perdew, Phys. Rev. Letters **55**, 1665 (1985).

64. J.P. Perdew, Physica B **172**, 1 (1991).

65. E.K.U. Gross and R.M. Dreizler, Z. Phys. A **302**, 103 (1981).

66. A.D. Becke, J. Chem. Phys. **85**, 7184 (1986).

67. A.D. Becke, Phys. Rev. A **38**, 3098 (1988).

68. S.H. Vosko and Engel, Phys. Rev. B **42**, 4940 (1990).

69. J.P. Perdew, J.A. Chevary, S.H. Vosko, K.A. Jackson, M.R. Pederson, D.J. Singh, and C. Fiolhais, Phys. Rev. B **46**, 6671 (1992).

70. D.M. Ceperley and B.J. Alder, Phys. Rev. Lett. **45**, 566 (1980).

71. X.J. Kong, C.T. Chan, K.M. Ho, and Y.Y. Ye, Phys. Rev. B **42**, 9357 (1990).

72. T.C. Leung, C.T. Chan, and B.N. Harmon, Phys. Rev. B **44**, 2923 (1991).

73. A. Garcia, Ch. Elsässer, Jing Zhu, S.G. Louie, and M.L. Cohen, Phys. Rev. B **46**, 9829 (1992).

74. B. Barbiellini, E.G. Moroni, and T. Jarlborg, J. Phys. Condens. Matter **2**, 7597 (1990).

75. M. Körling and J. Häglund, Phys. Rev. B **45**, 13293 (1992).

76. F.W. Kutzler and G.S. Painter, Phys. Rev. Lett. **59**, 1285 (1987).

77. A.D. Becke, J. Chem. Phys. **96**, 2155 (1992).

78. H. Stoll, C.M.E. Pavlidou, and H. Preuss, Theoret. Chim. Acta. **49**, 143 (1978).

79. J.A. Pople, M. Head-Gordon, D.J. Fox, K. Raghavachari, and L.A. Curtiss, J. Chem. Phys. **90**, 5622 (1989).

80. L.A. Curtiss, C. Jones, G.W. Trucks, K. Raghavachari, and J.A. Pople, J. Chem. Phys. **93**, 2537 (1990).

81. A.D. Becke, J. Chem. Phys. **97**, 9173 (1992).

82. A.D. Becke, J. Chem. Phys. **98**, 1372 (1993).

83. A.D. Becke, J. Chem. Phys. **98**, 5648 (1993).

84. A.D. Becke, J. Chem. Phys. **88**, 1053 (1987).

85. C.L. Lee, W. Yang, and R.G. Parr, Phys. Rev. B **37**, 785 (1988).

86. B. Miehlich, A. Savin, H. Stoll, and H. Preuss, Chem. Phys. Lett. **157**, 200 (1089).

87. B.G. Johnsson, P.M.W. Gill, and J.A. Pople, J. Chem. Phys. **98**, 5612 (1993).

88. R. Merkle, A. Savin, and H. Preuss, Chem. Phys. Lett. **194**, 32 (1992).

89. A.K. Theophilou, J. Phys. C **12**, 5419 (1979).

90. T. Ziegler, A. Rauk, and E.J. Baerends, Theor. Chim. Acta **43**, 262 (1977).

91. U. von Barth, Phys. Rev. A **20**, 1693 (1979).

92. V. Peuckert, J. Phys. C **11**, 4945 (1978).

93. E. Runge and E.K.U. Gross, Phys. Rev. Lett. **52**, 997 (1984).

94. C.-O. Almbladh and U. von Barth, Phys. Rev. B **31**, 3231 (1985).

95. M.S. Hybertsen and S.G. Louie, Comments Cond. Mat. Phys. **13**, 223 (1987).

96. L. Hedin and S. Lundqvist, in *Solid State Physics*, ed. by F. Seitz, D. Turnbull, and H. Ehrenreich, Vol. **23**, p. 1, Academic Press, New York (1969).

97. J.C. Wheatley, Rev. Mod. Phys. **47**, 415 (1975).

98. W. Kohn, Phys. Rev. Lett. **2**, 393 (1959).

99. D. Mearns, Phys. Rev. B **38**, 5906 (1988).

100. K. Schönhammer and O. Gunnarsson, Phys. Rev. B **37**, 3128 (1988).

101. O. Gunnarsson and K. Schönhammer, Phys. Rev. Lett. **56**, 1968 (1986).

102. R.W. Goodby and R.J. Needs, Phys. Rev. Lett. **62**, 1169 (1989).

103. J.C. Slater and J.H. Wood, Int. J. Quant. Chem. Suppl. **4**, 3 (1971).

104. C.-O. Almbladh and U. von Barth, Phys. Rev. B **33**, 3307 (1976).

105. J. Janak, Phys. Rev. B **18**, 7165 (1978).

106. L.J. Sham and M. Schlüter, Phys. Rev. Lett. **51**, 1888 (1983).

107. K. Schönhammer and O. Gunnarsson, J. Phys. C: Solid St. Phys. **20**, 3675 (1987).

108. J.P. Perdew, R.G. Parr, M. Levy, and J.L. Balduz, Phys. Rev. Lett. **49**, 1691 (1982).

109. J.P. Perdew and M. Levy, Phys. Rev. Lett. **51**, 1884 (1983).

110. J.P. Perdew, *What do the Kohn-Sham Orbital Energies Mean?* in *Density-Functional Methods in Physics*, edited by R.M. Dreizler and J. de Providencia, NATO ASI Series B: Physics, **123**, p. 265, Plenum, New York 1985.

111. N.D. Mermin, Phys. Rev. **137 A**, 1441 (1965).

112. In 1981 when Ref. 2 was initiated, we were well aware of the contradiction between this argument and that leading to the Eqs. (73) and (74) in the beginning of this subsection. In the general belief that physical quantities rarely show sigularities, we, at the time, decided to favor the latter - a choice which is reflected in Ref. 2. That this was the wrong choice became apparent when Perdew and his collaborators demonstrated (Ref. 108) that the xc-potential of the DFT is often a very unphysical and singular quantity (see also Ref. 7).

Exact Muffin-Tin Orbital Theory

O.K. Andersen, O. Jepsen, and G. Krier
Max-Planck-Institut für Festkörperforschung,
Heisenbergstr.1, D-70569 Stuttgart,
Federal Republic of Germany

1. INTRODUCTION

There is a need for a first-principles tight-binding electronic structure method which can be used as basis for many-electron as well as for molecular-dynamics calculations. In the present lecture notes we shall present such a method, a new version of the linear-muffin tin orbital (LMTO) method.

1.1 Overview

First of all, we shall explain how this method solves *Schrödingers equation,*

$$\left[-\nabla^2 + V(\mathbf{r})\right] \Phi_j(\mathbf{r}) = E_j \Phi_j(\mathbf{r}), \tag{1.1}$$

(in atomic Rydberg units) for one electron moving in the three-dimensional condensed-matter potential $V(\mathbf{r})$. This potential is local and energy independent, and will usually be the ground state potential of density functional theory. The potential may be a pseudo potential, but is usually the all-electron potential. Secondly, we shall show how the output *electron density,*

$$\rho(\mathbf{r}) = \sum_{j \in occ} |\Phi_j(\mathbf{r})|^2, \tag{1.2}$$

is obtained in such a form that it can be used to construct the input potential $V(\mathbf{r})$ for a next iteration towards charge (and spin) self-consistency. The main problem is here to solve *Poisson's equation,*

$$-\nabla^2 V_C(\mathbf{r}) = 8\pi \left[\rho(\mathbf{r}) - \sum_R Z_R \delta(r_R)\right] \tag{1.3}$$

for the electronic and protonic charge densities. $\mathbf{r}$ are the electronic- and $\mathbf{R}$ the nuclear positions. The relative electronic positions are $\mathbf{r}_R \equiv |\mathbf{r} - \mathbf{R}|$, with the distances $r_R \equiv |\mathbf{r}_R|$ from the nuclei.

The one-electron potential $V(\mathbf{r})$ in (1.1) is deep only inside the atoms, where it is nearly spherically symmetric, and it varies slowly between the atoms. This may be seen in Fig. 1.1 which gives equipotential contours in steps of 0.25 Ry for crystalline silicon in the (110)-plane. The lowest contour shown (inside the Si atom) is –4.25 Ry, the value at the saddle point between two Si nearest neighbors

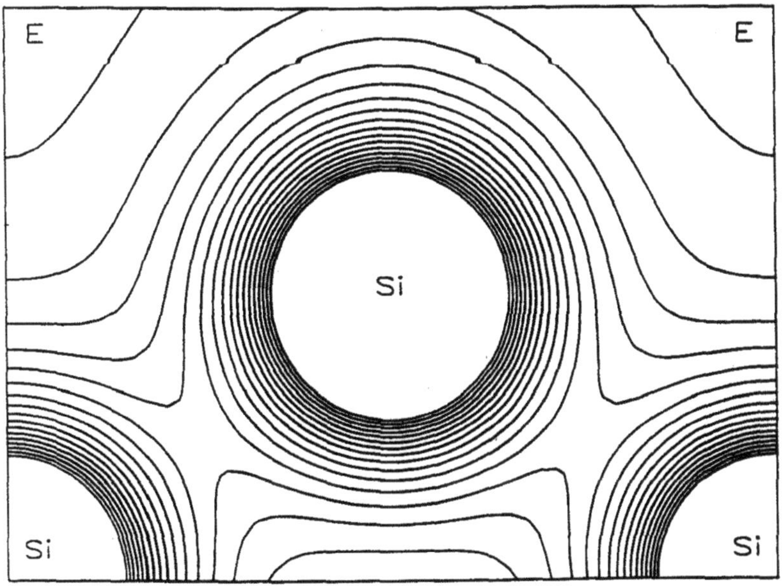

Figure 1.1 One electron potential for crystaline silicon

is −1.3 Ry, and the highest contour is −0.25 Ry. In Fig. 1.2 the potential is plotted along a straight line from the midpoint between the two nearest neighbors, through a Si atom, over an empty site (E), and to the midpoint between two empty sites. The dot-dashed curve is the *atomic-spheres approximation* (ASA), or *overlapping muffin-tin approximation,* to the full potential shown by the full curve. The ASA is the approximation we shall use for the construction of the *muffin-tin orbitals* (MTO's), and it consists of a superposition of the spherically symmetric Si- and E-potential wells, $v_{\mathrm{Si}}(r_{\mathrm{Si}})$ and $v_E(r_E)$, shown by the dashed curves. The Si well is attractive because the potential from the protons in the Si-nucleus is only incompletely screened by the repulsion from the electrons. The empty well is repulsive because it is empty of protons and contains some fraction of an electron. The common zero of these two types of MT-wells is V_0, and this muffin-tin floor is indicated by the horizontal dashed line in the figure. The MT-potential is thus

$$V_{mt}(\mathbf{r}) = \sum_R v_R(r_R), \tag{1.4}$$

with $V_0 \equiv 0$ and with $v_R(r_R)$ vanishing outside the MT-sphere centered at $\mathbf{R}$ with radius s_R (the zero of potential in the figures is chosen differently. From the figures we see that the ASA is a good approximation for the potential. For more complicated low- or no-symmetry structures it is not trivial to find those sphere

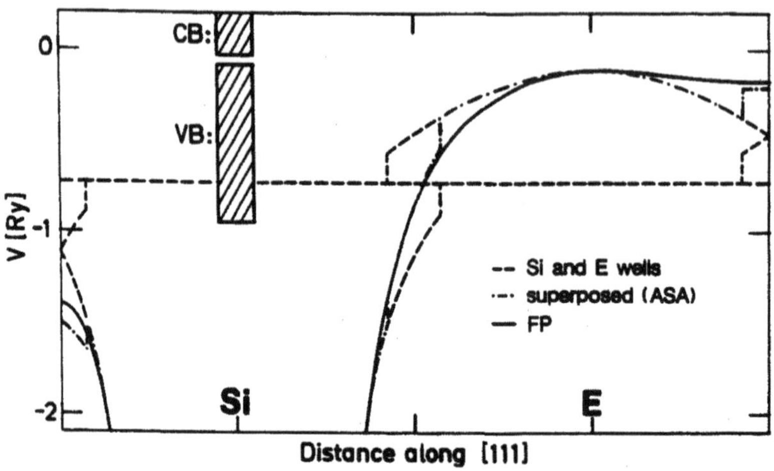

Figure 1.2 Silicon potential and its ASA.

radii and the empty-sphere positions which yield the best possible ASA to the full potential, but there exists an automatic procedure which works very well [1].

For the calculation of ground-state properties, the relevant range of one-electron energies E_j extends from the energy where an electron can move between atoms, and to the Fermi level. In silicon, this range is from the bottom of the valence band (denoted VB in Fig. 1.2) to the gap between the valence and conduction bands (CB). In general the energy range is at the order of a Rydberg, and extends on either side of the MT zero V_0. The center of gravity E_ν of the occupied bands is usually ~0.3 Ry *above* V_0. Several Rydbergs below the valence band are core states whose contribution to the charge density can usually be approximated by the core-density calculated for an atom. The valence charge density in crystalline silicon obtained from (1.2) as the output of a tight-binding LMTO-ASA calculation is shown in Fig. 1.3 and is indistinguishable from charge density obtained from the full potential.

LMTO's are one-electron basis functions and the key development in the present lecture notes (Section 2) is the construction of a complete set of solutions to the Helmholtz *wave equation*

$$\left[\nabla^2 + E\right] \Psi(E, \mathbf{r}) = 0 \tag{1.5}$$

in a region between non-touching hard-spheres, so-called *unitary spherical waves* $\psi_{Rlm}(E, \mathbf{r} - \mathbf{R})$. These functions are *localized* in r-space and in energy, but otherwise similar to plane waves $\exp\{i(\mathbf{k} + \mathbf{G}) \cdot \mathbf{r}\}$, the most common basis functions in solid-state physics. Like the plane-wave set the unitary spherical waves will be used to expand both the electronic wave functions, the charge density, and the

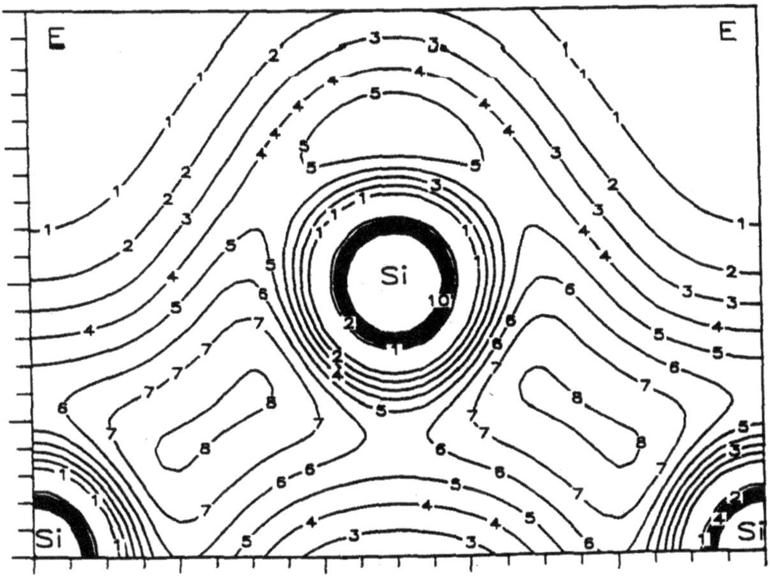

Figure 1.3 Silicon charge density.

potential (1.3) in the interstitial region. I contrast to the plane waves, the unitary waves diverge at the sites $\mathbf{R}$ and therefore have to be substituted (augmented) by numerical radial functions inside the spheres. The resulting one-electron functions are, first, the *kinky partial waves* $\phi_{Rlm}(E, \mathbf{r}-\mathbf{R})$, which are the localized equivalents of Slater's augmented plane waves (APW's). In our new formalism, they will play the same role as did the partial waves previously inside the MT-spheres. In Section 4 the kinky partial waves are then developed into smooth *muffin-tin orbitals* (MTO's) $\chi_{Rlm}(E, \mathbf{r}-\mathbf{R})$, whose energy denpendence is merely of second order from some chosen energy E_ν. Neglect of this second-order term in the basis set, finally leads to the *linear muffin-tin orbitals* (LMTO's) $\chi_{Rlm}(\mathbf{r}-\mathbf{R})$. The augmentation of the unitary spherical waves is tailor-made to the potential and, hence, the LMTO basis can be very small. Most importantly, the set is localized, as is necessary in order to obtain an electronic-structure method in which the effort increases linearly with the number of independent atoms (order-N method).

The LMTO method has been in use and under development for twenty years. These lecture notes are written for those who have little, or no previous knowledge of the method, and for those experts who are willing to learn about the new features at the expense of having to absorb yet another formulation. The following subsection briefly summarizes the existing versions of the method and characterizes the new one.

1.2 LMTO Versions

1.2.1 Standard LMTO-ASA

This original version [2, 3, 4, 5, 6] is simple and fast and is within the ASA closely related to the Korringa-Kohn-Rostoker (KKR) multiple-scattering method for MT-potentials [7]. It uses a small basis set, specifically *one* radial function per site ($\mathbf{R}$) and angular momentum ($lm \equiv L$) and therefore typically 9 LMTO's per site. With respect to the overlapping MT-potential used for its generation, the LMTO basis is complete to order $(E - E_\nu)^1$ inside the spheres and to order $(E - E_\nu)^0$ in the interstitial region. The overlap matrix $\mathrm{O}_{RL,R'L'}$ and the one-electron Hamiltonian in the ASA $\mathrm{H}_{RL,R'L'}$ are obtained from *potential parameters* and a *structure matrix* $\mathrm{S}^0_{RL,R'L'}$, plus its energy derivative for the so-called combined correction to the ASA. The potential parameters C_{Rl}, Δ_{Rl}, γ_{Rl}, and p_{Rl} may be calculated from the values and slopes of the radial wave functions $\varphi_{Rl}(E,r)$ at the spheres s_R, and for two neighboring energies E_ν and $E_\nu + dE_\nu$.

The energy-expansion parameters E_ν for the radial wave functions and for the wave-equation solution in the interstitial region are usually chosen to be different and are respectively named $E_{\nu Rl}$ and κ^2_ν. For the interstitial region one usually, but not always [3], chooses $\kappa^2_\nu \equiv 0$ whereby the wave equation degenerates to the Laplace equation. In this case, the LMTO envelopes behave simply like

$$\chi^0_{Rlm}(\mathbf{r} - \mathbf{R}) \propto |\mathbf{r} - \mathbf{R}|^{-l-1} Y_{lm}\left(\widehat{\mathbf{r} - \mathbf{R}}\right), \tag{1.6}$$

and the structure matrix like:

$$\mathrm{S}^0_{Rlm,R'l'm'} \propto |\mathbf{R} - \mathbf{R}'|^{-l-l'-1} Y_{l+l',m-m'}\left(\widehat{\mathbf{R} - \mathbf{R}'}\right). \tag{1.7}$$

Although analytically simple, the LMTO's (1.6) have *long range,* and this limits the method to crystals where the structure matrix must be evaluated in the Bloch representation,

$$\mathrm{S}^0_{Rlm,R'l'm'}(\mathbf{k}) = \sum_T \exp\left\{i\mathbf{k}\cdot(\mathbf{R} - \mathbf{R}' - \mathbf{T})\right\} \mathrm{S}^0_{Rlm,(R'+T)l'm'}\,. \tag{1.8}$$

using the Ewald summation procedure. In (1.8) R and R' label sites in the primitive cell and $\mathbf{T}$ are the lattice translations. However, in the Bloch representation the computational effort increases with the number of sites in the cell as $\sim N^3$. One might think that the way out of this problem of long range would be to use a negative κ^2_ν, but the value required to yield useful localization leads to an unacceptable $(E - \kappa^2_\nu)^0$-error.

An even weaker point of the standard LMTO-ASA version is that the output *charge density* is *neglected* between the spheres and *spheridized* in the spheres, that is, the straight ASA without combined correction is used for the charge density. As consequences, space must be filled with slightly overlapping spheres, and the calculated total energies are useful only for close-packed high-symmetry configurations; pressures can be calculated, but not forces.

1.2.2 Tight-Binding LMTO-ASA

It was realized that the LMTO $\kappa_\nu^2 \equiv 0$ basis set (1.6) can be ***transformed exactly*** into for instance *orthogonal* [8], *tight-binding* (TB) [9, 10], or *minimal* [11] basis-sets $|\chi^\alpha\rangle$. These transformations are of the form [12]:

$$|\chi^\alpha\rangle = |\chi^0\rangle (1 + \alpha S^\alpha), \text{ where } S^\alpha = S^0 \left(1 - \alpha S^0\right)^{-1}, \qquad (1.9)$$

and are called *screening* transformations. In the tight-binding representation, the structure matrix S^α can be generated *in real space* by for each site R inverting the positive definite matrix $\alpha_{Rl}^{-1}\delta_{RR'}\delta_{LL'} - S^0_{RL,R'L'}$ for the cluster of 20-40 nearest neighbors R'. The set of screening constants $\{\alpha_{Rl}\}$ specify the representation. Sets of screening constants which yield short range had to be found empirically.

With the choice $\alpha_{Rl} \equiv \gamma_{Rl}$ the LMTO's become orthonormal, albeit in the ASA and with neglect of the small parameters p_{Rl}. These orthonormal LMTO's are generalized Wannier functions and, of course, can have long range. However, the Hamiltonian in the orthonormal representation may be expressed as a power series in the following *two-center tight-binding Hamiltonian:*

$$h^\alpha_{RL,R'L'} \equiv (c^\alpha_{Rl} - E_{\nu Rl})\,\delta_{RR'}\delta_{LL'} + \sqrt{d^\alpha_{Rl}}S^\alpha_{RL,R'L'}\sqrt{d^\alpha_{R'l'}}, \qquad (1.10)$$

with coefficients proportional to $\alpha_{Rl} - \gamma_{Rl}$. In (1.10), c and d are potential parameters which may be derived from the slope and amplitude of the corresponding radial wave function $\varphi(E_\nu, r)$ at the sphere. The more terms are included in this power series in h^α, the wider is the energy range around E_ν in which the eigenvalues equal the one-electron energies E_j in (1.1). This power series makes *ab initio* calculations for structurally disordered condensed matter possible, for instance with Haydock's recursion method [13, 14, 15].

The orthonormal representation has proven useful for self-consistent Green-function calculations for point defects [16], for *ab initio* constructions of Anderson impurity models [17] and multi band Hubbard models (LDA+U) [18], and simply for analyzing complicated band structures by projection onto an orthonormal basis [19].

The tight-binding (TB) representation has been exploited for the development of a simple and accurate coherent-potential approximation for substitutionally disordered alloys [20] and to treat extended defects such as surfaces and interfaces [21, 22, 23]. For large systems, the normal self-consistency procedure becomes unstable due to the long range of the Coulomb interaction, but in the two-dimensional Green-function technique of Skriver and Rosengaard [22] this problem is solved by calculation of the linear response matrix. The TB and orthogonal representations also allows Hartree-Fock [24] and self-interaction corrected (SIC) density functional calculations [25, 26], as well as correlation-corrected band-structure calculations [27] to be performed. The TB-representation has finally been used for plotting the full, non-spheridized electron density resulting from a self-consistent LMTO-ASA calculation [12, 28] like in Fig. 1.3.

Minimal basis sets may be constructed by transformation to such representations where the tails of the LMTO's solve Scrödingers equation at E_ν for those Rl-channels which are to be deleted from the basis set [10, 11, 29, 19].

Although the LMTO's only constitute the basis set and any non-ASA perturbation to the Hamiltonian may in principle be included, it has turned out to be difficult to represent the full output charge density in a form suitable for calculating for instance the total energy [30]. As a consequence, the TB-LMTO method has been used exclusively with the straight ASA for the charge density. Most recently, *full-charge density* programs based on the TB-representation are under construction [31], but for molecular-dynamics calculations, where the sphere-packing cannot always be dense without causing large errors of the kinetic energy in the overlap regions, the κ_ν^2=0–approximation will hardly suffice.

1.2.3 Full-Potential LMTO

Some *full-potential* (FP) versions [32, 33, 34, 30] expand the smooth part of the charge density in plane waves and other versions [35, 36, 37] expand it in spherical waves with negative energies, that is, in decaying Hankel functions. Since the interstitial region has a complicated topology and the spherical waves do not constitute an orthogonal set, the expansion procedure may be cumbersome. Here, the techniques developed by Methfessel [36, 37] brought real advance. Poisson's equation (1.3) is now relatively simple to solve and the total energies, and even forces [37, 38] can be accurately calculated. A linear-response FP-LMTO technique for performing phonon calculations for general wave-vectors was developed by Savrasov [39].

All these full-potential versions use unscreened, *long-ranged* LMTO's. This prevents their use as order-N methods and, for molecular-dynamics calculations the time for calculating structure constants is prohibitively long except for small molecules [36]. The MT-spheres used to define the LMTO basis functions are chosen *not* to overlap in order to avoid uncertainty about what exactly happens in regions of overlap. This increases the importance of the interstitial region to such a degree that it becomes necessary to use large, *multiple-kappa* basis sets. Such a basis typically has 2-3 *negative* κ_ν^2-values per site and angular momentum.

Although highly accurate and computationally efficient for static calculations, existing FP-LMTO schemes lack the extreme efficiency and simplicity of the LMTO-ASA method with its close relation to the KKR formalism and its screening transformations which provide the possibility of constructing small, simple basis sets specific for the application. Moreover, the inclusion of the full-potential in the one-electron Hamiltonian, makes the form of the FP-Hamiltonian matrix far more complicated than (1.10).

1.2.4 New Version

Apart from combining the desirable features of the existing TB-LMTO and FP-LMTO versions, the new LMTO method has some additional advantages:

First of all, the new formalism no longer needs the ASA in order to be equivalent to the KKR multiple-scattering formalism. In fact, in Section 2 we derive an exact, generalized KKR formalism whose structure constants can be chosen to have *short-range* and to be *nearly independent of energy* for $E \equiv \kappa^2 < \kappa_1^2 \sim 1$ Ry. In a preliminary study [40] where the separation of the energy dependencies inside (E) and outside (κ^2) the spheres was maintained, as is customary in MTO-theory, the structure constants $S^{\alpha}_{RL,Rl'}(\kappa^2)$, the potential functions $P^{\alpha}_{Rl}(E,\kappa^2)$, and the screening constants $\alpha_{Rl}(\kappa^2)$ turned out to have far stronger κ^2-dependencies than expected. As a consequence, the screening constants could not be parameterized in a satisfactory way except for mono atomic close-packed structures. The new formalism shows that the $\alpha(\kappa^2)$'s can be interpreted in terms of energy-independent hard-spheres, which just have to be somewhat smaller than touching in order to yield structure constants with weak energy dependence and short range. The unitary spherical waves are the localized impurity solutions for this hard-sphere solid and κ_1^2 is the bottom of its continuum. The new structure matrix, which is a renormalized version of the one in Ref. [40], gives the slopes of the unitary spherical waves on the hard spheres.

Secondly, the new LMTO basis is complete to order $(E - E_\nu)^1$ inside *as well as outside* the spheres, with respect to a slightly overlapping MT-potential. Due to these developments, the new LMTO set is far more accurate in the interstitial region than the conventional set and, for this reason, a single-kappa set suffices (Section 4).

Thirdly, it was shown already in Ref. [40] that the LMTO method can deal accurately with up to about 20% linear overlap of the MT-spheres. This overlap is defined as:

$$\omega_{RR'} \equiv \frac{s_R + s_{R'}}{|\mathbf{R} - \mathbf{R}'|} - 1. \tag{1.11}$$

For reasons of space, we shall not repeat this proof and further developments [1] in the present notes.

Finally, in the interstitial region not only the LMTO's, but also the full charge density and potential are expanded in terms of the unitary spherical waves and their energy derivatives (Section 5).

2. LOW-ENERGY MULTIPLE-SCATTERING THEORY

In this section we shall derive a generalization of the exact Korringa-Kohn-Rostoker multiple-scattering formalism [7] in which the structure constants may be chosen to have *short range* when the energy is low.

2.1 Basic Concepts

What makes it possible to solve the Schrödinger equation (1.1) for a MT-potential *exactly,* is that Schrödingers differential equation *separates inside* the spheres and degenerates to the Helmholtz *wave equation* (1.5) in the *interstitial* region. For energies E_j where solutions Φ_j exist, these may therefore be expressed in terms of the *partial waves* φ_{Rlm} joined continuously and differentiably at the spheres of radii s_R by a *wave-equation solution* Ψ_j, that is,

$$\Phi_j(\mathbf{r}) = \Psi_j(E_j, \mathbf{r}) + \sum_{R=1}^{\infty}\sum_{l=0}^{\infty}\sum_{m=-l}^{l} \varphi_{Rlm}(E_j, \mathbf{r}_R)\, u_{Rlm,j}. \tag{2.1}$$

On the right-hand side of (2.1) and all such *one-center expansions* in the following, the wave-equation solution Ψ_j is supposed to be truncated outside the interstitial region and each partial wave φ_{Rlm} is supposed to be truncated outside its own MT-sphere.

2.1.1 Partial Waves

The partial wave $\varphi_{Rlm}(E, \mathbf{r})$ is the regular solution for energy E and angular-momentum lm of Schrödingers differential equation for the spherically symmetric potential-well $v_R(r)$. No boundary condition is installed at the MT-sphere, but E is a continuous parameter which vary in the range of interest. The solutions factorize into products of *radial* and *angular* functions, *viz.:*

$$\varphi_{Rlm}(E, \mathbf{r}) \equiv \varphi_{Rl}(E, r)\, Y_{lm}(\mathbf{r}/r), \tag{2.2}$$

where $Y_{lm}(\mathbf{r}/r)$ is a spherical harmonic (phase-convention of Condon and Shortley), because

$$\left[-\nabla^2 + v(r)\right]\varphi_l Y_{lm} = \left[-r^{-1}(r\varphi_l)'' + l(l+1)r^{-2}\varphi_l + v(r)\varphi_l\right] Y_{lm}. \tag{2.3}$$

Here, and in the following, a prime denotes a radial derivative, i.e., $' \equiv \partial/\partial r$. Moreover, we have dropped the subscript R which labels the well-type. $l(l+1)r^{-2}$ is the centrifugal potential. From (2.3) we see, that the radial function $\varphi_l(E,r)$ is that solution of the l'th *radial* Schrödinger equation

$$[r\varphi_l(E,r)]'' = \left[l(l+1)r^{-2} + v(r) - E\right] r\varphi_l(E,r) \tag{2.4}$$

which is regular at the origin. In actual calculations the potential $v(r)$ is known numerically on a radial mesh and the radial equations are solved by outwards numerical integration, starting with the function $r\varphi_l$ which behaves like r^{l+1}. In fact, not the radial Schrödinger equations, but the radial scalar-relativistic Dirac equations are usually solved.

In the following we shall use the notation (2.2) for all functions $f_L(\mathbf{r}) = f_l(r)\,Y_L(\hat{\mathbf{r}})$ which are products of a radial function and a spherical harmonic, and also for functions which are not. Here, L is short-hand for lm.

2.1.2 Matching Condition

If the partial waves are considered to be functions of energy, then the eigenvalues E_j are those energies for which coefficients $u_{RL,j}$ can be found such that the partial-wave expansions (2.1) from the various spheres join continuously and differentiably onto a wave-equation solution in the interstitial region. The information about the MT-well at $\mathbf{R}$ thus enters the equation determining E_j and $u_{RL,j}$ exclusively through the dimensionless *radial logarithmic derivative functions* at the MT-radius s_R,

$$D\left\{\varphi_{Rl}(E,s_R)\right\} \equiv \frac{s_R\varphi'_{Rl}(E,s_R)}{\varphi_{Rl}(E,s_R)}. \tag{2.5}$$

Said in another way: If we can find a solution $\Psi(E,\mathbf{r})$ at energy E of the wave equation in the interstitial region, which satisfies the *homogeneous* (Cauchy) *boundary condition*

$$\frac{s_R\Psi'_{RL}(E,s_R)}{\Psi_{RL}(E,s_R)} = D\left\{\varphi_{Rl}(E,s_R)\right\}, \quad \text{for all } RL, \tag{2.6}$$

then, this is a solution of Schrödingers equation in the interstitial region and E is an eigenvalue. Here, $\Psi_{RL}(E,s_R)$ denotes the projection of $\Psi(E,\mathbf{r})$ onto the spherical harmonic $Y_L(\hat{\mathbf{r}}_R)$ and MT-sphere, *viz.*

$$\Psi_{RL}(E,s_R) \equiv \int \delta(r_R - s_R)\,Y_L^*(\hat{\mathbf{r}}_R)\,\Psi(E,\mathbf{r})\,d^3r. \tag{2.7}$$

$\Psi'_{RL}(E,s_R)$ is its radial derivative,

$$\Psi'_{RL}(E,s_R) \equiv -\int \delta'(r_R - s_R)\,Y_L^*(\hat{\mathbf{r}}_R)\,\Psi(E,\mathbf{r})\,d^3r, \tag{2.8}$$

which may, of course, also be written as

$$\Psi'_{RL}(E,s_R) = \int Y_L^*(\hat{\mathbf{r}}_R)\,\frac{\partial\Psi(E,\mathbf{r})}{\partial \mathbf{r}_R}\cdot d^2\mathbf{r} \tag{2.9}$$

2.1.3 Orthonormalization

The normalization of the coefficients $u_{RL,j}$ in (2.1) follows from the orthonormality of the eigenfunctions and from splitting the integral over all space into integrals over the MT-spheres and over the interstitial region, that is,

$$\delta_{ij} = \langle \Phi_i | \Phi_j \rangle = \langle \Psi_i(E_i) \mid \Psi_j(E_j) \rangle + \sum_{RL} u^*_{RL,i} \langle \varphi_{RL}(E_i) \mid \varphi_{RL}(E_j) \rangle\, u_{RL,j}\,, \quad (2.10)$$

where we have used the orthonormality of the spherical harmonics, and where

$$\langle \varphi_L(E_i) \mid \varphi_L(E_j) \rangle = \int_0^s \varphi_l(E_i, r)\, \varphi_l(E_j, r)\, r^2 dr, \quad (2.11)$$

dropping the common subscript R.

In following sections 2.2 and 2.3.2 we shall relate the values of these integrals over the spheres and over the interstitial region to the values and normal gradients of respectively φ and Ψ *at* the spheres. At this place, let us just recall that the orthogonality of the eigenfunctions for two different energies follows from

$$0 = \langle \Phi_i | -\nabla^2 + V(\mathbf{r}) - E_j | \Phi_j \rangle =$$

$$\langle \Phi_j | -\nabla^2 + V(\mathbf{r}) - E_j | \Phi_i \rangle^* + \int \left[\Phi_j(\mathbf{r}) \vec{\nabla} \Phi_i^*(\mathbf{r}) - \Phi_i^*(\mathbf{r}) \vec{\nabla} \Phi_j(\mathbf{r}) \right] \cdot d^2\mathbf{r}$$

$$= (E_i - E_j) \langle \Phi_i | \Phi_j \rangle \quad (2.12)$$

where, for the first equation, we have used the Schrödinger equation, and for the second, Green's second theorem. The surface term vanishes because all eigenfunctions must satisfy the same boundary condition. This is elementary quantum mechanics.

2.1.4 Overview

Before proceeding further with the algebraic formulation of the matching condition (2.6), we shall in Sect. 2.2 recall the behaviors of the radial partial waves φ and their logarithmic derivative functions D. Next, in Sect. 2.3, we shall introduce the analogous for the interstitial region, namely the unitary spherical waves ψ and their radial derivative matrix S. The unitary spherical waves allows us to express the interstitial wave function as

$$\Psi_j(E_j, \mathbf{r}) = \sum_{Rlm} \psi_{Rlm}(E_j, \mathbf{r}_R)\, v_{Rlm,j}. \quad (2.13)$$

Armored with the partial waves and the unitary spherical waves, it turns out to be an easy task to bring the matching condition on algebraic form (Sect.s 2.4 and 2.5). This form is the generalized KKR equation.

2.2 Partial Waves

The partial waves were defined in (2.2-2.5). For the integrals (2.11) over a MT-sphere, the technique (2.12) yields:

$$0 = \langle \varphi_L(E_i) | -\nabla^2 + v(r) - E_j | \varphi_L(E_j) \rangle = \tag{2.14}$$

$$(E_i - E_j) \langle \varphi_L(E_i) \mid \varphi_L(E_j) \rangle + s^2 \left[\varphi_l(E_j, s) \varphi_l'(E_i, s) - \varphi_l(E_i, s) \varphi_l'(E_j, s) \right],$$

In terms of the radial logarithmic derivative functions (2.5), we thus have the desired result:

$$\frac{\langle \varphi_L(E_i) | \varphi_L(E_j) \rangle}{\varphi_l(E_i, s) \varphi_l(E_j, s)} = -s \frac{D\{\varphi_l(E_i, s)\} - D\{\varphi_l(E_j, s)\}}{E_i - E_j}, \tag{2.15}$$

which for $E_i \to E_j \equiv E$ becomes:

$$\frac{\int_0^s \varphi_l(E,r)^2 r^2 dr}{\varphi_l(E,s)^2} = -s\dot{D}\{\varphi_l(E,s)\}. \tag{2.16}$$

Here, and in the following, an energy derivative is denoted by an over-dot: $\dot{} \equiv \partial/\partial E$.

Let us next recall the typical behaviors of the radial partial wave $\varphi(E, r)$ as a function of r and with E as a continuous parameter. In Fig.s 2.1 and 2.2 we show the $5d$ and $6p$ radial functions $r\varphi$ for La in the compound LaI_2 [19] as functions of the radius, from the nucleus r=0 and to the Wigner-Seitz (WS) radius $s \equiv w \approx 4.0$ Bohr radii which corresponds to about 15% overlap (1.11). The functions are normalized to unity in the sphere, that is, the nominator on the left hand side of (2.16) has the value one. The energy is chosen at the center E_ν of the occupied part of the valence and conduction bands. This energy lies close to the bottom of the La $5d$-band and is somewhat below the bottom of the La $6p$-band. As we shall see, the La $5d$ LMTO's must thus be part of the basis set, but the La $6p$LMTO's need not, provided that the La $6p$ channel of the *tails* of the La $5d$'s, the I $5p$'s, and the other LMTO's included in the basis is adjusted to have the proper node position $c_{La\,6p} \sim 0.6w$. Partial waves which behave like La $5d$ in the energy region centered at E_ν, we shall call *active* and those behaving like La $6p$ we shall call *inactive*. Shown in the figures are also two different energy derivatives $r\dot{\varphi}^i$ and $r\dot{\varphi}^g$ of the radial functions.

$r\varphi$ starts out from the nucleus like r^{l+1}, then oscillates inside the atom where the effective kinetic energy, $E - [l(l+1)r^{-2} + v(r)]$, is large and positive, and goes through a total of $n - l - 1$ nodes. The radial Schrödinger equation (2.4) says that the curvature of $r\varphi$ towards the r-axis equals the effective kinetic energy times $r\varphi$. At a node, $r\varphi$ is therefore linear. In the outer part of the atom, the effective kinetic energy becomes numerically small and $r\varphi$ smooth; $r\varphi_{5d}$ is fairly constant and $r\varphi_{6p}$ slopes steeply away from its last node at about $0.6w$.

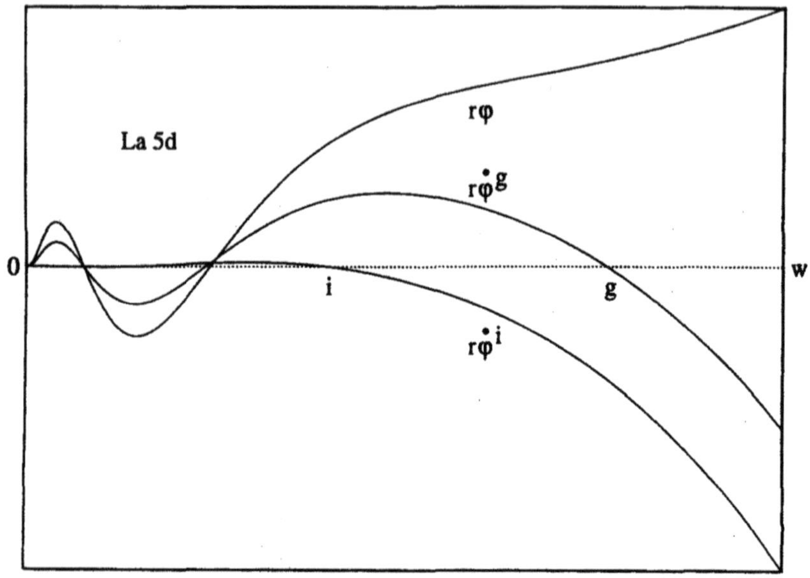

Figure 2.1 La radial $5d$ function and first energy derivatives.

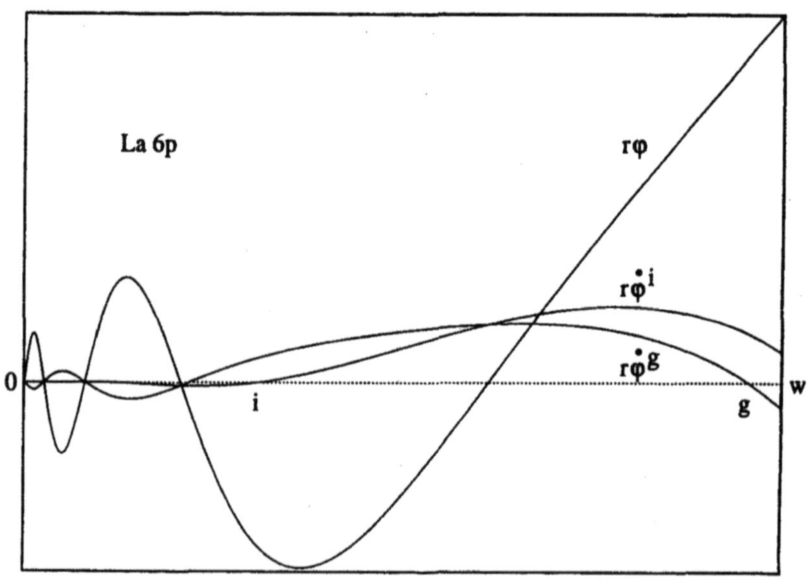

Figure 2.2 La radial $6p$ function and energy derivatives.

Let us now consider the energy dependence. The function $r\dot{\varphi}^i(r)$ is the energy derivative of $r\varphi(E,r)$ calculated with the initial condition being independent of energy, that is, near the nucleus we have chosen $r\varphi(E,r) \rightarrow \text{const.} \times r^{l+1}$ with an energy-independent constant. This energy derivative function is seen to be much smaller for La $6p$ than for La $5d$ in the outer part of the atom. As the energy is increased, $r\varphi(E,r) \sim r\varphi(r) + (E - E_\nu) r\dot{\varphi}^i(r)$ is relatively unaffected when r is smaller than the position i of the last node of $\dot{\varphi}^i$, but bends towards the axis in the outer part of the atom and, as we may imagine for the La $5d$, eventually pulls the $(n-l)$'th node into the WS-sphere. The energy at which this happens is near the top of the nl-band. For even higher energies this node "comes to rest" well inside the atom where the kinetic energy is large; now again, the radial function slopes steeply away from the last node. The bottom of the nl-band turns out to be near the energy where the slope of $\varphi(E,r)$ is horizontal at the WS-sphere, that is, where the tangent to $r\varphi(E,r)$ at the sphere passes through the origin. For La $5d$ this energy is seen to be quite close to E_ν.

As the energy is decreased from E_ν, $r\varphi(E,r)$ slopes steeply away from the last node. As may be imagined for La $6p$, when the energy is lowered this $(n-l-1)$'th node will slowly move outwards and eventually be pushed outside the WS-sphere. This happens approximately at the top of the La $5p$ semi-core band.

For high l-values the radial Schrödinger equation is dominated by the centrifugal potential such that $\varphi(E,r) \sim r^l$ for all r. This behavior resembles that of the La $6p$ function in the energy region of interest, to the extent that smooth (free-electron, $v(r) \equiv 0$) extrapolation of $\varphi(E,r)$ from $r \geq s$ to $r < s$ yields a node-position, a *scattering length* $c_l(E)$, which satisfies

$$0 \leq c_l(E) <\sim 0.7w \tag{2.17}$$

and is fairly independent of energy. Partial waves obeying these conditions do not give rise to states in the region of interest and need only be included as "tails", and not as "heads" in the orbitals of the basis. Such partial waves we refer to as *inactive*. Partial waves from *empty spheres* are usually inactive, because empty-sphere potentials are repulsive. For energies so small that the radial functions are centrifugal dominated near the sphere and the radial free-electron functions therefore become r^l and r^{-l-1}, the scattering length is related to the *radial logarithmic derivative function* $D\{\varphi(E,s)\}$ (2.5) through:

$$c_l(E) = \sqrt[2l+1]{\frac{D\{\varphi_l(E,s)\} - l}{D\{\varphi_l(E,s)\} + l + 1}} \cdot s, \quad \text{for} \quad |E| < \frac{l(l+1)}{s^2} \tag{2.18}$$

A definition valid for all energies will be given in (2.62).

The logarithmic derivative is an *ever decreasing function of energy*, as may be seen from (2.16). For inactive partial waves, it is positive ($\geq l$ for small E)

and decreases slowly. For *active* waves it decreases from a small positive value to zero, which is reached near the bottom of the corresponding band, it then falls rapidly to $-\infty \mid +\infty$, which is reached near the top of the band and, finally, it returns to an "inactive" positive value. With decreasing sphere radius the energy-dependence of $D\{\varphi(E,s)\}$ becomes less rapid, as seen from (2.16); the band structure, of course, does not depend on the sphere choice.

2.2.1 Energy Derivative Functions

It is important to note that the *energy derivative* of a radial wave function depends on the *normalization* of the latter because, changing from one normalization, $\varphi(E,r)$, to another, $n(E)\varphi(E,r)$, causes the energy-derivative function to change as follows:

$$\frac{1}{n(E)}\frac{\partial n(E)\varphi(E,r)}{\partial E} = \dot{\varphi}(E,r) + \frac{\dot{n}(E)}{n(E)}\varphi(E,r). \tag{2.19}$$

This means, that *any* linear combination of the partial wave and its energy-derivative function is *some* energy-derivative function. What is *invariant* to the normalization, is the *two-dimensional Hilbert space* spanned by the radial partial wave and an energy-derivative. This Hilbert space refers to *one* energy E, of course. We shall *specify* a particular function in this two-dimensional Hilbert space by the radius a of its (largest) *node*, and we shall denote this function $\dot{\varphi}^a(E,r)$, that is,

$$\dot{\varphi}^a(E,a) \equiv 0 \tag{2.20}$$

This function might therefore be generated by performing the energy derivative of the partial wave $\varphi(E,r)$ keeping its *value at* $r = a$ *constant.* From (2.19) with $n(E) \equiv 1/\varphi(E,a)$ this function is:

$$\dot{\varphi}^a(E,r) = \varphi(E,a)\frac{\partial\varphi(E,r)/\varphi(E,a)}{\partial E} = \dot{\varphi}(E,r) - \frac{\dot{\varphi}(E,a)}{\varphi(E,a)}\varphi(E,r), \tag{2.21}$$

in terms of any other energy-derivative function and the partial wave.

The second energy-derivative function shown in Fig.s 2.1 and 2.2, $r\dot{\varphi}^g$, is the one where the position g of the node is chosen such that the energy-derivative function is *orthogonal* to the partial wave *in the sphere*, that is:

$$\int_0^s r\dot{\varphi}^g(E,r)\, r\varphi(E,r)\, dr \equiv 0. \tag{2.22}$$

The position $g(E,s)$ of this node therefore depends on the sphere radius s and the energy E. Since the orthogonality equation (2.22) may be obtained by energy-differentiation of

$$\int_0^s \varphi(E,r)^2\, r^2 dr \equiv \text{constant}, \tag{2.23}$$

it is obvious that the orthogonal energy-derivative function might be generated by performing the energy derivative of the partial wave $\varphi(E,r)$ keeping its *normalization integral in the sphere constant.* From (2.19) with $n(E) \equiv 1/\sqrt{\langle\varphi^2(E)\rangle}$:

$$\dot{\varphi}^g(E,r) = \dot{\varphi}(E,r) - \frac{\langle\varphi(E)\mid\dot{\varphi}(E)\rangle}{\langle\varphi^2(E)\rangle}\varphi(E,r). \tag{2.24}$$

In (2.16) we expressed the *first* energy derivative of the logarithmic derivative function at the sphere in terms of the value of the radial function *at* the sphere and its normalization integral *in* the sphere. This sphere is of course arbitrary, so that in (2.16) we could substitute s by a. An alternative expression, in terms of the slope $\dot{\varphi}^a(E,a)'$ of the energy derivative function which vanishes at the chosen sphere, may be obtained by radial differentiation of the first equation (2.21). The result is:

$$\dot{D}\{\varphi(E,a)\} = \frac{a\dot{\varphi}^a(E,a)'}{\varphi(E,a)}, \tag{2.25}$$

and will turn out to have an important equivalent for the unitary spherical waves.

In Section 4 we shall make use of expressions like (2.16) which relate the n'th energy derivative of the radial logarithmic derivative function $\overset{(n)}{D}\{\varphi(E,a)\}$ to the integral in a sphere with radius a of the $(n-1)$'th energy derivative function $\overset{(n-1)}{\varphi^a}(E,r)$. Such relations may be derived by energy differentiation of (2.15) with E_i substituted by E, E_j by E_ν, and s by a, when we keep the value $\varphi(E,a)$ of the radial function at $r=a$ constant. The first energy derivative yields:

$$\langle\varphi(E)\mid\varphi\rangle + (E-E_\nu)\langle\dot{\varphi}^a(E)\mid\varphi\rangle = -a\varphi(E,a)\varphi(a)\dot{D}\{\varphi(E,a)\} \tag{2.26}$$

where an omitted energy argument means $E \equiv E_\nu$. The result of setting $E \equiv E_\nu$ in this equation is eq. (2.16). Differentiating (2.26) once more yields

$$2\langle\dot{\varphi}^a(E)\mid\varphi\rangle + (E-E_\nu)\langle\ddot{\varphi}^a(E)\mid\varphi\rangle = -a\varphi(E,a)\varphi(a)\ddot{D}\{\varphi(E,a)\} \tag{2.27}$$

and the result of setting $E \equiv E_\nu$ is:

$$a\ddot{D}\{\varphi(a)\} = -2\frac{\int_0^a r\varphi(r)\,r\dot{\varphi}^a(r)\,dr}{\varphi(a)^2} \tag{2.28}$$

The *turning point* of the logarithmic derivative function $D\{\varphi(E,a)\}$ thus occurs at the energy T where $\varphi(T,r)$ and $\dot{\varphi}^a(T,r)$ are orthogonal. This requires that the energy is so high (or low) that $\varphi(E,r)$ has pulled a node inside the a-sphere, to a position near the (last) maximum of $r\dot{\varphi}^a(T,r)$. The latter function is relatively independent of energy because its node is fixed at the sphere. At the turning point, the logarithmic derivative $D\{\varphi(T,a)\}$ is usually positive. From Fig. 2.2

we gather that for La $6p$ the logarithmic derivative evaluated slightly inside w has its turning point at E_ν. This confirms the inactive nature of this partial wave.

Differentiating (2.27) and setting $E \equiv E_\nu$ yields first of the following equations:

$$a\,\ddot{D}\,\{\varphi(a)\} = -3\frac{\int_0^a r\varphi(r)\,r\ddot{\varphi}^a(r)\,dr}{\varphi(a)^2} = -6\frac{\int_0^a r\dot{\varphi}^a(r)\,r\dot{\varphi}^a(r)\,dr}{\varphi(a)^2}. \tag{2.29}$$

The last equation may be obtained from the first, plus the result of differentiating (2.28) with respect to E_ν.

2.3 Unitary Spherical Waves

In order to formulate the homogeneous boundary value problem (2.6) algebraically, we need a *complete set of solutions* to the wave equation at energy E for the *interstitial* region, in analogy with the set of partial waves for the MT-spheres. The set that we shall use is denoted $\psi^a_{RL}(E, \mathbf{r}_R)$ where, as usual, R labels the MT centers and L the angular momenta. The set is characterized by the set of numbers $a_{R'l'}$. These $a_{R'l'}$ are radii of *non-touching* spheres centered at all the sites R' and which characterize the set in the following way:

$\psi^a_{RL}(E, \mathbf{r}_R)$ is *defined* as that wave-equation solution whose $Y_L(\hat{r}_R)$- projection on its *own* a_{Rl}-sphere is *unity*, and whose $Y_{L'}(\hat{r}_{R'})$-projections on all *other* $a_{R'l'}$-spheres *vanish*, *i.e.*:

$$\left[\nabla^2 + E\right]\psi^a_{RL}(E, \mathbf{r}_R) \equiv 0 \tag{2.30}$$

with

$$\int \delta\left(r_{R'} - a_{R'l'}\right) Y^*_{L'}\left(\hat{\mathbf{r}}_{R'}\right)\psi^a_{RL}(E, \mathbf{r}_R)\,d^3r \equiv \delta_{R'R}\delta_{L'L}. \tag{2.31}$$

$\psi^a_{RL}(E, \mathbf{r}_R)$ is thus the solution of the inhomogeneous Dirichlet boundary-value problem sketched in Fig. 2.3. In (2.31) we have allowed for the possibility that around each site, there may be a family of l-dependent a-spheres.

The $\psi^a(E)$-set always *exists*. This assumes that the right-hand side of (2.31) contains a factor $1 - \delta(E, E^a_i)$ where E^a_i are the eigenvalues for the system of *impenetrable-*, or hard a-spheres, because for those energies the solutions of the wave equation have nodes at *all* a-spheres. We shall, however, mostly be interested in those energies for which the $\psi^a(E)$-functions are localized; these are below the bottom E^a_1 of the *hard-sphere continuum* and for this reason $1 - \delta(E, E^a_i)$ has been dropped in (2.31).

The $\psi^a(E)$-set is clearly *unique;* for example $\psi^a_{RL}(0, \mathbf{r}_R)$ is proportional to the electrostatic potential from a 2^l-pole at $\mathbf{R}$, surrounded by conducting, earthed (zero-potential) a-spheres at all other sites.

The $\psi^a(E)$-set is finally *complete,* because an arbitrary solution $\Psi(E, \mathbf{r})$ of the wave-equation is obviously given by the linear combination

$$\Psi(E, \mathbf{r}) = \sum_{RL}\psi^a_{RL}(E, \mathbf{r}_R)\,\Psi_{RL}(E, a_{Rl}). \tag{2.32}$$

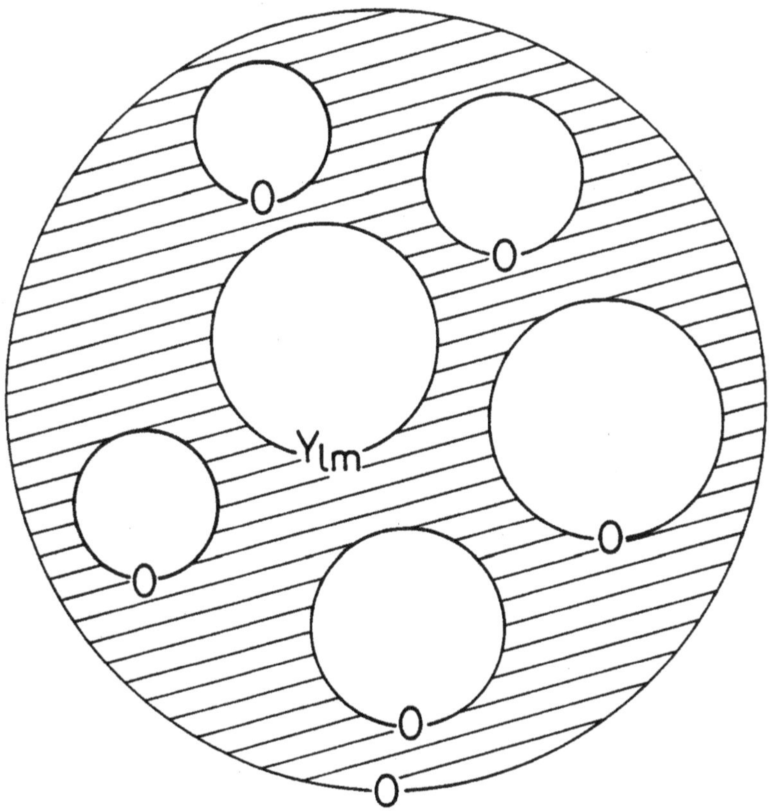

Figure 2.3 Boundary condition for a unitary spherical wave.

where the projections $\Psi_{RL}(E, a_{Rl})$ were defined in (2.7). Since any $\psi^a(E)$-set is complete, all sets must be linearly dependent, that is, they span the same Hilbert space.

The ψ^a-set is a set of *spherical waves* which may be said to be *unitary* with respect to the $a_{R'l'}$-spheres. It is obvious why the a-spheres are not allowed to touch or cut: at a hypothetical common point of the a_R and $a_{R'}$-spheres, the ψ^a_{RL}-functions should be both zero and non-zero. When the a-radii are chosen to depend on angular momentum, it is not the entire ψ^a_{RL}-function but merely its l'-projection on the $a_{R'l'}$-sphere which vanishes or is unity. The set of radii a_{Rl} thus specifies the nodal structure of the unitary spherical waves.

In order that the matching condition (2.6) be *automatically* satisfied in the centrifugal dominated *high-l* limit, it is necessary that the high-l components of the ψ^a-functions be *regular* at all the sites. This means, that we must choose

$$a_{Rl} = 0 \quad \text{for } l > \lambda_R. \tag{2.33}$$

It is most economic to take the low-l cut-off λ_R as the highest l-value for which the logarithmic derivative function deviates substantially from the constant value l in the energy range of interest. In practice, $\lambda \sim 2$.

Since the low-l components of the unitary waves are forced to vanish at *all* surrounding a-spheres, these waves are *localized* when all a_{Rl} are positive and E is below the bottom of the hard-sphere continuum. It is our experience that the nodes should satisfy

$$0.3t_R < a_{Rl} < 0.9t_R \;\; \text{for} \;\; l \leq \lambda_R, \tag{2.34}$$

in terms of the radii t_R of *touching* spheres, in order that the unitary spherical waves decrease *smoothly* in the a-interstitial and at the same time remain well *localized* when E is below the hard-sphere continuum. The best localization is obtained with $a_{Rl} \sim 0.5t_R$ for $\lambda = 0$, and increasing to $\sim 0.85t_R$ for λ increasing to ~ 5. When the packing of the a-spheres is close, the continuum usually starts well above the top of the first free-electron ($a \equiv 0$) band, that is, well above $E = \sqrt[3]{9\pi/2}^2 w^{-2} \approx 5.85w^{-2}$. Here, w is the Wigner-Seitz radius (w^2 ~5-20 Bohr radii), which is 10-15% larger than the touching-sphere radius t in close-packed structures.

The figure at the front of this book shows the nodal surfaces of the unitary s-wave $\psi^a_{l=0}(E, \mathbf{r})$ in the body centered cubic structure for the following parameters: E=0, λ=2, and a_{Rl}=0.70w=0.80t. The shape of the central sheet resembles that of the Wigner-Seitz cell. In contrast to the partial wave $\varphi_{RL}(E, \mathbf{r}_R)$, the unitary wave $\psi^a_{RL}(E, \mathbf{r}_R)$ does not have pure L−character and it diverges at the sites, as the donut-shaped nodal sheets surrounding the eight nearest-neighbors indicate. These divergencies which are due to the image charges from the 2^l-poles ($l \leq \lambda$) at the centers of the hard spheres will not disturb, because the unitary spherical waves will only be used in the interstitial region. The divergencies are

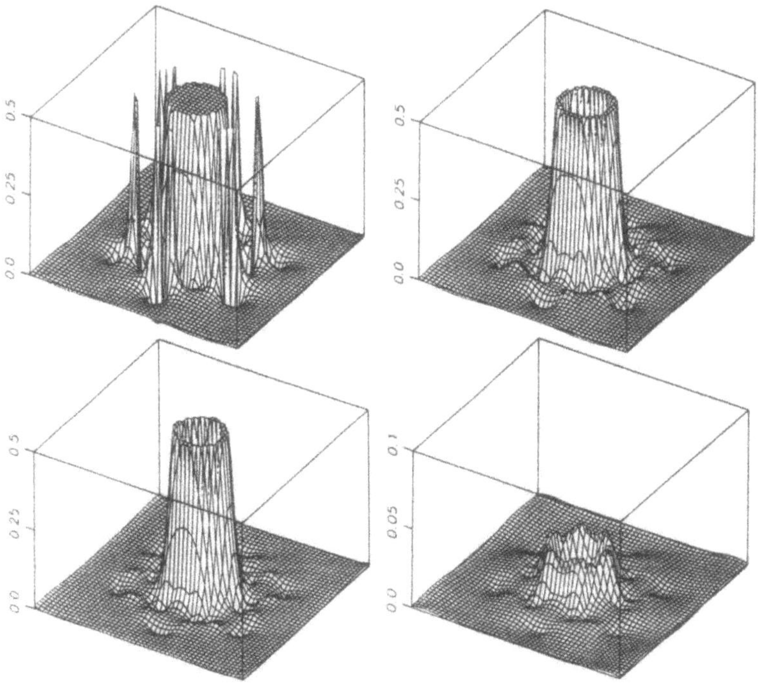

Figure 2.4 Unitary s-wave and its energy derivative (see text).

more easily seen in the upper left-hand part of Fig. 2.4 which shows the unitary s-wave in the (111)-plane of the face-centered cubic structure for Ew^2=3, λ=2, and a_{Rl}=0.70w=0.77t. The upper right-hand part shows the same function, but now with the low-l (i.e. $\leq \lambda$) projections removed inside the a-spheres. The remaining high-l contributions are regular. It is these low-l projections which we shall later substitute by partial waves to form MTO's and eventually LMTO's. Although the unitary spherical wave in the upper part of Fig. 2.4 has positive energy, it is seen to be well localized around the central site. In lower left-hand part of the figure the energy has been changed to Ew^2=–3. This has hardly any effect on the unitary spherical wave! The reason is, that as long as the wavelength $2\pi/\sqrt{|E|}$ ($\approx 3.6w$ in Fig. 2.4) is much larger than the linear extent of the interstitial holes between the a-spheres, the energy-dependence of the unitary waves is small in the interstitial region. As discussed in connection with Fig. 1.2, this range of weak energy dependence includes the range of the valence bands for real systems. We shall later discuss the energy derivative of the unitary spherical wave which is shown in the lower right-hand part of Fig. 2.4.

As will become clear in Section 3 when we explain how to compute the unitary waves, they have been called screened Neumann or Hankel functions. With only the *low-l* averages being required to take the values 1 or 0 at the a-spheres, it *is* possible to let the spheres touch, or even overlap . The resulting unitary waves will however then be long ranged, and consequently useless for our immediate purpose. Similarly, the unitary spherical waves of course exist also for energies above the bottom of the hard-sphere continuum and the formalism to be developed in the following remains valid for such energies, but the functions are delocalized and cannot be generated in real space.

2.3.1 Slope Matrix

The unitary spherical waves were specified by their *values* at the a-spheres (2.31) and now we need their *radial derivatives*. These we characterize by a dimensionless *matrix* $S^a_{R'L',RL}(E)$ which is $a_{R'l'}$ times the L'-component of the radial derivative at the $a_{R'l'}$-sphere (with the positive direction taken from $\mathbf{R}'$ towards the interstitial region) of the unitary spherical wave $\psi^a_{RL}(E, \mathbf{r}_R)$, that is:

$$\underline{S^a_{R'L',RL}(E)} \equiv a_{R'l'} \int Y^*_{L'}(\hat{\mathbf{r}}_{R'}) \left. \frac{\partial \psi^a_{RL}(E, \mathbf{r}_R)}{\partial \mathbf{r}_{R'}} \right|_{r_{R'}=a_{R'l'}} \cdot d^2\mathbf{r}_{R'} \qquad (2.35)$$

$$\underline{= -a_{R'l'} \int \delta'(r_{R'} - a_{R'l'}) Y^*_{L'}(\hat{\mathbf{r}}_{R'}) \psi^a_{RL}(E, \mathbf{r}_R) d^3r.}$$

This is illustrated in the upper part of Fig. 2.5. The radial derivative- or slope matrix is the counterpart to the radial logarithmic derivative functions (2.5) and it depends on the structure, the energy, and the choice of nodes, but *not* on the MT potential, unless the a 's have been chosen to do so. Its diagonal elements are the logarithmic derivatives of the unitary waves at their own a-spheres. As mentioned above, the radial derivative matrix depends little on E in the long wave-length limit and, for E=0, it will turn out to be the usual LMTO screened structure matrix, apart from normalizations and diagonal elements.

The *ssσ*-element $S^a_{d0,00}(E)$ of the slope matrix for the fcc structure and with a=0.7w and λ=2 is shown in Fig. 2.6 for the different shells (at distance d from the central site) as a function of energy. The on-site term is negative and the off-site terms positive, as expected from Fig. 2.5a. The good localization follows from the fact that the slope matrix decays by an order of magnitude for each of the first three shells. The relatively small decrease from the third to the fourth shell is presumably caused by the fact that we used a cluster of only four shells to compute S (by a method to be described in Section 3). The very weak energy dependence up to $Ew^2 \approx 6$ may be appreciated. At higher energies a resonance in the more remote shells heralds the beginning of the continuum and here, our cluster method becomes useless.

The behavior of the slope matrix can be understood in a semi-quantitative manner from the fact that the unitary wave decreases from 1 to 0 over a distance

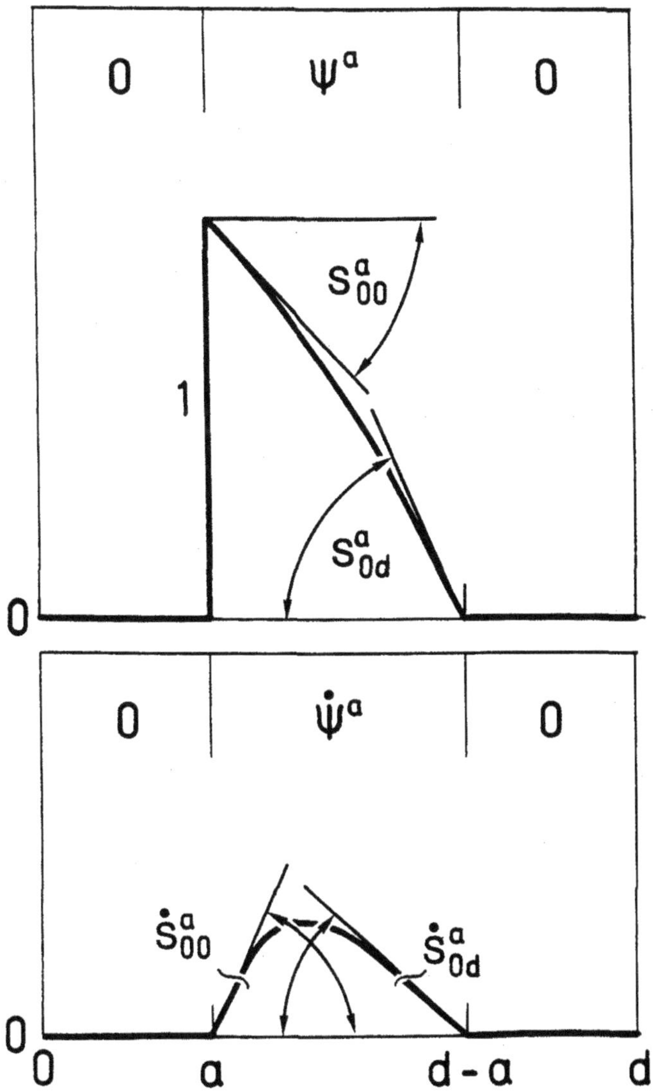

Figure 2.5 Unitary spherical wave and its energy derivative.

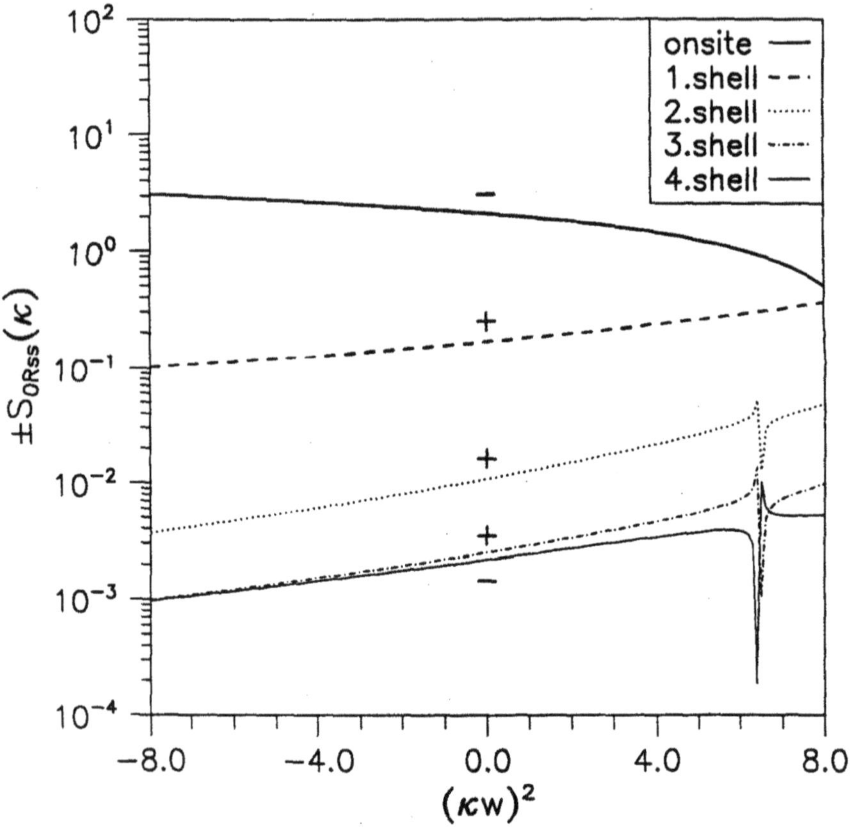

Figure 2.6 $ss\sigma$ slope matrix as a function of energy.

$\sim 2t - 2a$ when going from the surface of its own sphere to the surface of a nearest-neighbor sphere, which is centered at the distance $d = 2t$ (Fig. 2.5). As usual, t is the radius of touching spheres. For close packing, we thus have that $S^a_{RL',RL}(E) \sim -1/[2(t/a-1)]\,\delta_{LL'}$ for the on-site elements. In order to estimate the magnitude of the nearest-neighbor elements we use the following argument: On that part of a near-neighbor sphere (R') which directly faces the central sphere (R), the slope is numerically about the same, and on the remaining part it is much smaller. In order to obtain the near-neighbor l'=0 elements, we must take the average of the slope over the surface of the R'-sphere, that is, we divide by the number n of its nearest neighbors. We thus have the crude estimate: $S^a_{R'0,RL}(E) \sim -S^a_{RL',RL}(E)/n$, further details of course depending on the orientation of the central L-spherical harmonic to the nearest-neighbor direction.

We may improve a bit on these very crude straight-line estimates and, among other things, include the energy dependence, by using the radial wave equation (2.4) with $v(r) \equiv 0$ and $\varphi \equiv \psi$ for a spherical model of the nearby interstitial region. This model consists of the spherical shell between the own a-sphere and a larger sphere with radius $a+\delta$ representing all other spheres. δ should thus be a bit larger than $2t-2a$. From the values $r\psi(E,r) = a$ and 0 at the two radii, the radial wave equation yields $[r\psi(E,r)]'' = [l(l+1) - Ea^2]/a$ and 0 for the respective curvatures. A third-order polynomial approximation to $r\psi(E,r)$ can fit these four values and will be valid as long as δ/a and Ea^2 are not much larger than 1. For the slopes at the end points we then obtain:

$$S^a_{RL,RL}(E) \approx -\left[1 + \frac{a}{\delta} + \left\{l(l+1) - Ea^2\right\}\frac{\delta}{3a}\right] \tag{2.36}$$

for the on-site element, and

$$S^a_{R'0,RL}(E) \sim \frac{1}{n}\left[\frac{a}{\delta} - \left\{l(l+1) - Ea^2\right\}\frac{\delta}{6a}\right]\left[1 + \frac{\delta}{a}\right]^{-1} \tag{2.37}$$

for a nearest-neighbor element. We see that the absolute value of the on-site term decreases with energy and that of the off-site term increases. This is in agreement with the computed results in Fig. 2.6. For $a \approx 0.7t$, $\delta \approx a$, so that $S^{0.7t}_{on}(E) \approx -[2 + \{l(l+1) - Ea^2\}/3]$ and $S^{0.7t}_{off}(E) \sim \frac{1}{24}[1 - \{l(l+1) - Ea^2\}/6]$, where we have taken n=12. The highest energies that we shall be interested in, are at about the top of the first free-electron band, which is at $E = 5.85w^{-2} \approx 2.4a^{-2}$. This means that at the highest energies, the energy dependence of the radial derivative matrix starts to be noticeable, as may also be seen in Fig. 2.6. For E=0 and l=0, the expressions yield: $S^{0.7t}_{on}(0) \approx -2$, and $S^{0.7t}_{off}(0) \sim \frac{1}{24}$, in rough agreement with the figure.

2.3.2 Overlap Integral

We shall often need the value of the integral over the interstitial region of the product of two unitary spherical waves, centered at the same or at different sites.

This overlap integral is for instance needed in (2.10) and for calculation of the total Coulomb energy. The integral over the a-interstitial (in contrast to the one over the MT-interstitial) will turn out to be particularly simple, and this we shall now evaluate. The remaining integrals in the spherical shells between the a- and the s-spheres can easily be included in a different part of the formalism (see Sect. 2.5.1). In analogy with (2.12) and (2.14), we obtain for the integral over the a-interstitial:

$$0 = \langle \psi^a_{RL}(E_i) | -\nabla^2 - E_j | \psi^a_{R'L'}(E_j) \rangle = (E_i - E_j) \langle \psi^a_{RL}(E_i) \mid \psi^a_{R'L'}(E_j) \rangle -$$

$$\sum_{\bar{R}} \int_{a_{\bar{R}l}} \left[\psi^a_{R'L'}(E_j, \mathbf{r}_{R'}) \frac{\partial \psi^{a*}_{RL}(E_i, \mathbf{r}_R)}{\partial r_{\bar{R}}} - \psi^{a*}_{RL}(E_i, \mathbf{r}_R) \frac{\partial \psi^a_{R'L'}(E_j, \mathbf{r}_{R'})}{\partial r_{\bar{R}}} \right] \cdot d^2 \mathbf{r}_{\bar{R}}. \tag{2.38}$$

Here, the minus sign comes about because the normals pointing out of the interstitial region are in the negative radial directions. We may now expand both unitary spherical waves in spherical harmonics $Y_{\bar{l}\bar{m}}(\hat{\mathbf{r}}_{\bar{R}})$around site $\bar{R}$. The value at $r_{\bar{R}} = a_{\bar{R}\bar{l}}$ is given by (2.31) and the slope is given by (2.35). The negative of the surface integral in (2.38) therefore simplifies to:

$$\sum_{\bar{R}\bar{L}} a_{\bar{R}\bar{l}} \left[\delta_{R'\bar{R}} \delta_{L'\bar{L}} S^{a*}_{\bar{R}\bar{L},RL}(E_i) - \delta_{R\bar{R}} \delta_{L\bar{L}} S^a_{\bar{R}\bar{L},R'L'}(E_j) \right] =$$

$$a_{R'l'} S^{a*}_{R'L',RL}(E_i) - a_{Rl} S^a_{RL,R'L'}(E_j),$$

that is, the *three-center integrals* which have $\bar{R} \neq R$ *and* $\bar{R} \neq R'$ *vanish.* Note, that this holds even for the higher partial waves $\bar{l} > \lambda_{\bar{R}}$, because these are regular at $\bar{\mathbf{R}}$. Since, according to (2.38), the surface integral vanishes for $E_i = E_j$, we realize that $a_{Rl} S^a_{RL,R'L'}(E)$ *is hermitian.* The result for the overlap integral in the a-interstitial is thus:

$$\langle \psi^a_{RL}(E_i) \mid \psi^a_{R'L'}(E_j) \rangle = a_{Rl} \frac{S^a_{RL,R'L'}(E_i) - S^a_{RL,R'L'}(E_j)}{E_i - E_j} \tag{2.39}$$

which for $E_i \to E_j \equiv E$ becomes

$$\langle \psi^a_{RL}(E) \mid \psi^a_{R'L'}(E) \rangle = a_{Rl} \dot{S}^a_{RL,R'L'}(E) \tag{2.40}$$

These simple results relating the overlap integrals of the unitary spherical waves to their radial derivative matrix, are completely analogous to (2.16) for the partial waves and their logarithmic derivative functions.

As a corollary: Whereas the logarithmic derivative of the partial wave is an ever decreasing function of energy, the logarithmic derivative of the unitary wave at its own sphere $S^a_{RL,RL}(E)$ is an ever *increasing* function of energy. The larger the interstitial volume, the more rapid the energy dependence. The results

(2.36) and (2.37) for the spherical model clearly illustrate this: The normalization integral (2.40) is approximately

$$\left\langle \psi^{a}_{RL}(E)^{2} \right\rangle = a\dot{S}^{a}_{RL,RL}(E) \approx \frac{1}{3}a^{2}\delta, \tag{2.41}$$

and the nearest-neighbor overlap integral is roughly

$$\langle \psi^{a}_{RL}(E) \mid \psi^{a}_{R'0}(E) \rangle = a\dot{S}^{a}_{RL,R'0}(E) \sim \frac{1}{72}a^{2}\delta \left[1+\frac{\delta}{a}\right]^{-1}. \tag{2.42}$$

(The fact that (2.41) is considerably smaller than the volume $4\pi a^2\delta$ of the near-interstitial although the average of the integrand over the central sphere is one, is a consequence of the strong localization.)

2.3.3 Energy Derivative Functions

By taking the energy-derivative of (2.31) we realize that the *energy derivative functions* $\dot{\psi}^a$ for fixed, energy independent a's have *nodes at all* the a-spheres, including the own sphere, *i.e.* that

$$\int \delta(r_{R'} - a_{R'l'})\, Y^{*}_{L'}(\hat{\mathbf{r}}_{R'})\, \dot{\psi}^{a}_{RL}(E, \mathbf{r}_R)\, d^3r \equiv 0. \tag{2.43}$$

This is illustrated in the lower part of Fig. 2.5 and is completely equivalent to eq. (2.20) for the energy derivative of the radial wave function, normalized to have a constant value for $r = a$.

Similarly to (2.25) and to (2.16), the dimensionless radial derivatives of the $\dot{\psi}^a$-functions at the a-spheres are the energy derivatives of the slope matrix and, hence, a^{-1} times the overlap matrix for the a-interstitial:

$$a_{R'l'} \int Y^{*}_{L'}(\hat{\mathbf{r}}_{R'}) \left.\frac{\partial \dot{\psi}^{a}_{RL}(E,\mathbf{r}_R)}{\partial \mathbf{r}_{R'}}\right|_{r_{R'}=a_{R'l'}} \cdot d^2\mathbf{r}_{R'}$$

$$= -a_{R'l'} \int \delta'(r_{R'} - a_{R'l'})\, Y^{*}_{L'}(\hat{\mathbf{r}}_{R'})\, \dot{\psi}^{a}_{RL}(E, \mathbf{r}_R)\, d^3r$$

$$= \dot{S}^{a}_{R'L',RL}(E) = (a_{R'l'})^{-1} \langle \psi^{a}_{R'L'}(E) \mid \psi^{a}_{RL}(E) \rangle. \tag{2.44}$$

This is obtained by energy-differentiation of (2.35) and by use of (2.40), and is illustrated in the lower part of Fig. 2.5.

In complete analogy with the development for the partial waves in eq.s (2.26-2.29), we can relate the energy derivatives of the slope matrix in terms of integrals over the a-interstitial of the unitary spherical waves and their energy derivative functions. These relations are thus derived by energy differentiations of eq. (2.39) and the results in addition to (2.44) are:

$$a_{R'l'}\ddot{S}^{a}_{R'L',RL} = 2\left\langle \dot{\psi}^{a}_{R'L'} \mid \psi^{a}_{RL} \right\rangle = 2\left\langle \psi^{a}_{R'L'} \mid \dot{\psi}^{a}_{RL} \right\rangle \tag{2.45}$$

and

$$a_{R'l'} \dddot{S}^{a}_{R'L',RL} = 3\left\langle \ddot{\psi}^{a}_{R'L'} \mid \psi^{a}_{RL} \right\rangle = 3\left\langle \psi^{a}_{R'L'} \mid \ddot{\psi}^{a}_{RL} \right\rangle$$
$$= 6\left\langle \dot{\psi}^{a}_{R'L'} \mid \dot{\psi}^{a}_{RL} \right\rangle \tag{2.46}$$

An important task will be to calculate the radial derivative matrix and its energy derivatives, but let us for the moment assume that it is known and proceed with the algebraic formulation of the Cauchy matching problem (2.6).

2.4 Matching Equations in the MT-Representation

Let us first assume that the MT-spheres are slightly smaller than touching so that we can take $a_{Rl} \equiv s_R$. This does seldomly yields a good approximation for the potential, nor does it yield the shortest range for the slope matrix, or the most linear energy depence of the resulting equations, but the algebraic result is conceptually the most simple.

Our task is to find coefficients

$$v^{s}_{RL} = \Psi_{RL}(E, s_R) \tag{2.47}$$

such that the linear combination $\Psi(E, \mathbf{r})$ of unitary waves (2.13) or (2.32) have the proper logarithmic derivatives $D\{\varphi_{Rl}(E, s_R)\}$ at the MT-spheres, *i.e.* that it satisfies the matching conditions (2.6). From (2.32) and the definition (2.35) of the radial derivative matrix, the matching conditions are simply:

$$[v^{s}_{R'L'}]^{-1} \sum_{RL} S^{s}_{R'L',RL}(E)\, v^{s}_{RL} = D\{\varphi_{R'l'}(E, s_{R'})\} \quad \text{for all} \quad R'L', \tag{2.48}$$

which are a set of *linear, homogeneous equations.* If we consider $\mathbf{v}^{s}$ as a column-*vector* with components v^{s}_{RL}, and $D\{\varphi(E, s)\}$ as a *diagonal matrix* with elements $\delta_{RR'}\delta_{LL'}D\{\varphi_{Rl}(E, s_R)\}$, these equations may be written in vector form as:

$$K^{s}(E)\,\mathbf{v}^{s} \equiv s\left[D\{\varphi(E, s)\} - S^{s}(E)\right]\mathbf{v}^{s} = \mathbf{0}, \tag{2.49}$$

where we have multiplied each equation by the MT-radius $s_{R'}$ in order that the *kink matrix* $K^{s}(E)$ be *hermitian.* The linear, homogeneous equations have solutions for those *energies* E_i where the determinant of the kink matrix vanishes, and the *eigenfunctions* are then given by the one-center expansions (2.1) with

$$u_{RL,i} = \frac{v^{s}_{RL,i}}{\varphi_{Rl}(E_i, s_R)}. \tag{2.50}$$

The subscript i has been added on the coefficient vector $\mathbf{v}^{s}_{i}$, first of all in order to have the possibility of enumerating different solutions with the same energy. *E.g.* for a crystal, the states with a given energy (outside gaps) are enumerated by a two-dimensional surface, with various sheets, in three-dimensional k-space.

In this case, we would take i to be the combination of the three-dimensional Bloch vector $\mathbf{k}$ and the band index, and hence, to completely enumerate all the solutions, not only the degenerate ones. In the following, we therefore drop the energy argument in the coefficient vector.

The normalization of these coefficients is obtained from (2.10) by use of (2.16) for the integrals in the MT-spheres, of (2.32) for the interstitial functions, and of (2.40) for their integral. As a result, the overlap matrix (2.10) reduces to:

$$\langle \Phi_i | \Phi_j \rangle = \delta_{ij} = \mathbf{v}_i^{s\dagger} \frac{K^s(E_j) - K^s(E_i)}{E_i - E_j} \mathbf{v}_j^s, \tag{2.51}$$

which is automatically satisfied for non-degenerate solutions of the matching equations (2.49). For degenerate eigenstates, $E_i = E_j = E$, the eigenvectors should be orthonormalized according to:

$$\langle \Phi_i | \Phi_j \rangle = \delta_{ij} = -\mathbf{v}_i^{s\dagger} \dot{K}^s(E) \mathbf{v}_j^s. \tag{2.52}$$

Note, that the matching equations (2.48) do not become ill behaved at an energy where one of the logarithmic derivatives diverge. For such an Rl, no only the corresponding components $v_{Rlm,i}^s$ of the eigenvector, but also the denominator of the right-hand side of (2.50) vanishes so that the coefficients $u_{Rlm,i}$ usually remain finite.

This algebraic formulation of the matching problem (2.6) is exceedingly simple. The equations (2.49) are equivalent with the original Korringa-Kohn-Rostoker (KKR) equations [7] which, however, are expressed in terms of phase shifts, specifically $\cot \eta_{Rl}(E)$, and a *long-ranged* structure matrix $B_{R'L',RL}(E)$, *both* of which can be wildly energy dependent. As we shall explain later, the comparatively well-behaved matching equations (2.49) may be viewed as the KKR equations in the s-representation.

2.4.1 Interpretation in Terms of Kinky Partial Waves

The solution of the matching problem just derived may be interpreted in terms of the following set of *kinky partial waves* (Fig. 2.7). These are defined in *all* space as

$$\phi_{Rlm}^s(E, \mathbf{r}_R) \equiv \frac{\varphi_{Rlm}(E, \mathbf{r}_R)}{\varphi_{Rl}(E, s_R)} + \psi_{Rlm}^s(E, \mathbf{r}_R) \quad \text{with} \quad l \leq \lambda_R. \tag{2.53}$$

Here, like in (2.1), the partial wave is supposed to be truncated outside the s-sphere and the unitary spherical wave outside the interstitial region, except for its high-l components which penetrate into the spheres. We see that a kinky partial wave is a low partial wave inside its own MT-sphere, and is matched continuously, but not differentiably *at* the sphere to a unitary spherical wave in the interstitial region. The wave has kinks also at the neighboring MT-spheres, inside which its low-l components vanish. Except for the higher partial waves, the kinky partial

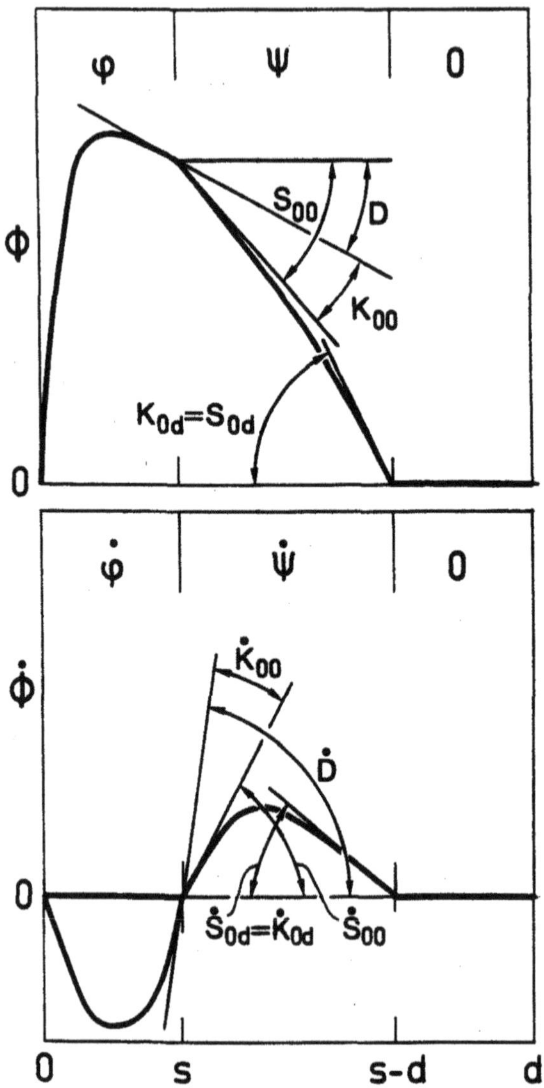

Figure 2.7 Kinky partial wave and its energy derivative.

wave ϕ^s_{RL} thus has pure L-character inside its own sphere and vanishes inside all other spheres. The high-l components extend smoothly into all spheres, and only the low partial waves have discontinuous slopes. The kinky partial wave gives the following meaning to the kink matrix: $(s_{R'l'})^{-1} K^s_{R'L',RL}(E)$ is the kink of $\phi^s_{RL}(E, \mathbf{r}_R)$ in the $R'L'$-channel at the $s_{R'l'}$-sphere.

These kinky partial waves are completely analogous to Slaters ***augmented plane waves*** (APW's) and might therefore also be called augmented unitary spherical waves. Each kinky wave is the solution of Schrödinger's equation for its own MT-well and the flat interstitial potential, albeit with the addition of a shell of infinitely high potential at the surface of the own MT, and of infinitely high potentials inside the other MT's.

What is achieved at a solution of the matching equations (2.49) is that for the *linear combination* of kinky partial waves $\sum_{RL} \phi^s_{RL}(E, \mathbf{r}_R)\, v^s_{RL}$, the kink

$$D\{\varphi_{R'l'}(E, s)\} - S^s_{R'L',R'L'}(E)$$

in the logarithmic derivative of each kinky wave $\phi^s_{R'L'}(E, r_{R'})$ at its central sphere R', is *cancelled* by the sum of the kinks

$$\sum_{RL \neq R'L'} S^s_{R'L',RL}(E)\, v^s_{RL}$$

in the *tails* of the kinky wave's coming from the other sites. When there are off-diagonal elements in the on-site block of the radial derivative matrix (as is the case for sites of low symmetry), then the other kinky partial waves at the same site contribute with their off-diagonal "head-kinks" to this cancellation. In addition to being a solution of Schrödingers equation in the sub-regions and being continuous, the resulting linear combination is also differentiable and, hence, is a solution of Schrödingers equation in all space:

$$\Phi_i(\mathbf{r}) = \sum_{R=1}^{\infty} \sum_{l=0}^{\lambda} \sum_{m=-l}^{l} \phi^s_{Rlm}(E_i, \mathbf{r}_R)\, v^s_{Rlm,i}. \tag{2.54}$$

An advantage of this representation of the wave function as a superposition of kinky partial waves, over that of the truncated one-center expansions (2.1), is that the l-sums in (2.54) only include the low partial waves whereas those in (2.1) go to infinity.

2.4.2 Energy linearization of the kink matrix

If now, instead of using the matching conditions (2.49), we use the Raleigh-Ritz variational principle for the Hamiltonian $H_{mt} \equiv -\nabla^2 + V_{mt}(\mathbf{r})$ with a linear combination like (2.54) of kinky partial waves $\phi^s_{RL}(E_\nu, \mathbf{r}_R)$ with a *fixed* energy E_ν as trial function, then standard theory tells us that the best linear combination is that one whose coefficients satisfy the linear, homogeneous equations

$$\langle \phi^s(E_\nu) | H_{mt} - E | \phi^s(E_\nu) \rangle\, \mathbf{v}^s \equiv [H^s - EO^s]\, \mathbf{v}^s = 0. \tag{2.55}$$

Here, v is a column-vector with components v_{RL}, $|\phi\rangle$ is a row-vector of functions $|\phi_{RL}\rangle$, and $\langle\phi|$ is a column-vector of functions $\langle\phi_{R'L'}|$, such that $\langle\phi|H_{mt}-E|\phi\rangle$ is a matrix. This matrix is *linear in E* and the linear, homogeneous equations (2.55) consequently lead to an ordinary *eigenvalue problem* with a *Hamiltonian matrix* $H^s \equiv \langle\phi^s|H_{mt}|\phi^s\rangle$ and an *overlap matrix* $O^s \equiv \langle\phi^s|\phi^s\rangle$. This problem can be solved by a diagonalization procedure which yields the eigenvalues E_i and eigenvectors $\mathbf{v}_i^s$ *simultaneously,* rather than by the cumbersome two-step procedure needed to solve the *secular* problem (2.49).

Since the error of the trial function is of order $E_i - E_\nu$, the eigenvalues will have errors of order $(E_i - E_\nu)^2$. Now, like in (2.52) the overlap matrix is seen to be:

$$O^s \equiv \langle\phi^s(E_\nu)|\phi^s(E_\nu)\rangle = -\dot{K}^s(E_\nu) \tag{2.56}$$

Insertion of this equation in (2.55) and comparing the result with (2.49) then leads to the conclusion that the linear matrix in (2.55) must equal the kink matrix $K^s(E)$ to *first* order in $E - E_\nu$. Consequently,

$$H^s - EO^s \equiv \langle\phi^s|H_{mt}-E|\phi^s\rangle = K^s + (E-E_\nu)\dot{K}^s \tag{2.57}$$

$$= s\left[(D\{\varphi(s)\} - S^s) + (E-E_\nu)\left(\dot{D}\{\varphi(s)\} - \dot{S}^s\right)\right] \tag{2.58}$$

Here, and often in the following, an omitted energy argument means that $E \equiv E_\nu$. We have thus learned, that we can regard the kink matrix $K^s(E_\nu)$ at a fixed energy E_ν as the Hamiltonian matrix $\langle\phi^s(E_\nu)|H_{mt}-E_\nu|\phi^s(E_\nu)\rangle$ relative to that energy, in the representation of the kinky partial waves for that energy. Similarly, the negative energy derivative of the kink matrix $-\dot{K}^s(E_\nu)$ is the overlap matrix. In (2.58) we have expressed both the overlap matrix (the coefficient of $E - E_\nu$) and the Hamiltonian matrix with respect to E_ν as the sum of a MT and an interstitial term. The MT-terms are *diagonal* ($\propto \delta_{RR'}\delta_{LL'}$). The higher the point symmetry, the more diagonal are also the interstitial terms. Note, that these Hamiltonian and overlap matrices have the *two-center form,* a property that can be traced back to the derivation of the overlap integral (2.40) of two unitary spherical waves.

Over the energy range where the matrix $\frac{1}{2}(E-E_\nu)\ddot{K}^s\left(\dot{K}^s\right)^{-1}$ is small, this Hamiltonian formulation is highly efficient. However, although for the slope matrix the linear approximation $S^s(E) \approx S^s + (E-E_\nu)\dot{S}^s$ is good when the s-spheres are nearly touching, this is less so for the logarithmic derivative functions: $\ddot{D}\{\varphi_l(E,w)\}$ evaluated at the Wigner-Seitz radius is usually large for all E inside the l-band as was mentioned in connection with Fig. 2.1 and, although $\ddot{D}\{\varphi_l(E,s)\}$ evaluated at the ~15% smaller touching spheres is smaller, the useful energy range is too small for most purposes. The numbers which should be much smaller than unity in order that the eigenvalues can be trusted, are $E - E_\nu$ times

$$\frac{\ddot{K}^s_{RL,RL}}{2\dot{K}^s_{RL,RL}} = \frac{1}{2}\frac{\ddot{D}\{\varphi_{Rl}(s_R)\} - \ddot{S}^s_{Rl,RL}}{\dot{D}\{\varphi_{Rl}(s_R)\} - \dot{S}^s_{Rl,RL}} \tag{2.59}$$

$$=\frac{\frac{1}{2}\left\{\int_0^s \varphi(r)\dot{\varphi}^s(r)\,r^2dr+\varphi(s)\left\langle\psi^s|\dot{\psi}^s\right\rangle\varphi(s)+h.c.\right\}}{\int_0^s \varphi(r)^2\,r^2dr+\varphi(s)\,s\dot{S}^s\varphi(s)}=\frac{\frac{1}{2}\left\{\left\langle\phi^s|\dot{\phi}^s\right\rangle+h.c.\right\}}{\left\langle\phi^s|\,\phi^s\right\rangle},$$

as obtained by use of (2.28), (2.45), and (2.53). The problem with the energy linearization of the kink matrix $K^s(E)$ for nearly touching s-spheres is thus, that $\left\langle\phi^s_{RL}|\dot{\phi}^s_{RL}\right\rangle$ is numerically large because the (radial) nodal surface of ϕ^s_{RL} is fairly close to the (radial) nodal surface of $\dot{\phi}^s_{RL}$, the former being essentially the WS-cell boundary (see book front), and the latter being just a bit smaller than the inscribed sphere. In order that $\left\langle\phi_{RL}|\dot{\phi}_{RL}\right\rangle$ be small, the nodal surface of each function should run close to the maximum of the other.

Now, in earlier sections we have seen that the radial node of the energy derivative function $\dot{\varphi}^a_{RL}$ and the central radial node of the interstitial energy derivative function $\dot{\psi}^a_{RL}$ can be adjusted at will. By choosing a_{Rl} smaller than s_R, we can therefore shift the central node of $\dot{\phi}^a_{RL}$ to smaller radii and thereby reduce the size of $\left\langle\phi^a_{RL}|\dot{\phi}^a_{RL}\right\rangle$. By choosing a_{Rl} as small as $g_{Rl}(E_\nu, s_R)$ (see eq. (2.22) and Fig. 2.1), the integral of $\phi^g_{RL}\dot{\phi}^g_{RL}$ over the WS-cell will essentially vanish. Since the integral $\left\langle\phi^g_{RL}|\dot{\phi}^g_{RL}\right\rangle$ extends a bit further than the WS-cell, approximately to a cell touching the g-spheres centered on the neighboring sites, its sign will be opposite to that of $\left\langle\phi^s_{RL}|\dot{\phi}^s_{RL}\right\rangle$. Therefore, by adjusting the a_{Rl}'s to values slightly larger than the g_{Rl}'s, but smaller than the radii t_R of touching spheres, we can achieve that all diagonal elements $\left\langle\phi^a_{RL}|\dot{\phi}^a_{RL}\right\rangle$ vanish, or even that the entire matrix $\left\langle\phi^a|\dot{\phi}^a\right\rangle\propto\ddot{K}^a$ vanishes. We may thus reduce the second-order error of the energy eigenvalues arbitrarily.

Finally, we should mention that although the matching equations work at an energy where one of the logarithmic derivative functions diverge, the energy linearization does *not:* If, for instance $\varphi_{Rl}(s)=0$, then the linearized equations with the matrix (2.58) yield a bound state of Rl-character at E_ν. The error given by (2.59) is, however infinitely large because the normalization of $\dot{\varphi}^s_{Rl}(r)$ diverges when $\varphi_{Rl}(r)$ and $\dot{\varphi}^s_{Rl}(r)$ are linearly dependent, as may seen from eq. (2.21) or (2.25). The bound state is thus a "ghost" state. What should be done, is to remove the corresponding kinky partial waves $\phi^s_{Rlm}(\mathbf{r}_R)$'s (which diverge) from the basis set and augment all the unitary spherical waves $\psi^s_{R'L'}(\mathbf{r}_{R'})$ with the $\varphi_{Rlm}(\mathbf{r}_R)$'s. Since both $\psi^s_{R'L'}(\mathbf{r}_{R'})$ and $\varphi_{RL}(\mathbf{r}_R)$ vanish on the s_R-sphere, the augmentation can be made continuous *and* differentiable. *In the formalism, the Rl-channel will then appear as high.* Since the Rl-kinky partial waves have been removed from the basis, the linearized formalism will deliver no states of predominantly Rl-character, but this is all right because with the MT-spheres being touching or smaller, the energy E_ν for which the node of the Rl-partial wave is *at* the sphere, is far above the top of the Rl-band. In the following sections, we shall derive the matching formalism for a *general* a-representation and show how to transform between representations. This will allow us to get rid of all the inactive partial waves (2.17), not only the high ones. The way this is done, is by choosing a_{Rl} equal to the scattering length $c_{Rl}(E_\nu)$, and then delete the corresponding kinky partial wave.

2.5 Matching Equations in a General Representation

In practice, we wish to use MT-spheres which are as large as possible, because this gives the best representation of the potential, and in Ref. [40] its has been shown that we may even use MT-spheres which overlap by up to about 15% ($s_R \sim 1.15t_R$). The a-spheres are, however, not allowed to touch and the shortest range of the unitary spherical waves is, in fact, obtained for $a_R \sim 0.7t_R$, as described before. The minimal second-order energy dependence of the kink matrix $K^a(E)$, that we shall now define is achieved for $g_{Rl} \leq a_{Rl} \leq t_R$, as was explained in the preceding section. Finally, in order to obtain small matrix dimensions and in order to avoid ghost bands in the linearized version of the theory, we would like to fold down the inactive partial waves by choosing the corresponding hard-sphere radii equal to the scattering lengths, $a_{Rl} = c_{Rl}(E_\nu)$. Hence, there are four reasons for choosing the a_{Rl}'s different from the MT-radii (short range, small dimension, and energy-linearity of the kink matrix and a good overlapping MT-representation of the potential). Therefore, we now generalize the matching equations (2.49) to an arbitrary choice of the a_{Rl}-parameters, and show how to transform from one set of a-parameters to another.

2.5.1 Partial Waves seen from the Outside

When introducing the scattering length in (2.17) we used a concept of continuing the partial wave smoothly from the outer part of the sphere. This concept will now be specified and generalized.

The (numerical) integration of the radial Schrödinger equation (2.4) starts at r=0 with the initial condition $\varphi(E,r) \propto r^l$ and then proceeds through the MT-potential $v(r)$ to the MT-radius s. From the MT-radius, we may *continue* the integration *backwards* and *forwards,* into and out of the MT-sphere, through the *constant* potential $V_{mtz} \equiv 0$ and define the radial function *seen from the outside* $\varphi(E,r)^o$ as that solution of the radial wave equation [(2.4) with $v(r) \equiv 0$] which *touches* $\varphi(E,r)$ at the MT-radius. Since the matching of $\varphi(E,r)^o$ and $\varphi(E,r)$ at s is continuous and differentiable, and since the change of curvature,

$$[s\{\varphi(E,s)^o - \varphi(E,r)\}]'' = -v(s)\, s\varphi(E,s), \tag{2.60}$$

is small, $\varphi(E,r)^o$ remains close to $\varphi(E,r)$ and depends little on s, as long as r is close to s.

We shall mostly be interested in extrapolating the partial wave to the hard a-sphere, which is smaller than the MT-sphere, and therefore in using the function extrapolated backwards into the MT-sphere. Note, that we can still *think* about $\varphi(E,a)^o$ and $D\{\varphi(E,a)^o\}$ as respectively the radial amplitude and logarithmic-derivative function *at a*.

All the previously derived relations between partial waves, energy derivatives of the logarithmic derivatives, and energy derivatives of the partial waves hold for the back-extrapolated quantities provided that the integrals are interpreted as, for example:

$$\int_0^a \varphi(E,r)^2 r^2 dr \equiv \int_0^s \varphi(E,r)^2 r^2 dr - \int_a^s [\varphi(E,r)^o]^2 r^2 dr. \tag{2.61}$$

The scattering length is defined by the equation

$$\varphi_{Rl}(E, c_{Rl}(E))^o \equiv 0 \tag{2.62}$$

which may or may not have solutions. If it does, and if one of these solutions satisfy the conditions (2.17), then the corresponding partial wave is inactive.

2.5.2 Matching Equations

It is now obvious, that the wave-equation solution (2.32) which matches the backwards extrapolated logarithmic derivative functions $D\{\varphi(E,a)^o\}$ at all the a-spheres *will also match* the logarithmic derivatives $D\{\varphi(E,s)^o\} \equiv D\{\varphi(E,s)\}$ at all the MT-spheres. To find this solution, we can therefore use the procedure (2.47)-(2.50), but for the radii a. As a result, we obtain the generalized matching- or KKR equations

$$0 = \underline{\sum_R \sum_l^{\lambda} \sum_{m=-l}^{l} a_{Rl}\left[D\{\varphi_{Rl}(E,a)^o\}\delta_{R'R}\delta_{L'L} - S^a_{R'L',RL}(E)\right] v^a_{RL}}$$

$$\equiv \underline{\sum_R \sum_l^{\lambda} \sum_{m=-l}^{l} K^a_{R'L',RL}(E)\, v^a_{RL} \quad \text{for all} \quad R'l'm'} \tag{2.63}$$

in terms of the kink (or generalized KKR) matrix $K^a(E)$. In vector form, these equations are:

$$K^a(E)\,\mathbf{v}^a \equiv a\left[D\{\varphi(E,a)^o\} - S^a(E)\right]\mathbf{v}^a = \mathbf{0}.$$

These equations look as if the matching were performed at the a-spheres.

The equations have solutions for those energies E_i where

$$|D\{\varphi^o(E,a)\} - S^a(E)| = 0,$$

and the eigenfunctions are then given by the one-center expansions (2.1) with

$$u_{RL,i} = \frac{v^a_{RL,i}}{\varphi_{Rl}(E_i, a_{Rl})^o}. \tag{2.64}$$

The orthonormalization is given by eqs. (2.51-2.52) with the superscripts s substituted by superscripts a.

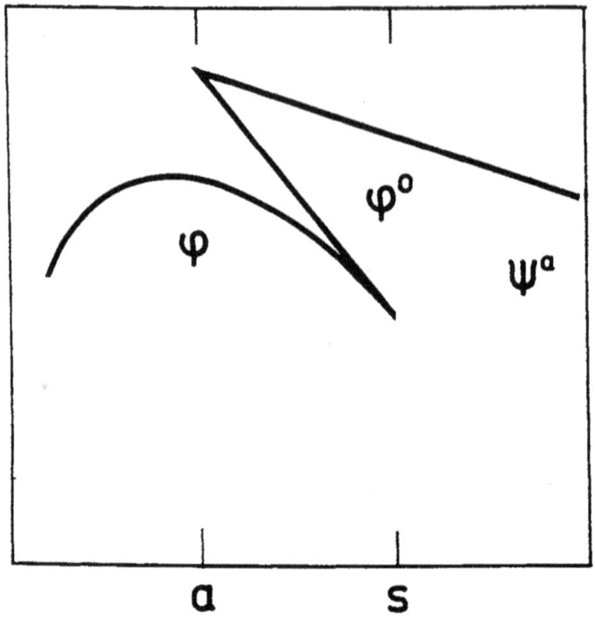

Figure 2.8 Triple valuedness of the kinky partial wave (schematic).

2.5.3 Kinky Partial Waves and their Energy Derivatives

The kinky partial wave in the a-representation is:

$$\underline{\phi^a_{RL}(E,\mathbf{r}_R) \equiv \varphi^a_{RL}(E,\mathbf{r}_R) + \psi^a_{RL}(E,\mathbf{r}_R), \quad \text{with} \quad l \in \lambda_R.} \tag{2.65}$$

where we have used the normalization

$$\varphi^a_{RL}(E,\mathbf{r}_R) \equiv \frac{\varphi_{RL}(E,\mathbf{r}_R)}{\varphi_{Rl}(E,a_{Rl})^o}, \tag{2.66}$$

where the truncations of $\varphi(E,\mathbf{r})$ and $\psi^a(E,\mathbf{r})$ are supposed to be with respect to the a-partitioning of space (in the sense of eq. (2.61)), and where λ_R from now on denotes the set of active partial waves at site R. As shown in Fig. 2.8, for r_R increasing from the origin, $\varphi^a(E,r)$ in (2.65) should be taken as the Schrödinger equation solution $\varphi^a(E,r)$ up to s, and then as $\varphi^a(E,r)^o$ for r running backwards from s to a, where it is truncated. At this central a-sphere, the kinky partial wave is now matched continuously, but not differentiably to the ψ^a-function. The latter decays in the interstitial region and goes continuously, but not differentiably to zero at the neighboring a-spheres. The inactive partial waves supplied by the ψ^a-function penetrate all spheres. This kinky partial wave thus has ***kinks at the*** *a-spheres* with a kink matrix, i.e. the matrix whose $(R'L', RL)$-element is the kink of $\phi^a_{RL}(E, \mathbf{r}_R)$ at the $a_{R'l'}$-sphere in the $R'L'$-channel, *is* $a^{-1}K^a(E)$. The kinky partial wave is *tripple-valued* in the central $a \to s$ spherical shell as shown in the figure. This tripple-valuedness may seem a bit artificial, although mathematically correct, and we shall later modify the kinky partial wave to become a linear muffin-tin orbital which is both smooth, single-valued, and energy-independent .

Like previously, the *matching equations* (2.63) express that for the linear combination of kinky partial wave's,

$$\underline{\Phi_i(\mathbf{r}) = \sum_{R=1}^{\infty} \sum_{l=0}^{\lambda} \sum_{m=-l}^{l} \phi^a_{Rlm}(E_i, \mathbf{r}_R)\, v^a_{Rlm,i}.} \tag{2.67}$$

the kinks should cancel. When this happens, the linear combination of the $\varphi_{RL}(E, \mathbf{r}_R)^{\circ}$-functions coincide with that of the $\psi^a_{RL}(E, \mathbf{r}_R)$-functions throughout the $a \to s$ shells, and this brings the ψ^a linear combination into smooth matching with linear combination of partial waves *at the MT-spheres.*

For the construction of MTO's we shall, in addition to the kinky partial waves, need their energy-derivative functions

$$\dot{\phi}^a_{RL}(E, \mathbf{r}_R) \equiv \dot{\varphi}^a_{RL}(E, \mathbf{r}_R) + \dot{\psi}^a_{RL}(E, \mathbf{r}_R), \quad \text{with} \quad l \in \lambda_R. \tag{2.68}$$

These vanish at all the a-spheres [see eq.s (2.20) and (2.43)] and their kink matrix is $a^{-1}\dot{K}^a(E)$, as seen from (2.25), (2.44), and (2.63).

The energy derivatives of the Kink matrix may be related to integrals in all space of the kinky partial waves and their energy derivatives. From (2.26-2.29) for the partial waves and from (2.44-2.46) for the unitary spherical waves, we find:

$$-\dot{K}^a_{R'L',RL} = \langle \phi^a_{R'L'} \mid \phi^a_{RL} \rangle = O^a_{R'L',RL}, \tag{2.69}$$

$$-\ddot{K}^a_{R'L',RL} = 2\left\langle \dot{\phi}^a_{R'L'} \mid \phi^a_{RL} \right\rangle = 2\left\langle \phi^a_{R'L'} \mid \dot{\phi}^a_{RL} \right\rangle, \tag{2.70}$$

and

$$\begin{aligned} -\dddot{K}^a_{R'L',RL} &= 3\left\langle \ddot{\phi}^a_{R'L'} \mid \phi^a_{RL} \right\rangle = 3\left\langle \phi^a_{R'L'} \mid \ddot{\phi}^a_{RL} \right\rangle \\ &= 6\left\langle \dot{\phi}^a_{R'L'} \mid \dot{\phi}^a_{RL} \right\rangle. \end{aligned} \tag{2.71}$$

2.5.4 Energy Linearization of the Kink Matrix

The Hamiltonian variational formalism expressed through eqs. (2.55-2.59) holds, also when the superscripts s are substituted by a, $D\{\varphi(E,s)\}$ is substituted by $D\{\varphi(E,a)^{\circ}\}$, $\varphi(E,s)$ is substituted by $\varphi(E,a)^{\circ}$, and $\psi(E,s)$ is substituted by $\psi(E,a)$. In particular:

$$H^a - EO^a \equiv \langle \phi^a | H_{mt} - E | \phi^a \rangle = K^a + (E - E_\nu)\dot{K}^a \tag{2.72}$$

where we should use (2.61) for the radial integrals, and where an omitted energy argument, as usual, means $E \equiv E_\nu$. This energy linearization now becomes very useful, because by decreasing the radius a, the logarithmic derivatives become weaker functions of energy and, at the same time, the energy dependence of the slope matrix increases.

Specifically, with decreasing a, $\left|\dot{D}\left\{\varphi(E,a)^{\circ}\right\}\right|$ decreases and $\dot{S}^{a}(E)$ increases with the result that $O^{a}_{RL,RL}$ remains rather constant because the a-dependence mostly reflects the change in the range of the unitary spherical wave.

The leading error caused by the energy linearization is due to the diagonal elements of the *second* energy derivative matrix (2.70). These elements are controlled by the values of a, because a_{Rl} is the backwards extrapolated position of the *central* node in $\dot{\phi}^{a}_{RL}(E,\mathbf{r}_R)$, and such a node is *not* present in $\phi^{a}_{RL}(E,\mathbf{r}_R)$ for energies inside a band with Rl-character, as was mentioned in connection with Fig. 2.1 for La 5d. Here, we had $g\sim 0.8w\sim 0.9t$. It is therefore conceivable that values $0<g_{Rl}\leq t_R$ can be found for a_{Rl} such that

$$\left\langle \phi^{g}_{RL}\mid \dot{\phi}^{g}_{RL}\right\rangle = 0 \quad \text{for all } RL \tag{2.73}$$

This was discussed at the end of the previous section. The g-values defined by (2.73) differ somewhat from those defined explicitly in (2.22-2.24) which refer to $\varphi(E,r)$ in the WS-sphere. The modified definition (2.73) refers to the kinky partial wave in all space and accordingly, it is the energy derivative of the back-extrapolated partial wave $\dot{\varphi}^{go}_{Rl}$, and not $\dot{\varphi}^{g}_{Rl}$, which has a node at the central g_{Rl}-sphere. Since the extent of the kinky partial wave depends on the g's on the neighboring sites, definition (2.73) merely gives an implicit prescription for finding the g's.

In conclusion, the representation $\{a\}$ specifies the nodes of the $\dot{\phi}^{a}$-functions, and when these are adjusted such as to make $\dot{\phi}^{a}_{RL}$ orthogonal to ϕ^{a}_{RL}, the corresponding kink matrix $K^{g}(E)$ becomes *linear to third order in energy* (apart from negligibly small non-diagonal terms). In this particular, so-called ***nearly orthogonal representation*** ($a\equiv g$), the KKR matching equations (2.63) with good accuracy reduce to eigenvalue equations whose overlap and Hamiltonian matrices are given by $\langle\phi^{g}|H_{mt}-E|\phi^{g}\rangle = K^{g}+(E-E_{\nu})\dot{K}^{g}$. Since the Hamiltonian and overlap matrices are linear in the slope matrix, they have the *two-center* form. The reason for the name ***nearly orthogonal representation*** will become evident when we derive the LMTO method.

It should be remembered, that whereas the *exact* $K^{a}(E)$ matrix yields *all* the energies and wave functions, a linearization of $K^{a}(E)$ can only yield a number of states equal to its dimension. What can be achieved by choice of the nearly orthogonal representation is that, around a chosen expansion point E_{ν}, this ***finite number of energies*** are the best possible approximation to the exact ones.

One drawback of the nearly orthogonal representation is, that the nodes are adjusted to make ϕ^{g} and $\dot{\phi}^{g}$ orthogonal rather than to achieve the shortest possible range and, hence, a two-center ***tight-binding*** formalism. Nevertheless, by judicious choice of the a-parameters a good compromise between linearity, short range, and small matrix dimensions can usually be obtained without having to go to the LMTO formalism which, as we shall see in Section 4, gives energies correct to *third* rather than merely to first order at the expense of containing *three-center* integrals.

We shall now see, how to transform between representations, for instance, from a tight-binding representation to the potential-dependent nearly orthogonal representation, from the tight-binding to a down-folded minimal-basis representation, or from the so-called *bare* representation to a tight-binding representation. This latter transformation is needed for the actual calculation of the slope matrix in real space, and will be dealt with in the following section.

2.6 Transformation between Representations

The solutions (2.67) of the Schrödinger equation are independent of the *representation*, that is, of the nodes chosen for the unitary spherical waves. The set $|\psi^a(E)\rangle$ of wave-equation solutions is a linear combination of any other set $\left|\psi^b(E)\right\rangle$ with the same energy. The nodes determine the shapes of the individual ψ^a-functions (2.74) and this, for instance, allows us to use a tight-binding representation instead of the conventional, long-ranged KKR representation, which will turn out to correspond to $a_{Rl}=0$. We shall now figure out how to *transform* ψ-functions and radial derivatives, D and S, from one arbitrary representation (a) to another (b).

It will prove useful to express the ψ^a_{RL}-function in *all* space as one-center expansions like (2.1):

$$\psi^a_{RL}(E,\mathbf{r}_R)=f^a_{RL}(E,\mathbf{r}_R)+\sum_{R'L'}g^a_{R'L'}(E,\mathbf{r}_{R'})\,S^a_{R'L',RL}(E). \tag{2.74}$$

The functions f^a and g^a are products of a radial function and a spherical harmonic as in eq.(2.2). In principle, the radial functions should be truncated outside their a-sphere and there should be a $\psi^a_{RL}(E,\mathbf{r}_R)$ truncated outside the a-interstitial on the right-hand side of (2.74). Since, however, the radius of convergence of any such one-center expansion is as large as the nearest-neighbor distance, the separation of space into non-overlapping spheres and interstitial can be chosen at convenience. For this reason, we have omitted the interstitial. (The function $g(E,r)$ should not be confused with the position $r=g$ of the node of that energy-derivative function $\dot{\phi}^g$ which is orthogonal to ϕ).

From the definitions of ψ^a and S^a [see eq.s (2.31) and (2.35) and Fig.2.5a] it follows that the radial part of the f^a-function must have *value one* and *slope zero* at a. Similarly, the radial part of the g^a-function must have *value zero* and *slope* $1/a$ at its sphere. That is,

$$\begin{aligned} f^a_{Rl}(E,a_{Rl})&=1, \quad f^a_{Rl}(E,a_{Rl})'=0,\\ g^a_{Rl}(E,a_{Rl})&=0, \quad g^a_{Rl}(E,a_{Rl})'=1/a_{Rl}. \end{aligned} \tag{2.75}$$

$f_{Rl}^{a}(E,r_R)$ and $g_{Rl}^{a}(E,r_R)$ are thus the two linearly independent solutions of the l'th radial wave equation (eq. (2.4) with $v_R(r) \equiv 0$) defined by the initial conditions (2.75). The latter are E-independent and a_{Rl}-dependent . In the following, we shall refer to f and g as the "value" and "slope" functions. Note, that f occurs only in the head of the ψ-function and that the tail of a ψ-function contains only g-functions. The head of a ψ-function usually contains both f and g-functions.

For one-center expansions like (2.74), the following matrix-bra-ket notation will prove convenient:

$$|\psi^a(E)\rangle = |\mathbf{f}^a(E)\rangle + |\mathbf{g}^a(E)\rangle\, S^a(E) \tag{2.76}$$

Here, the ket $|\psi\rangle$ is a row-vector of functions defined in all space, $|\mathbf{f}\rangle$ and $|\mathbf{g}\rangle$ are row-vectors of functions truncated outside their own spheres (conveniently chosen), and S is a matrix. Similarly, the bra's such as $\langle\psi|$ will be column-vector functions.

According to (2.32), the coefficients in the expansion of $\psi_{R'L'}^{b}(\mathbf{r}_{R'})$ in terms of the set $|\psi^a\rangle$ are simply the angular-momentum projections of $\psi_{R'L'}^{b}(\mathbf{r}_{R'})$ on the a-spheres, *viz.*

$$\begin{aligned}\psi_{R'L'}^{b}(\mathbf{r}_{R'}) &= \sum_{RL}\psi_{RL}^{a}(\mathbf{r}_R)\int\delta(r_R-a_R)\,Y_L^{*}(\hat{\mathbf{r}}_R)\,\psi_{R'L'}^{b}(\mathbf{r}_{R'})\,d^3r_R\\ &= \sum_{RL}\psi_{RL}^{a}(\mathbf{r}_R)\left[f_{Rl}^{b}(a_R)+g_{Rl}^{b}(a_R)\,S_{RL,R'L'}^{b}\right].\end{aligned} \tag{2.77}$$

In order to keep the notation simple, we have dropped the common energy E. If we now, in the analogous way, expand the ψ^a-functions in terms of the ψ^b-functions and insert the result on the right-hand side of (2.77), we should regain $\psi_{R'L'}^{b}(\mathbf{r}_{R'})$. This means that

$$\left[f^a(b)+g^a(b)\,S^a\right]\left[f^b(a)+g^b(a)\,S^b\right] = 1, \tag{2.78}$$

in matrix notation where $f^b(a)$ a.s.o. are to be treated as diagonal matrices with elements $f_{Rl}^{b}(a)\,\delta_{RR'}\delta_{LL'}$ a.s.o.. This is the transformation for slope matrices. It is seen to have the form of an *inhomogeneous Dyson's equation.* In order to find S^b, we must thus invert the matrix $\left[f^a(b)+g^a(b)\,S^a\right]$, or better, a hermitian matrix. The result is:

$$S^b + \frac{f^b(a)}{g^b(a)} = \frac{1}{ag^b(a)}\left[\frac{f^a(b)}{ag^a(b)} + S^a a^{-1}\right]^{-1}\frac{1}{g^a(b)}. \tag{2.79}$$

The logarithmic derivative functions $D\{\varphi^o(E,a)\}$ of course transform in the same way as the slope matrices, except that the corresponding equation factorizes in scalar equations because the D's are diagonal matrices. The analogue to (2.76) is, for instance, simply

$$|\varphi^{o}(E)\rangle\,\varphi(E,a)^{-1} = |\mathbf{f}^{a}(E)\rangle + |\mathbf{g}^{a}(E)\rangle\, D\{\varphi^{o}(E,a)\} \tag{2.80}$$

Primarily, we know $f^{a}(b)$ and $g^{a}(b)$ as obtained by integrating the radial Helmholtz equations from a with the "10" and "01" initial conditions (2.75) and to b. One may of course compute $f^{b}(a)$ and $g^{b}(a)$ by the reciprocal procedure, or one may find them from the derivatives of f^{a} and g^{a} at b in the following way:

$|\mathbf{f}^{b}\rangle$ and $|\mathbf{g}^{b}\rangle$ are the solutions of the following 2×2 equation obtained by matching at b:

$$\{|\mathbf{f}^{a}\rangle\,|\mathbf{g}^{a}\rangle\} = \{|\mathbf{f}^{b}\rangle\,|\mathbf{g}^{b}\rangle\}\left\{\begin{array}{cc} f^{a}(b) & g^{a}(b) \\ bf^{a}(b)' & bg^{a}(b)' \end{array}\right\} \tag{2.81}$$

and matrix inversion yields

$$\{|\mathbf{f}^{b}\rangle\,|\mathbf{g}^{b}\rangle\} \equiv \tag{2.82}$$

$$\{|\mathbf{f}^{a}\rangle\,|\mathbf{g}^{a}\rangle\}\left\{\begin{array}{cc} f^{b}(a) & g^{b}(a) \\ af^{b}(a)' & ag^{b}(a)' \end{array}\right\} = \{|\mathbf{f}^{a}\rangle\,|\mathbf{g}^{a}\rangle\}\left\{\begin{array}{cc} bg^{a}(b)' & -g^{a}(b) \\ -bf^{a}(b)' & f^{a}(b) \end{array}\right\}\frac{b}{a}.$$

These are the four desired relations. We have made use of the fact that the value of the determinant of the matrix on the right-hand side of (2.81) is a/b because it equals $W\{b; f^{a}, g^{a}\}/b$, where the Wronskian,

$$W\{r; f, g\} \equiv r^{2}\left[f(r)g'(r) - f'(r)g(r)\right] = rf(r)\,g(r)\left[D\{g(r)\} - D\{f(r)\}\right], \tag{2.83}$$

is independent of r if f and g are solutions of the same second-order differential equation. Its value is therefore

$$W\{f^{a}, g^{a}\} = a. \tag{2.84}$$

We can now insert the expression (2.82) on the right-hand side of (2.79) and obtain the result:

$$bS^{b} - bD\{g^{a}(b)\} = -\frac{1}{g^{a}(b)}\left[\frac{f^{a}(b)}{ag^{a}(b)} + S^{a}a^{-1}\right]^{-1}\frac{1}{g^{a}(b)}. \tag{2.85}$$

The analogous equation for the logarithmic derivative function could be used to find $D\{\varphi^{o}(E,a)\}$ from $D\{\varphi(E,s)\}$, provided that we have found the functions $f^{s}(E,r)$ and $g^{s}(E,r)$. Otherwise, continued backwards integration of the radial Schrödinger equation is the more direct approach.

2.7 Joining onto the conventional KKR and LMTO formalisms

This last sub-section merely serves the purpose of relating to previous work. It is not necessary for the logical development of the formalism.

Even within one and the *same* global a-representation, the *appearance* of the matching equations (2.63) change if the bases in the 2-dimensional (2D) radial Hilbert spaces are chosen differently than in (2.75). We shall now first of all show that with a specific choice of these 2D-bases, the matching equations become the screened KKR equations [40]

$$\sum_{RL} \left[\kappa \cot \eta^a_{Rl}(E)\, \delta_{R'R}\delta_{L'L} + B^a_{R'L',RL}(E) \right] v^a_{RL,j} = 0 \quad \text{for all} \quad R'L'. \tag{2.86}$$

The conventional KKR equations correspond to the *bare* representation ($a \to 0$) which we shall return to in Section 3. Secondly, we shall specify the additive and multiplicative renormalization of the slope matrix $S^a(E)$ which makes it coincide with the conventional [12, 41] LMTO structure matrix S^α when $E = 0$, and which has the least possible E-dependence (note the difference between S^a and S^α). It is worth while noting that only the conventional LMTO structure matrix can be derived by change of the 2D representations; the conventional KKR-ASA formalism cannot, because it treats the energy dependencies inside (E) and outside (κ^2) the spheres on different footings (E is variable, but κ^2 is fixed). The present LMTO tail-cancellation is *exact* for the MT-potential and equivalent with the KKR equations.

Instead of the value and slope functions, f and g in (2.75), we could use so-called n and j^a-functions whose slopes, as well as the value of n at $r = a$ are left unspecified at the moment:

$$\begin{aligned} n_{Rl}(E, a_{Rl}) &= \text{arbitrary}, & n_{Rl}(E, a_{Rl})' &= \text{arbitrary}, \\ j^a_{Rl}(E, a_{Rl}) &= 0, & j^a_{Rl}(E, a_{Rl})' &= \text{arbitrary}. \end{aligned} \tag{2.87}$$

From the invariance of $|\psi^a(E)\rangle$ to a change of basis in each of the 2D sub-spaces, we find by expressing $|\mathrm{f}^a\rangle$ and $|\mathrm{g}^a\rangle$ in (2.76) in terms of $|\mathrm{n}\rangle$ and $|\mathrm{j}^a\rangle$:

$$|\psi^a(E)\rangle = \left(|\mathrm{n}(E)\rangle - |\mathrm{j}^a(E)\rangle \frac{n(E)'}{j^a(E)'} \right) n(E)^{-1} + |\mathrm{j}^a(E)\rangle \left[a j^a(E)' \right]^{-1} S^a(E) \tag{2.88}$$

Here, we have used the same notation as in (2.74) and (2.76). An omitted argument, r, means that $r \equiv a$, and $n(E)'/j^a(E)'$ and $n(E)$ should be considered as diagonal matrices with the respective elements $n_{Rl}(E, a_R)'/j^a_{Rl}(E, a_R)'$ and $n_{Rl}(E, a_R)$. The same expression is valid for the partial waves seen from the outside, $|\varphi^o(E)\rangle / \varphi^o(E, a)$, except that $S^a(E)$ is substituted by the diagonal matrix $D\{\varphi(E, a)^o\}$ (2.80). The one-center expansions (2.88) of the unitary spherical waves may now be compared with the equivalent expansions for the screened Neumann (or Hankel) functions used in the screened KKR and LMTO formalisms [12, 40]:

$$|\mathbf{n}^a(E)\rangle = |\mathbf{n}(E)\rangle + |\mathbf{j}^a(E)\rangle\, \kappa^{-1} B^a(E), \tag{2.89}$$

and the expansion (2.80) of the partial waves seen from the outside may be compared with

$$|\varphi(E)^{\circ}\rangle \propto |\mathbf{n}(E)\rangle - |\mathbf{j}^a(E)\rangle \cot\eta^a(E), \tag{2.90}$$

Here we have chosen KKR notation, but with suitable definitions this and the following expressions are quite general. Comparison with (2.88) now reveals that $|\mathbf{n}^a(E)\rangle = |\psi^a(E)\rangle\, n(E)$, and that the relation between the conventional screened structure matrix and the slope matrix is:

$$\kappa^{-1} B^a(E) = \left[a j^a(E,a)'\right]^{-1} \left[S^a(E) - D\{n(E,a)\}\right] n(E,a) \tag{2.91}$$

The same relation holds between $-\cot\eta^a(E)$ and $D\{\varphi(E,a)^{\circ}\}$. What may be achieved by change of 2D basis vectors and renormalizations of the unitary spherical waves are thus *multiplicative* and *additive* renormalizations of the slope matrix.

The three parameters $n(E,a)$, $n(E,a)'$, and $j^a(E,a)'$ are free to be chosen. In our opinion, one should *not* choose them to be energy dependent, because this would introduce unnecessary energy dependencies of the structure matrix $B^a(E)$. Our choice (2.75) is consistent herewith, but the choices in conventional KKR theory and to some extend in the previous formulation [40] of the general-κ^2 LMTO theory are not.

2.7.1 KKR Equations

In the screened KKR formalism [40]

$$n_{Rl}(E,r) \equiv n_l(\kappa r), \quad \text{and} \quad j^a_{Rl}(E,r) \equiv j_l(\kappa r) - n_l(\kappa r)\tan\beta^a_{Rl}(E) \tag{2.92}$$

with $n_l(\kappa r)$ and $j_l(\kappa r)$ being the conventional spherical Neumann and Bessel functions. They will be discussed in Section 3 in (3.9) and (3.10). The global representation is characterized by the hard-sphere phase shifts $\beta^a_{Rl}(\kappa)$ which are related to the radii of the hard spheres by:

$$\tan\beta^a_{Rl}(E) \equiv \frac{j_l(\kappa a_{Rl})}{n_l(\kappa a_{Rl})}. \tag{2.93}$$

$B^a(E)$ is the screened KKR structure matrix as defined in Ref. [40] (apart from the factor κ^{-1} neglected there) and $\eta^a(E)$ are the screened phase shifts, related to the unscreened ones $\eta(E)$ through

$$\tan\eta^a(E) = \tan\eta(E) - \tan\beta^a(E). \tag{2.94}$$

2.7.2 LMTO Equations

For the initial conditions (2.87) we shall now take the values used in the conventional κ^2=0 LMTO formalism because with this choice, $\mathrm{S}^\alpha(E) \equiv -\kappa^{-1}B^a(E)$ in (2.89) will coincide with the conventional structure matrix S^α when $E=0$ and, otherwise, it will have the weakest possible E–dependence. In the conventional formalism [12] we use:

$$n(E,a) = \left(\frac{a}{w}\right)^{-l-1} \quad \text{and} \quad an(E,a)' = (-l-1)\left(\frac{a}{w}\right)^{-l-1}, \tag{2.95}$$

where the global length scale w is usually taken to be the average WS-radius, and where we have dropped the subscripts R and l, as usual. Moreover,

$$0 \equiv j^a(E,a) \equiv j(E,a) - n(E,a)\alpha(E) \equiv \frac{1}{2(2l+1)}\left(\frac{a}{w}\right)^l - \left(\frac{a}{w}\right)^{-l-1}\alpha(E)$$

which yields the *relation between the screening parameter* α *and the node* a:

$$\underline{\alpha_{Rl} = \frac{1}{2(2l+1)}\left(\frac{a_{Rl}}{w}\right)^{2l+1}}. \tag{2.96}$$

We see that with this choice of initial conditions, the screening parameter is independent of energy. This is very different from the LMTO formalism presented in Ref. [40]. Finally,

$$aj^a(E,a)' \equiv \frac{l}{2(2l+1)}\left(\frac{a}{w}\right)^l - an(E,a)'\alpha = \frac{1}{2}\left(\frac{a}{w}\right)^l \tag{2.97}$$

Using these initial values in (2.91) we obtain the desired relation between the LMTO structure matrix for general energy $\mathrm{S}^\alpha(E)$ and the slope matrix $S^a(E)$:

$$\underline{\mathrm{S}^\alpha_{R'L',RL}(E) \equiv -2\left(\frac{w}{a_{R'l'}}\right)^{l'}\left[S^a_{R'L',RL}(E) + (l+1)\delta_{R'R}\delta_{L'L}\right]\left(\frac{w}{a_{Rl}}\right)^{l+1}} \tag{2.98}$$

This continuation away from zero energy of the conventional LMTO structure matrix is the smoothest possible. Note that $\mathrm{S}^\alpha(E)$ is hermitian and is made dimensionless by introduction of a global length scale w. The slope matrix $S^a(E)$ is not hermitian, but $aS^a(E)$ is. For practical purposes $\mathrm{S}^\alpha(E)$ and $aS^a(E)$ are equally simple to work with; their difference consists of an energy-independent diagonal element and energy-independent scalings. The small-a dependence is different. As will be derived in Section 3, $S^a_{on}(E) \to (-l-1)\delta_{LL'}$ and $S^a_{off}(E \leq 0)$ go to zero like $a^{l+l'+1}$. Eq. (2.98) then shows that $\mathrm{S}^\alpha(E \leq 0)$ tends to a constant.

The corresponding relation between the potential function $\mathrm{P}^\alpha(E) = \cot\eta^a(E)$ in (2.90) and the radial logarithmic derivative is:

$$\underline{\mathrm{P}^\alpha_{Rl}(E) \equiv -\frac{D\{\varphi_{Rl}(E,a_{Rl})^o\} + l + 1}{(2l+1)\alpha_{Rl}}}. \tag{2.99}$$

This expression is different from and more accurate than the usual expression [12]. The matching (tail-cancellation) equations is this conventional LMTO representation are

$$\sum_{RL}\left[\mathrm{P}_{Rl}^{\alpha}(E)\,\delta_{R'R}\delta_{L'L}-\mathrm{S}_{R'L',RL}^{\alpha}(E)\right]v_{RL,j}^{\alpha}=0\quad\text{for all}\quad R'L'. \tag{2.100}$$

Comparing the well-known homogeneous Dyson's equation (1.9) for transforming the conventional structure matrix $\mathrm{S}^{\alpha}(0)$ from the bare to any screened *global* representation with the new, inhomogeneous one (2.78), (2.79), or (2.85), one might fell that it is simpler to work with $\mathrm{S}^{\alpha}(E)$ than with $aS^{a}(E)$. This is, however not the case because for $E\neq 0$ the Dyson's equation for $\mathrm{S}^{\alpha}(E)$ becomes inhomogeneous as well. From (3.26) in Section 3, this equation is found to be:

$$\mathrm{S}^{\alpha}(E)=\frac{\alpha^{-1}}{\tilde{j}(\kappa a)}\left[\alpha^{-1}\frac{\tilde{n}(\kappa a)}{\tilde{j}(\kappa a)}-\mathrm{S}^{0}(E)\right]^{-1}\frac{\alpha^{-1}}{\tilde{j}(\kappa a)}-\alpha^{-1}\left[1+\frac{D\left\{\tilde{j}(\kappa a)\right\}}{2l+1}\right]. \tag{2.101}$$

For $Ea^{2}\rightarrow 0$ this reduces to

$$\mathrm{S}^{\alpha}(0)=\alpha^{-1}\left[\alpha^{-1}-\mathrm{S}^{0}(0)\right]^{-1}\alpha^{-1}-\alpha^{-1}, \tag{2.102}$$

because $\tilde{n}(\kappa a)\rightarrow 1$, $\tilde{j}(\kappa a)\rightarrow 1$, and $D\left\{\tilde{j}_{l}(\kappa a_{Rl})\right\}\rightarrow 0$, as seen from (3.11) and (3.13). Eq.s (2.101-2.102) are the computationally useful version of the Dyson's equation, but only (2.102) depends only *one* parameter, α, is homogeneous, and may be cast in the form (1.9).

3. CALCULATION OF THE STRUCTURE MATRIX

We now evaluate the slope- or structure matrix $S_{RL,R'L'}(E)$ explicitly. This, we do by first choosing a representation in which we know the unitary spherical waves analytically so that we can derive the corresponding slope matrix. Thereafter, we use (2.85) to transform to a tight-binding representation, and this transformation may be carried out by computation in real-space.

3.1 The Bare Representation

The representation in which we know the $\psi_{RL}(E, \mathbf{r}_R)$'s is the so-called *bare-*, or unscreened representation where the a-spheres are vanishingly small $(a \equiv \varepsilon \to 0)$, because in this case $\psi^0_{RL}(E, \mathbf{r}_R)$ is a wave-equation solution, *regular* in *all* space, except at its own site and therefore factorizes into the product of a radial function and a spherical harmonic, $\psi^0_{Rl}(E, r_R) Y_L(\hat{\mathbf{r}}_R)$. Here, the radial function is some linear combination of a spherical Neumann and Bessel function. This we now prove and work out the details.

A bit of care is needed when taking the limit $\varepsilon \to 0$. In particular, we must smooth out the rapid variations of $\psi^\varepsilon_{RL}(E, \mathbf{r}_R)$ on length-scale ε near the other sites (ψ vanishes at the ε–spheres and diverges at their centers). Similarly, the radial $(f^\varepsilon, g^\varepsilon)$-basis also exhibits this rapid, unphysical variation and even becomes linearly dependent in the limit $\varepsilon = 0$, as may be gathered from the vanishing of the Wronskian (2.84).

3.1.1 r small and a small

For small r, two linearly independent solutions of the radial Helmholtz equation are r^l and r^{-l-1}. In terms of these functions, we see that for small ε,

$$f_l^\varepsilon(E, r) = \frac{1}{2l+1}\left[(l+1)\left(\frac{r}{\varepsilon}\right)^l + l\left(\frac{r}{\varepsilon}\right)^{-l-1}\right] \tag{3.1}$$

and

$$g_l^\varepsilon(E, r) = \frac{1}{2l+1}\left[\left(\frac{r}{\varepsilon}\right)^l - \left(\frac{r}{\varepsilon}\right)^{-l-1}\right] \tag{3.2}$$

These, as well as expressions (3.3-3.7) in the following, are valid if both ε^2 and r^2 are much smaller than $2\left|(2l-1)/E\right|$. For the special case $E = 0$, they are valid for all ε and r. The radial part of the L -component of the *head* of $\psi^{\varepsilon}_{RL}(E, \mathbf{r}_R)$ is

$$f^{\varepsilon}_l(E,r) + g^{\varepsilon}_l(E,r)S^{\varepsilon}_{RL,RL}(E) =$$

$$\frac{1}{2l+1}\left[\left\{l+1+S^{\varepsilon}_{RL,RL}(E)\right\}\left(\frac{r}{\varepsilon}\right)^{l} + \left\{l - S^{\varepsilon}_{RL,RL}(E)\right\}\left(\frac{r}{\varepsilon}\right)^{-l-1}\right] \tag{3.3}$$

and the radial part of the L'-component of the *tail* expanded around site R' is

$$g^{\varepsilon}_{l'}(E,r)\, S^{\varepsilon}_{R'L',RL}(E)\ , \tag{3.4}$$

with g given by (3.2).

3.1.2 r small and a vanishing

If we now keep r fixed and let $\varepsilon \to 0$, we see that the components of the tail-expansion become $S^{0}_{R'L',RL}(E)$ times

$$\lim_{\varepsilon\to 0} g^{\varepsilon}_{l'}(E, r \gg \varepsilon) \equiv g^{0}_{l'}(E,r) = \frac{1}{2l'+1}\left(\frac{r}{\varepsilon}\right)^{l'}, \tag{3.5}$$

which are the *regular* solutions. There is only one head component (3.3), and it becomes either

$$\lim_{\varepsilon\to 0} \psi^{\varepsilon}_l(E, r \gg \varepsilon) = (r/\varepsilon)^l \quad \text{with} \quad S^{0}_{RL,RL}(E) = l, \tag{3.6}$$

or

$$\lim_{\varepsilon\to 0} \psi^{\varepsilon}_l(E, r \gg \varepsilon) \equiv \psi^{0}_l(E,r) = (r/\varepsilon)^{-l-1} \quad \text{with} \quad S^{0}_{RL,RL}(E) = -l-1. \tag{3.7}$$

Since *all* tail-components are *regular* solutions, $\psi^{0}_{RL}(E, \mathbf{r}_R)$ is a solution which is regular in all space of the Helmholtz equation, except possibly at its own site $\mathbf{R}$. At this site, ψ^0 is regular in case (3.6) and singular in case (3.7). Since *for* $\varepsilon = 0$ *the boundary conditions have spherical symmetry,* we can write

$$\psi^{0}_{RL}(E, \mathbf{r}) = \psi^{0}_l(E,r)\, Y_L(\hat{\mathbf{r}}), \quad \text{for all} \quad r, \tag{3.8}$$

i.e., the ψ^0-functions are angular-momentum eigenfunctions, and this is what we have used in (3.6-3.7).

Now, the set consisting exclusively of *regular* ψ^0-functions (3.6), is clearly not complete because it can never produce an r^{-l-1}-function. The set of merely *irregular* functions (3.7) *is* complete, as we shall see, because from this set we can construct any set characterized by finite-sized nodal spheres, provided that there is an infinite number of sites. In eq.(2.32) we showed that such sets *are* complete. If the system is finite, the set consisting of the irregular functions at *all* the sites plus the regular functions at *one* site is complete. From now on, we shall therefore let ψ^0 denote only the irregular functions, which for small r are given by eq. (3.7).

3.1.3 Spherical Bessel and Neumann Functions

The two standard, linearly independent solutions of the radial wave-equations are the spherical Bessel and Neumann functions, $j_l(\kappa r)$ and $n_l(\kappa r)$, which are respectively regular and irregular at the origin. Their *normalization* is dictated by a boundary condition at *infinity:*

$$j_l(\kappa r) \rightarrow \begin{cases} \frac{(\kappa r)^l}{(2l+1)!!}\left(1 - \frac{(\kappa r)^2}{2(2l+3)}\cdots\right) & \text{for } r \rightarrow 0 \\ \frac{\sin(\kappa r - l\pi/2)}{\kappa r} & \text{for } r \rightarrow \infty \end{cases}, \tag{3.9}$$

and

$$n_l(\kappa r) \rightarrow \begin{cases} -\frac{(2l-1)!!}{(\kappa r)^{l+1}}\left(1 + \frac{(\kappa r)^2}{2(2l-1)}\cdots\right) & \text{for } r \rightarrow 0 \\ -\frac{\cos(\kappa r - l\pi/2)}{\kappa r} & \text{for } r \rightarrow \infty \end{cases}. \tag{3.10}$$

Here, $\kappa^2 \equiv E$, $(2l+1)!! \equiv (2l+1)\cdot(2l-1)\cdot \ldots \cdot 3\cdot 1$, and $(-1)!! \equiv 1$. The Wronskian has the value

$$W\{j_l(\kappa r), n_l(\kappa r)\} = \kappa^{-1}. \tag{3.11}$$

Whereas the spherical Bessel function is *the* solution which is regular at the origin, the spherical Neumann function is that irregular solution which contains no Bessel function (*i.e.*, it contains no powers of r with the same parity as l). There is a *branch-cut at* κ=0, and when $E < 0$, both functions diverge exponentially for $r \rightarrow \infty$. For negative E, it is therefore convenient to take the irregular solution as the exponentially decreasing Hankel function $n_l(\kappa r) - ij_l(\kappa r)$.

In order to avoid the inconvenient κ-dependent normalizations of the spherical Bessel and Neumann functions, we shall use

$$\tilde{j}_l(\kappa r) \equiv \frac{(2l+1)!!}{(\kappa r)^l} j_l(\kappa r) = 1 - \frac{(\kappa r)^2}{2(2l+3)} + \ldots \tag{3.12}$$

and

$$\tilde{n}_l(\kappa r) \equiv -\frac{(\kappa r)^{l+1}}{(2l-1)!!} n_l(\kappa r) = 1 + \frac{(\kappa r)^2}{2(2l-1)} - \ldots \tag{3.13}$$

to express the *relevant* energy dependence, which is of $\kappa^2 = E$ and not of κ.

3.1.4 a vanishing

For $r \gg \varepsilon \rightarrow 0$, expression (3.7) for the radial part of the ψ^0-function should therefore be substituted by:

$$\psi_l^0(E,r) = (r/\varepsilon)^{-l-1}\,\tilde{n}_l(\kappa r). \tag{3.14}$$

Similarly, the radial part of the expansion of $\psi_{RL}^0(E, \mathbf{r}_R)$ in $Y_{L'}(\mathbf{r}_{R'})$, which for small r was given by (3.5), now becomes

$$g_{l'}^0(E,r) = \frac{1}{2l'+1}\left(\frac{r}{\varepsilon}\right)^{l'} \tilde{j}_{l'}(\kappa r). \tag{3.15}$$

For completeness, we finally mention that

$$f_l^0(E,r) \equiv -g_l^0(E,r)S^0_{RL,RL}(E) + \psi_l^0(E,r)$$
$$= \frac{l+1}{2l+1}\left(\frac{r}{\varepsilon}\right)^l \tilde{j}_l(\kappa r) + \left(\frac{r}{\varepsilon}\right)^{-l-1} \tilde{n}_l(\kappa r) \tag{3.16}$$

Our *bare* ψ^0 and g^0-functions therefore essentially equal the weakly energy-dependent parts, $\tilde{n}_l(\kappa r)$ and $\tilde{j}_l(\kappa r)$, of the spherical Neumann and Bessel functions. The vanishing and diverging prefactors, ε^{l+1} and ε^{-l}, will cancel out of the formalism. Besides, we shall not often use the bare representation, but merely need an analytical expression for the bare slope matrix $S^0(E)$ to be used in (2.85) for calculating the screened matrix $S^a(E)$ corresponding to good-sized nodal spheres.

3.2 Bare Slope Matrix

The spherical-harmonics expansion around the origin of the Neumann function $n_L(\kappa \mathbf{r}_R) \equiv n_l(\kappa r_R) Y_L(\hat{\mathbf{r}}_R)$ centered at a different site $\mathbf{R}$ is known *analytically* [42]. This basic *one-center expansion* is:

$$i^l n_L(\kappa(\mathbf{r}-\mathbf{R})) = \sum_{L'}\sum_{L''} 4\pi C_{LL'L''}\, i^{l'} j_{L'}(\kappa \mathbf{r}) \left[i^{l''} n_{L''}(\kappa \mathbf{R})\right]^*, \quad for \quad r < R, \tag{3.17}$$

where $j_{L'}(\kappa \mathbf{r}) \equiv j_{l'}(\kappa r) Y_{L'}(\hat{\mathbf{r}})$ and

$$C_{LL'L''} \equiv \int Y_L(\hat{r}) Y^*_{L'}(\hat{r}) Y_{L''}(\hat{r}) d\hat{r} = \sqrt{\frac{2l''+1}{4\pi}}\, c^{l''}(L,L') \tag{3.18}$$

are the Gaunt coefficients. These coefficients vanish unless the vector addition rules are obeyed, *i.e.* $m'' = m' - m$ and the l'' -summation in (3.17) only includes the few terms with $l'' = |l'-l|,\ |l'-l|+2,\ \dots,\ l'+l$. The factor $i^{-l+l'-l''}$ is therefore real. The same expansion holds for any linear combination of the Neumann and the Bessel function, that is, with the substitutions: $n_L(\kappa(\mathbf{r}-\mathbf{R})) \to n_L(\kappa(\mathbf{r}-\mathbf{R}))\alpha + j_L(\kappa(\mathbf{r}-\mathbf{R}))\beta$ and $n_{L''}(\kappa \mathbf{R}) \to n_{L''}(\kappa \mathbf{R})\alpha + j_{L''}(\kappa \mathbf{R})\beta$. In terms of the conventional *KKR structure constants* (in real space) [7],

$$B_{R'L',RL}(\kappa) \equiv \sum_{l''} 4\pi\, i^{-l+l'-l''} C_{LL'L''}\, \kappa n^*_{L''}(\kappa(\mathbf{R}-\mathbf{R}')) \tag{3.19}$$

for $\mathbf{R} \neq \mathbf{R}'$ and

$$B_{RL',RL}(\kappa) \equiv 0, \tag{3.20}$$

the expansion (3.17) of the Neumann function at site $\mathbf{R}$ in Bessel functions at site $\mathbf{R}'$ is:

$$\kappa n_L(\kappa \mathbf{r}_R) = \sum_{L'} j_{L'}(\kappa \mathbf{r}_{R'})\, B_{R'L',RL}(\kappa), \tag{3.21}$$

and is valid for $\mathbf{r}$ inside the sphere centered at $\mathbf{R}'$ passing through $\mathbf{R}$. These structure constants form a *hermitian* matrix.

By renormalizing the basic one-center expansion (3.17) to our bare head and tail functions, (3.14) and (3.15), we obtain the desired result for the *bare slope matrix*:

$$\frac{S^0_{R'L',RL}(E)}{\varepsilon^{l+l'+1}} = \frac{-\kappa^{l+l'}}{(2l-1)!!(2l'-1)!!} B_{R'L',RL}(\kappa) = \tag{3.22}$$

$$\sum_{l''} 4\pi C_{LL'l''} \frac{(2l''-1)!!\, i^{-l+l'-l''}}{(2l-1)!!(2l'-1)!!} \frac{Y^*_{l''(m'-m)}\left(\widehat{\mathbf{R}-\mathbf{R}'}\right)}{|\mathbf{R}-\mathbf{R}'|^{l''+1}} \kappa^{l+l'-l''} \tilde{n}_{l''}(\kappa|\mathbf{R}-\mathbf{R}'|)$$

for $\mathbf{R} \neq \mathbf{R}'$, and

$$S^0_{RL',RL}(E) + (l+1)\,\delta_{L'L} = 0. \tag{3.23}$$

for $\mathbf{R} = \mathbf{R}'$. Expression (3.22) is what we use for evaluation in real space. This bare matrix has long range when $E > 0$; it oscillates and falls off like $|\mathbf{R}-\mathbf{R}'|^{-1}$ as may be seen from (3.10). For $E < 0$, it falls off like $|\mathbf{R}-\mathbf{R}'|^{-1} \exp(-|\kappa|\,|\mathbf{R}-\mathbf{R}'|)$, and for $E = 0$, the behavior is $|\mathbf{R}-\mathbf{R}'|^{-l-l'-1}$ for all distances. Note, that $l+l'-l''$ and $-l+l'-l''$ are *even*.

A warning as to the interpretation of $S^0(E)$ in terms of radial derivatives: The $S^\varepsilon(E)$-matrix expresses $\varepsilon\nabla\psi^\varepsilon_{RL}(E, \mathbf{r} = \varepsilon\hat{\mathbf{r}}_{R'})$ *at* the ε–spheres, but this is irrelevant because $\psi^\varepsilon_{RL}(E, \mathbf{r}_R)$ has ε-sized "pinholes" at all other sites.

3.3 Screened Representations

Since we now *explicitly* know the bare ψ^0- and g^0-functions and the bare slope matrix S^0, we can transform to a *short-ranged* (tight-binding) representation, characterized by a set of a-spheres which are somewhat smaller than touching (see eq.(2.34)). For the transformation (2.85) we need $g^0(E,a)$, which we get from (3.15), and

$$D\left\{g^0(E,a)\right\} = D\{j(\kappa a)\} = l + D\left\{\tilde{j}(\kappa a)\right\}, \tag{3.24}$$

as well as

$$\frac{f^0(E,a)}{g^0(E,a)} = \frac{\psi^0(E,a)}{g^0(E,a)} - S^0(E) = (2l+1)\left(\frac{\varepsilon}{a}\right)^{2l+1} \frac{\tilde{n}(\kappa a)}{\tilde{j}(\kappa a)} + l + 1, \tag{3.25}$$

obtained from (3.16). The resulting transformation is:

$$a\left[S^a(E) - D\{j(\kappa a)\}\right] =$$

$$-\frac{2l'+1}{a^{l'}\tilde{j}(\kappa a)} \left[\frac{2l+1}{a^{2l+1}} \frac{\tilde{n}(\kappa a)}{\tilde{j}(\kappa a)} + \frac{S^0(E) + l + 1}{\varepsilon^{l+l'+1}}\right]^{-1} \frac{2l+1}{a^l \tilde{j}(\kappa a)}. \tag{3.26}$$

Note that $[S^0(E)+l+1]\varepsilon^{-l-l'-1}$ is a matrix whose diagonal elements vanish and whose off-diagonal elements equal those of $S^0(E)\varepsilon^{-l-l'-1}$. Transformation (3.26) together with (3.22-3.23) is the main result of the present section.

As discussed before, $S^a(E)$ is well localized when the a-spheres have a good size (2.34) and the energy is not too high. In this case, we can therefore compute $S^a_{RL,R'L'}(E)$ by performing the *matrix inversion* required by (3.26) *in real space*.

For *positive* energies the unitary spherical waves for a *finite* cluster of hard spheres exhibit surface resonances. For positive energies the slope matrix for an *infinite* system of hard spheres can therefore not be calculated reliably from (3.26) using merely a finite cluster. Increasing the size of the cluster so that the resonances sharpen up, is not an efficient cure because the number of resonances increases with the surface area. What we must do, is to mimic the effects in the central region of all the spheres outside the cluster by a hard cavity wall, that is, a concave hard sphere (Watson sphere) which encloses the cluster. The modification of (3.26) needed for this may be found in Ref. [43] or may be derived from Ref. [3].

The behavior of the fcc structure matrix as a function of E shown in Fig. 2.6 was calculated with this cavity technique using a cluster consisting of the central site plus the 4 nearest shells. The surface of the concave sphere had the distance $0.1w$ from the surface of the 4'th shell of convex spheres. For the convex spheres, $\lambda = 2$ and for the concave sphere, $\lambda = 8$. Usually such a high accuracy is not needed and 2-3 shells inside a concave sphere with $\lambda = 4-6$ suffices. The energy dependence of the structure matrix is seen to be very weak, and even weaker than that of the structure matrix with the previous definition [40], not to speak of the bare KKR structure matrix which has a branch cut at $E = 0$ and poles at the free-electron parabola $E = k^2$. The poles of the screened structure matrix start at the bottom of the hard-sphere continuum, which is above the energy range of interest for ground state properties.

In Fig. 3.1 we show the dependence of the $ss\sigma$-elements of the fcc structure matrix $S^a(E=0)$ on the hard sphere radius. Note, that for small a the on-site term goes to -1 and the off-site terms vanish proportional to a or faster. It may be seen that there is a comfortable range of hard-sphere radii in which the localization is good.

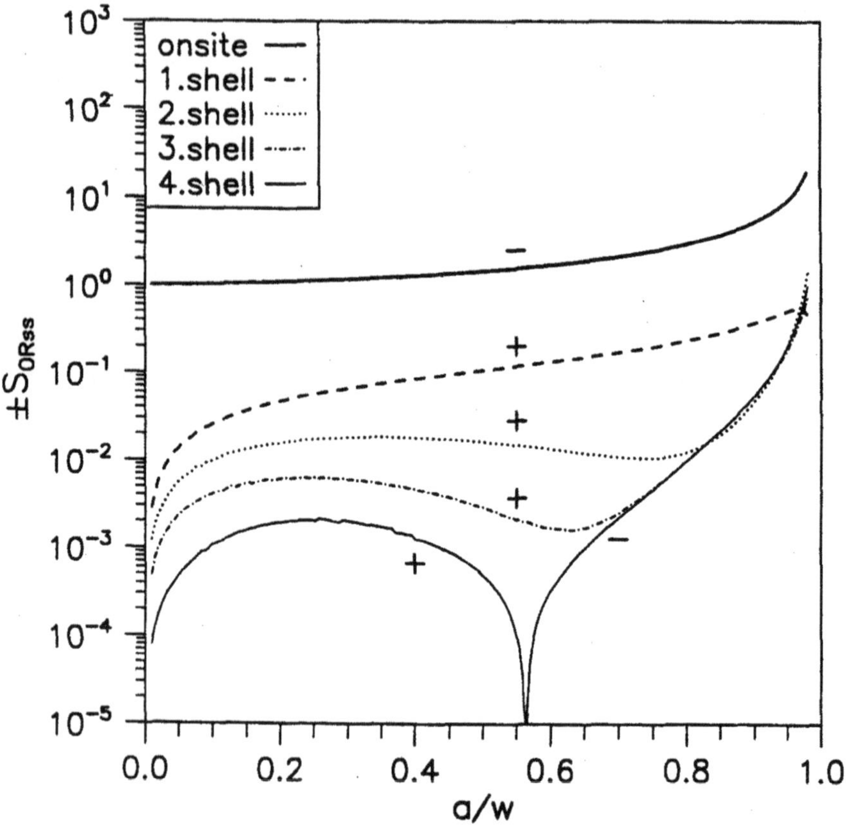

Figure 3.1 $ss\sigma$ slope matrix as a function of a/w.

4. MUFFIN TIN ORBITALS

The kinky partial waves $|\phi(E)\rangle$ defined in (2.65) give exact solutions for slightly overlapping [40] MT-potentials. We now want to modify these energy-dependent, discontinuous basis functions in such a way that they become *differentiable in all space* and *independent of energy* to *first* order in $E - E_\nu$. This gives the advantages that the resulting *muffin-tin orbitals* (MTO's) are simple to work with and can be used for *fixed* energy ($E \equiv E_\nu$) in connection with the Raleigh-Ritz variational principle for the MT-Hamiltonian to yield eigenvalues which equal the one-electron energies plus an error term proportional to $\left[(E - E_\nu)^2\right]^2$. This means that the eigenvalues are correct to third order. These smooth MTO's at fixed energy are what we call *linear* muffin-tin orbitals (LMTO's). This is a new development, to the extend that the LMTO's defined previously [40] had errors of first, rather than of second order in the *interstitial* region and therefore gave rise to energy errors of order $(E - E_\nu)^2$ with a pre factor depending on the relative size of the interstitial region. With LMTO's we can use standard Hamiltonian techniques to include any kind of perturbations to the overlapping MT potential.

4.1 Muffin Tin Orbitals

The modification of the set of kinky partial waves (2.65) is:

$$\underline{|\chi(E)\rangle \equiv |\phi(E)\rangle - |\dot{\phi}\rangle \dot{K}^{-1} K(E)}, \tag{4.1}$$

which is named the MTO set. As usual, an omitted energy argument means $E = E_\nu$. Moreover, we have dropped the superscripts a labelling the (arbitrary) representation. The MTO set has the following properties:

It is *complete* with respect to the MT-potential used for its generation because the linear combination $|\chi(E_j)\rangle \mathbf{v}_j$ of MTO's specified by a solution $\mathbf{v}_j$ of the KKR equations, $K(E_j)\mathbf{v}_j = \mathbf{0}$ (2.63), solves Schrödingers equation:

$$|\chi(E_j)\rangle \mathbf{v}_j = \left[|\phi(E_j)\rangle - |\dot{\phi}\rangle \dot{K}^{-1} K(E_j)\right] \mathbf{v}_j = |\phi(E_j)\rangle \mathbf{v}_j = |\Phi_j\rangle . \tag{4.2}$$

In the last equation we have used (2.67). Eq.(4.2) expresses the MTO "tail-cancellation condition" [12].

The MTO is everywhere *continuous and differentiable* because the kink matrix for the MTO, i.e. the matrix whose $(R'L', RL)$-element is the kink of the RL-MTO at the $a_{R'L'}$-sphere in the $R'L'$-channel, is a^{-1} times

$$K(E) - \dot{K}\dot{K}^{-1}K(E) = 0, \tag{4.3}$$

as may be seen from (4.1).

The MTO is *independent of energy to order* $(E - E_\nu)^1$ because energy differentiation of (4.1) yields:

$$|\dot{\chi}\rangle = \mathbf{0}. \tag{4.4}$$

The Taylor series of the MTO is therefore:

$$|\chi(E)\rangle \equiv \left(|\phi\rangle - \left|\dot{\phi}\right\rangle \dot{K}^{-1}K\right) + \left(\left|\ddot{\phi}\right\rangle - \left|\dot{\phi}\right\rangle \dot{K}^{-1}\ddot{K}\right)\frac{1}{2}(E - E_\nu)^2 - .... \tag{4.5}$$

In the nearly orthogonal representation $\ddot{K}^g$=0, so that the second-order term only contains the function $\left|\ddot{\phi}^g\right\rangle$ which essentially vanishes outside its own sphere due to the weak energy dependence of the unitary partial waves.

We can easily understand the behavior of the MTO (4.1) if we use the conventional approximation [40] of neglecting the energy dependence in the interstitial, that is, if we neglect the small functions$\left|\dot{\psi}\right\rangle$ [see Fig. 2.4d], $\dot{S}$, and that energy-dependence of $D\{\varphi(E, a)^o\}$ which arises through the backwards extrapolation from the MT- to the hard sphere. With this approximation the MTO is a unitary spherical wave at energy E in the interstitial region and inside the spheres each low partial wave is substituted by that linear combination of the partial wave at E and its energy derivative function at E_ν which makes the MTO continuous and differentiable. Since the MTO-envelope is the unitary partial wave, the MTO is *localized.* In the formal definition (4.1) the multiplication with $\dot{K}^{-1}$ does not cause any delocalization because, in this approximation, $-\dot{K}^{-1} = -\left[a\dot{D}\right]^{-1} = \langle\varphi^2\rangle^{-1}$, which is diagonal and just normalizes the partial waves. In this approximation, the definition (4.1) thus reduces to the conventional one [12, 40].

Without this approximation and with E_ν well below the hard-sphere continuum, the positive definite matrix $-\dot{K}$ remains nearly diagonal and can easily be inverted. When E too, is well below the hard-sphere continuum, the multiplication with $\dot{K}^{-1}K(E)$ in (4.1) causes little delocalization. The role of $\left|\dot{\psi}\right\rangle$ in $\left|\dot{\phi}\right\rangle \equiv |\dot{\varphi}\rangle + \left|\dot{\psi}\right\rangle$ is to make the new MTO set complete in the interstitial region. Note, that in this new formulation the interstitial region is treated on exactly the same footing as the intra-sphere regions. Apart from increasing the accuracy by one order in $E - E_\nu$, this makes the formalism more elegant because the only matrix we need to work with is the kink-, or KKR-matrix $K(E)$.

In view of the triple-valuedness illustrated in Fig. 2.8 of the kinky partial wave in the spherical $a \rightarrow s$ shell, we would like to point out that the MTO (4.1) is *single-valued* to first order in $E-E_\nu$ and, for non-overlapping MT's, it takes the value $\varphi(E,r)-\dot{\varphi}^a(r)\dot{K}^{-1}K(E)$ given by the *forwards* integrated radial function and its energy derivative. The radial function $\varphi(E,r)^\circ - \dot{\varphi}^a(r)^\circ \dot{K}^{-1}K(E)$ integrated backwards from s to a, matches continuously *and differentiably* onto the appropriate L-projection of the interstitial function $\psi(E,\mathbf{r}) - \dot{\psi}(\mathbf{r})\dot{K}^{-1}K(E)$ at *both* end-points, s and a. The difference in curvature of the two functions is of order $(E-E_\nu)^2$. All contributions from integrals over these two latter functions are therefore small and cancel to order $(E-E_\nu)^2$.

Here we are always using the normalization (2.66) which has become possible after the inactive partial waves have been removed from the basis by setting the corresponding hard-sphere radii equal to the scattering lengths. All inactive partial waves are then treated implicitly, as tails of the unitary spherical waves.

4.2 Linear Muffin Tin Orbitals

The set of LMTO's is obtained by dropping the second and higher order terms in (4.5):

$$|\chi\rangle = |\phi\rangle - \left|\dot{\phi}\right\rangle \dot{K}^{-1}K = |\phi\rangle + \left|\dot{\phi}\right\rangle O^{-1}(H - E_\nu O), \tag{4.6}$$

and is strictly *single-valued* in the $a \to s$ shells. In the second equation we have used the facts that $-\dot{K}$ is the overlap matrix and K the Hamiltonian matrix with respect to E_ν in the basis of the *kinky partial waves* (2.72). Also, remember that O and H both have *two-center* form with hopping integrals specified by the slope matrix S and its energy derivative matrix $\dot{S}$. Note, that we have not written $O^{-1}E_\nu O = E_\nu$ because in practice we may choose Rl-dependent energy expansion parameters such that E_ν is a diagonal matrix with different elements $E_{\nu Rl}$, and therefore does not commute with O. Finally, we stress, that H and O refer to the kinky partial-wave basis rather than to the LMTO basis. The latter matrices we shall now evaluate.

4.2.1 LMTO Overlap Matrix and MT-Hamiltonian

For the integrals in all space of a kinky partial wave and one of its energy derivatives, we previously derived the equations (2.69-2.71):

$$\langle \phi \mid \phi \rangle = -\dot{K},$$

$$\left\langle \phi \mid \dot{\phi} \right\rangle = \left\langle \dot{\phi} \mid \phi \right\rangle = -\frac{1}{2}\ddot{K},$$

$$\left\langle \dot{\phi} \mid \dot{\phi} \right\rangle = -\frac{1}{6}\dddot{K}.$$

For the LMTO *overlap matrix*,

$$\langle \chi \mid \chi \rangle \equiv \langle \phi \mid \phi \rangle - \left\langle \phi \mid \dot{\phi} \right\rangle \dot{K}^{-1}K - K\dot{K}^{-1}\left\langle \dot{\phi} \mid \phi \right\rangle + K\dot{K}^{-1}\left\langle \dot{\phi} \mid \dot{\phi} \right\rangle \dot{K}^{-1}K, \tag{4.7}$$

we then find the following important result:

$$\underline{\langle \chi \mid \chi \rangle = -\dot{K} + \frac{1}{2}\left\{\ddot{K}\dot{K}^{-1}K + K\dot{K}^{-1}\ddot{K}\right\} - \frac{1}{6}K\dot{K}^{-1}\,\dddot{K}\,\dot{K}^{-1}K} \tag{4.8}$$

The LMTO overlap matrix is thus expressed solely in terms of the kink (KKR) matrix at the chosen energy E_ν and its first three energy derivatives. In practice it may be more convenient to evaluate the *radial* integrals than to differentiate the logarithmic derivative functions $D\{\varphi(E,a)^o\}$ three times. For the unitary spherical waves, however, energy differentiation of the slope matrix is the only efficient way of calculating the integrals over the a-interstitial.

Now we understand why the g-representation in which $\ddot{K}$ vanishes is called the nearly orthogonal one; it makes the LMTO's nearly orthogonal.

The matrix elements of the MT-Hamiltonian used to generate the LMTO set may be found in a similar way. Since the LMTO is continuous and differentiable, there are no problems with hermiticity like those occurring for the matrix elements between kinky partial waves *alone*. What we mean is that the result (2.72): $\langle \phi | H_{mt} - E_\nu | \phi \rangle = K$ cannot be obtained by naively taking the matrix element of an equation like

$$(H_{mt} - E_\nu)\,|\phi\rangle = 0\,; \tag{4.9}$$

this would yield zero. For matrix elements between continuously differentiable *linear combinations* of kinky partial waves, like LMTO's,

$$\begin{aligned}\langle \chi | H_{mt} - E_\nu | \chi \rangle &= \langle \phi | H_{mt} - E_\nu | \phi \rangle - \left\langle \phi | H_{mt} - E_\nu | \dot{\phi} \right\rangle \dot{K}^{-1}K \\ &- K\dot{K}^{-1}\left\langle \dot{\phi} | H_{mt} - E_\nu | \phi \right\rangle + K\dot{K}^{-1}\left\langle \dot{\phi} | H_{mt} - E_\nu | \dot{\phi} \right\rangle \dot{K}^{-1}K,\end{aligned} \tag{4.10}$$

such procedures are however correct when used consistently for all terms. By use of (4.9) and its energy derivative:

$$(H_{mt} - E_\nu)\left|\dot{\phi}\right\rangle = |\phi\rangle\,, \tag{4.11}$$

but not simultaneously with (2.72), we evaluate (4.10)as follows:

$$\begin{aligned}\langle \chi | H_{mt} - E_\nu | \chi \rangle &= -\left\langle \phi | H_{mt} - E_\nu | \dot{\phi} \right\rangle \dot{K}^{-1}K + K\dot{K}^{-1}\left\langle \dot{\phi} | H_{mt} - E_\nu | \dot{\phi} \right\rangle \dot{K}^{-1}K \\ &= -\langle \phi \mid \phi \rangle \dot{K}^{-1}K + K\dot{K}^{-1}\left\langle \dot{\phi} \mid \phi \right\rangle \dot{K}^{-1}K.\end{aligned}$$

The final result for the matrix elements of the *MT-Hamiltonian* is therefore simply:

$$\underline{\langle \chi | H_{mt} - E_\nu | \chi \rangle = K - \frac{1}{2}K\dot{K}^{-1}\ddot{K}\dot{K}^{-1}K.} \tag{4.12}$$

In the nearly orthogonal representation, this reduces to K^g, and it is thus the overlap matrix (4.8) which provides the third-order correction.

4.2.2 *Conventional Normalization of the Kinky Partial Waves and the LMTO's*

The normalization of the kinky partial waves and the LMTO's which corresponds to the conventional one [12] is:

$$\left|\bar{\phi}(E)\right\rangle \equiv |\phi(E)\rangle \langle\phi \mid \phi\rangle^{-\frac{1}{2}} \tag{4.13}$$

such that, by energy differentiation,

$$\left|\dot{\bar{\phi}}\right\rangle = \left|\dot{\phi}\right\rangle \langle\phi \mid \phi\rangle^{-\frac{1}{2}}, \tag{4.14}$$

and

$$|\bar{\chi}\rangle \equiv |\chi\rangle \langle\phi \mid \phi\rangle^{-\frac{1}{2}} = \left|\bar{\phi}\right\rangle + \left|\dot{\bar{\phi}}\right\rangle h. \tag{4.15}$$

Here,

$$h \equiv \langle\phi \mid \phi\rangle^{-\frac{1}{2}} \langle\phi | H_{mt} - E_\nu | \phi\rangle \langle\phi \mid \phi\rangle^{-\frac{1}{2}},$$

that is,

$$\underline{h = O^{-\frac{1}{2}} (H - E_\nu O) O^{-\frac{1}{2}} = \left[-\dot{K}\right]^{-\frac{1}{2}} K \left[-\dot{K}\right]^{-\frac{1}{2}}} \tag{4.16}$$

is the new accurate definition of the two-center Hamiltonian (1.10). The integrals over the renormalized Kinky waves are:

$$\left\langle\bar{\phi} \mid \bar{\phi}\right\rangle = 1, \tag{4.17}$$

$$\left\langle\bar{\phi} \,|\dot{\bar{\phi}}\right\rangle = \left\langle\dot{\bar{\phi}}| \,\bar{\phi}\right\rangle = -\frac{1}{2}\left[-\dot{K}\right]^{-\frac{1}{2}} \ddot{K} \left[-\dot{K}\right]^{-\frac{1}{2}} \tag{4.18}$$

and

$$\left\langle\dot{\bar{\phi}}|\dot{\bar{\phi}}\right\rangle = -\frac{1}{6}\left[-\dot{K}\right]^{-\frac{1}{2}} \dddot{K} \left[-\dot{K}\right]^{-\frac{1}{2}} \tag{4.19}$$

The notation is now exactly like in conventional LMTO theory. The new meaning given to the symbols is that $\left|\bar{\phi}\right\rangle$ and $\left|\dot{\bar{\phi}}\right\rangle$ are no longer the normalized partial wave and its $\left|\dot{\bar{\phi}}^{a}\right\rangle$-derivative *truncated* outside the MT-sphere, but they now have (localized) tails, have become kinky partial waves. Here again, the multiplication of the set of kinky partial waves with the matrix $\langle\phi \mid \phi\rangle^{-1/2} = \left[-\dot{K}\right]^{-1/2}$ is essentially a renormalization rather than a Löwdin orthogonalization because $\dot{K}$ is essentially diagonal.

With this conventional normalization the LMTO overlap matrix and Hamiltonian matrices are given by the conventional expressions:

$$\langle\bar{\chi} \mid \bar{\chi}\rangle \equiv 1 + \left\langle\bar{\phi} \,|\dot{\bar{\phi}}\right\rangle h + h\left\langle\dot{\bar{\phi}}| \,\bar{\phi}\right\rangle + h\left\langle\dot{\bar{\phi}}|\dot{\bar{\phi}}\right\rangle h \tag{4.20}$$

and

$$\langle \bar{\chi} | H_{mt} - E_\nu | \bar{\chi} \rangle = h + h \left\langle \bar{\phi} \,|\, \dot{\bar{\phi}} \right\rangle h = h + h \left\langle \dot{\bar{\phi}} \right| \bar{\phi} \Big\rangle h. \tag{4.21}$$

Finally, in the nearly orthonormal representation,

$$\underline{\langle \bar{\chi}^g \mid \bar{\chi}^g \rangle \equiv 1 + h^g \left\langle \dot{\bar{\phi}}^g \,|\, \dot{\bar{\phi}}^g \right\rangle h^g} \tag{4.22}$$

and

$$\underline{\langle \bar{\chi}^g | H_{mt} - E_\nu | \bar{\chi}^g \rangle = h^g}. \tag{4.23}$$

5. REPRESENTING THE CHARGE DENSITY

In this final section we shall briefly indicate how the output charge density (1.2) is represented such that Poisson's equation can be solved and the Coulomb integrals evaluated.

The clue to this problem is to expand the charge density in the interstitial region in the same unitary spherical waves and their first energy derivatives as used for the wave functions. Since these waves satisfy the Helmholtz equation, Poisson's equation can be solved with due course given to the solutions of the Laplace equation. The continuous and differentiable fit to ψ^a and $\dot{\psi}^a$ at the a-spheres is performed for the charge density $\sim \varphi^o_{RL}\varphi^o_{RL'}$ seen from the outside. The remaining charge density is localized to inside the spheres and is treated numerically via the radial Poisson's equations [35, 37].

The representation of the interstitial charge density is thus

$$\rho(\mathbf{r}) \approx \sum_{RL} \left[\psi^a_{RL}(\mathbf{r}_R)\, \rho_{RL}(a_{Rl})^o + \dot{\psi}^a_{RL}(\mathbf{r}_R) \sum_{R'L'} X_{RL,R'L'}\, \rho_{R'L'}(a_{R'l'})^o \right] \tag{5.1}$$

The matrix X which makes the fit differentiable is easily found from the condition that the kink matrix for $|\rho\rangle - |\psi\rangle\,\rho - |\dot{\psi}\rangle\,X\rho$ vanishes, that is, X must solve the matrix equation

$$D\{\rho\} - S - \dot{S}X = 0$$

whose formal solution is:

$$X = \dot{S}^{-1}\left(D\{\rho\} - S\right). \tag{5.2}$$

This can all be made to work in real space. The resulting decomposition of the charge density in localized contributions are shown for silicon in Fig. 5.1. The upper part of the figure shows the Si contribution and the lower the empty-sphere contribution. Summing up the localized contributions yields the periodic charge density shown in Fig. 1.3 in the Introduction.

With this decomposition, we may solve Poisson's equation and fit the interstitial potential in exactly the same way as we did for the charge density. Finally, we compute the Coulomb energy using $\dot{S}$, $\ddot{S}$, and $\dddot{S}$ for the interstitial integrals.

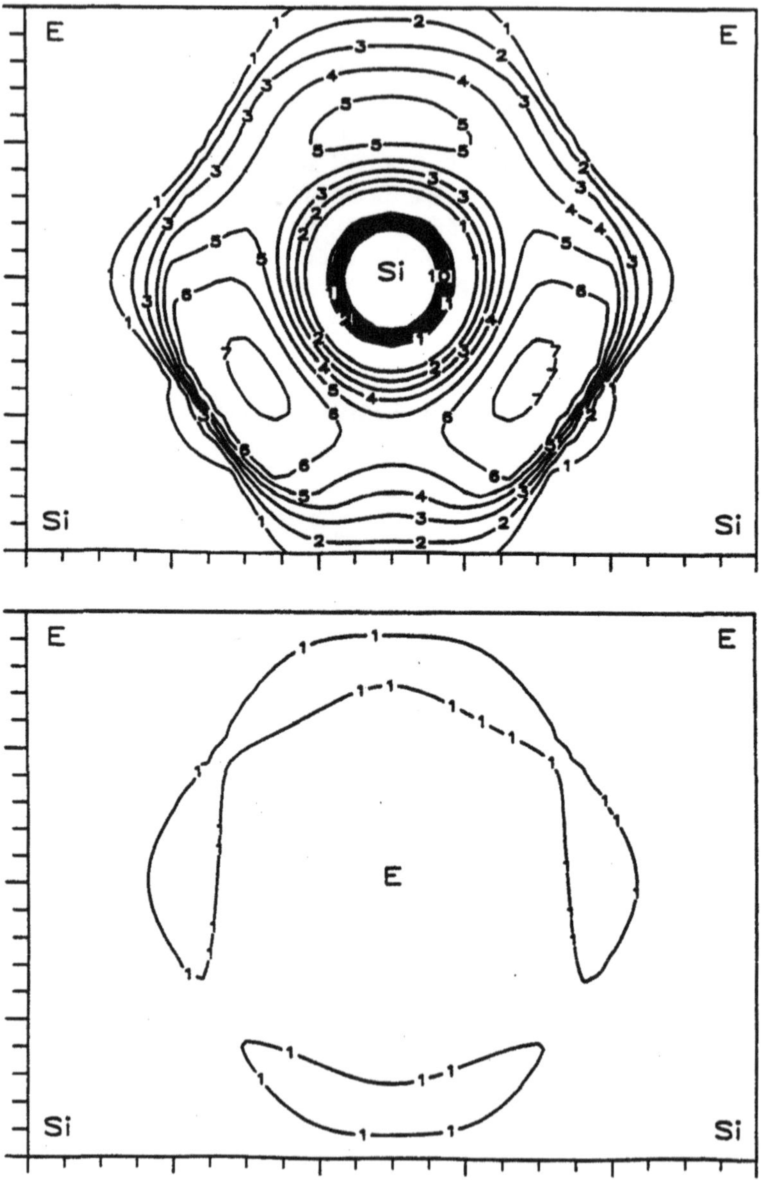

Figure 5.1 Localized Si and E centered charge density contributions.

1. G. Krier, O.K. Andersen, and O. Jepsen (to be published)
2. O.K. Andersen, *Solid State Commun.* **13** (1973) 133.
3. O. K. Andersen and R. G. Wolley, *Mol. Phys.* **26** (1973) 905.
4. O.K. Andersen, *Phys. Rev.* **B12** (1975) 3060.
5. A. R. Williams, J. Kübler, and C. D. Gelatt, *Phys. Rev.* **B19** (1979) 6094.
6. H. L. Skriver, *The LMTO Method* (Springer, New York, 1984).
7. J. Korringa, Physica (Utrecht) **13**, (1947) 392; W. Kohn and J. Rostoker, Phys. Rev. **94** (1954) 1111.
8. O.K. Andersen, in *The Electronic Structure of Complex Systems*, eds. W. Temmerman and P. Phariseau (Plenum, 1984).
9. O. K. Andersen and O. Jepsen, *Phys. Rev. Lett.* **53** (1984) 2571.
10. O. K. Andersen, O. Jepsen, and D. Glötzel, in *Highlights of Condensed-Matter Theory* eds. F. Bassani, F. Fumi, and M. P. Tosi (North-Holland, New York, 1985).
11. W.R.L. Lambrecht and O.K. Andersen, *Phys. Rev.* **B34** (1986) 2439; O.K. Andersen, T. Paxton, M. Schilfgaarde, and O. Jepsen (unpublished).
12. O. K. Andersen, Z. Pawlowska, and O. Jepsen, *Phys. Rev.* **B34** (1986) 5253.
13. R. Haydock, V. Heine, and M. J. Kelly, *J. Phys. C: Solid St. Phys.* **8** (1975) 2591; and R. Haydock in the present volume.
14. H.J. Nowak, O.K. Andersen, T. Fujiwara, O. Jepsen, and P. Vargas, *Phys. Rev.* **B44** (1991) 3577.
15. P. Vargas, this volume.
16. O. Gunnarsson, O. Jepsen, and O. K. Andersen, *Phys. Rev.* **B27** (1983) 7144.
17. O. Gunnarsson, O.K. Andersen, O. Jepsen, and J. Zaanen, *Phys. Rev.* **B39** (1989) 1708
18. V.I. Anisimov, J. Zaanen, and O.K. Andersen, *Phys. Rev.* **B44** (1991) 943
19. O. Jepsen and O.K. Andersen, Z. Phys. (submitted).
20. J. Kudrnovsky, V. Drchal, and J. Masek, *Phys. Rev.* **B35** (1987) 2487; J. Kudrnovsky, S.K. Bose, and O.K. Andersen, *Phys. Rev.* **B43** (1991) 4613.
21. W. R. L. Lambrecht and O. K. Andersen, *Surface Science* **178** (1986) 256.
22. H.L. Skriver and N.M. Rosengaard, *Phys. Rev.* **B43** (1991) 9538.
23. B. Velicky and J. Kudrnovsky, *Surf. Sci.* **64** (1977) 411; J. Kudrnovsky, P. Weinberger, and V. Drchal, *Phys. Rev.* **B44** (1991) 6410.
24. A. Svane and O. K. Andersen, *Phys. Rev.* **B34** (1986) 5512.
25. A. Svane and O. Gunnarsson, *Phys. Rev. Lett.* **65** (1990) 1148.
26. Z. Szotek, W.M. Temmerman, and H. Winter, *Phys. Rev. Lett.* **72** (1994) 1244.
27. F. Aryasetiawan, *Phys. Rev.* **B46** (1992) 13051
28. Z. Pawlowska, N. E. Christensen, S. Satpathy, and O. Jepsen, *Phys. Rev.* **B34** (1986) 7080.

29. O.K.Andersen, O. Jepsen, A.I. Liechtenstein, and I.I. Mazin, *Phys. Rev.* **B49** (1994) 4145.

30. P. Blöchl, *Thesis, University of Stuttgart* (1989).

31. J. Kollar and H. L. Skriver (to be published)

32. G.W. Fernando, B.R. Cooper, M.V. Ramana, H. Krakauer, and C.Q. Ma, *Phys. Rev. Lett.* **56** (1986) 2299.

33. K.H. Weyrich, *Phys. Rev.* **B37** (1988) 10269.

34. J.M. Wills (unpublished); J.M. Wills and B.R. Cooper, *Phys. Rev.* **B36** (1987) 3809.

35. M. Springborg and O.K. Andersen, *J. Chem. Phys.* **87** (1987) 7125.

36. M. Methfessel and M. Schilfgaarde (to be published).

37. M. Methfessel, *Phys. Rev.* **B38** (1988) 1537; M. Metfessel, C.O. Rodriguez, and O.K. Andersen, *Phys.Rev.* **B40** (1989) 2009.

38. S. Savrasov and D. Savrasov, *Phys. Rev.***B46** (1992) 12181. In this full-potential LMTO method the integrals over all space are performed as sums of integrals over cells. Each cell integral is performed as a surface integral and thus avoids the slowly converging spherical-harmonics expansion of the cellular step-function $\theta_R(\mathbf{r})$.

39. S.Yu. Savrasov, *Phys. Rev. Lett.* **69** (1992) 2819

40. O.K. Andersen, A.V. Postnikov, and S.Yu. Savrasov, in *Applications of Multiple Scattering Theory to Materials Science,* eds. W.H. Butler, P.H. Dederichs, A. Gonis, and R.L. Weaver. (Materials Research Society symposium proceedings, Pittsburgh, 1992); A.V. Postnikov and O.K. Andersen, (preprint).

41. O. K. Andersen, O. Jepsen, and M. Sob, in *Electronic Band Structure and its Applications*, ed. M. Yussouff (Springer Lecture Notes, 1987).

42. W. Hobson, *The Theory of Spherical and Ellipsoidal Harmonics* (New York, N.Y. 1955)

43. G. Krier and O.K. Andersen (to be published)

SOLVING ELECTRONIC STRUCTURE PROBLEMS WITH THE RECURSION METHOD

ROGER HAYDOCK
Department of Physics and Materials Science Institute
University of Oregon
Eugene, Oregon 97403, USA

ABSTRACT

The goal of these lectures is to achieve a working knowledge of how to apply the recursion method to problems in electronic structure. The first lecture introduces the basic issues of the electronic structure of condensed matter and explains how the recursion method addresses them. In a macroscopic system it is only possible to compute local quantities which converge exponentially with computational effort due to a quantum analogue of the black body radiation theorem. The second lecture develops the recursion method from quantum path counting, leading to the chain model and its physical interpretation. The third lecture is about the calculation of electronic energies, bounds on them, and differences in total energies. The last lecture is on the recursion method in finite precision, ghosts and how to eliminate their consequences. In summary, the contents of the lectures may be described as: 1. Local environment and the quantum black body theorem, 2. Quantum path counting and The Recursion Method, 3. Orbital peeling-microscopic differences of macroscopic energies, 4. Finite precision and ghosts.

1. Local Environment and the Quantum Black Body Theorem

a. Adiabatic Electrons

The first step in applying the recursion method is to choose a model for the electronic equations of motion. Like the other lecturers, I adopt the independent, or one-electron model, which I prefer to call the adiabatic model because its minimum assumptions are that there is no flow of entropy between electrons or between the electrons and the lattice. Another way of saying this is that there are states for each electron with infinite lifetimes. These stationary states are the ones we seek to study. Of course, this model is approximate because single-electron states have only finite lifetimes in real materials, although these lifetimes are often very long with respect to the quantum mechanical period of the state.

We further assume that the adiabatic electrons obey the Schroedinger Equation:

$$-i\hbar\hat{\Psi} = \left[-\left(\hbar^2/2m\right)\nabla^2 + V(r)\right]\Psi \tag{1}$$

where V(r) is some real potential and Ψ is one of these adiabatic states ($\hbar$ is Planck's Constant and m is the mass of the electron which need not be its free mass).

b. Sparse Representation

In addition we assume that the equations of motion can be written in a sparse form. Suppose that there are atomic-like orbitals φ_α such that

$$\Psi = \sum_\alpha \Psi_\alpha \varphi_\alpha \tag{2}$$

and that

$$-i\hbar\dot{\Psi}_\alpha = \sum_\beta t_{\alpha\beta}\Psi_\beta \tag{3}$$

where the parameters $\{t_{\alpha\beta}\}$ form a matrix H representing the Hamiltonian operator. We do not require that the $\{\varphi_\alpha\}$ be orthogonal, only that the overlap matrix S with elements $S_{\alpha\beta}$ be known and that its diagonal elements be unity. The property of sparseness is that the sums $\sum_\alpha |t_{\alpha\beta}|$ and $\sum_\beta |t_{\alpha\beta}|$ all converge exponentially — there are only a finite number of significant elements in each row and column of H (and S). This sparseness is the essence of "tight-binding" but is sufficient to represent a completely general Hamiltonian. It is also the property on which convergence of the recursion method depends.

c. Strong Coupling

Solving the above equations of motion is difficult because in most solids the Hamiltonian H is degenerate rather than diagonally dominant. The problem is that the off diagonal elements of H are bigger than the differences between diagonal elements. What is worse is that this degeneracy is of (infinite) macroscopic dimension.

The consequence of degeneracy is that the equations cannot be solved by perturbation theory because it does not converge. If H were diagonally dominant then perturbation theory would be appropriate.

If the system has a complete set of translation symmetries (is crystalline) then a basis of Bloch waves reduces H to a block diagonal form in which each diagonal block is also diagonally dominant. So, for crystals, perturbation theory can be applied to the band structure.

The problem which we address with the recursion method is the case when, even after symmetries have been taken into account, there are degenerate matrices of macroscopic

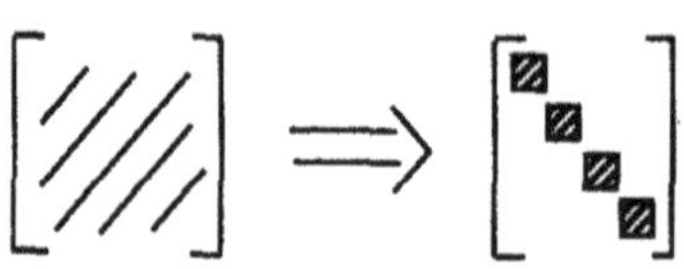

Fig. 1 Reduction of the Hamiltonian to block diagonal form.

dimension which describe electronic motion. This is the case when surfaces or interfaces are present, and more so for line or point defects in crystals as well as for amorphous and quasi-crystalline materials.

d. Instability of Global Solutions

One might think that for practical purposes, we should take a large submatrix of H and diagonalize it on our biggest computer. Out of such a calculation would come a long list of eigenstates and their energies. The problem with these eigenstates is that if we change so much as one matrix element in the submatrix of H, they change completely. This is a consequence of degeneracy, namely that the eigenstates are spaced so closely in energy and are so delocalized over all the basis elements that even a small change in one matrix-element completely changes the eigenstates. A simple example is of two very different but almost degenerate eigenstates which hybridize and mix significantly when a matrix element is changed.

The problem is that eigenstates are global solutions of the equations of motion and are unstable because of (near) degeneracy. In order to solve the electronic motion in solids we need a stable quantity which describes electronic motion. This is the projected density of states explained below.

e. Projected Density of States

The projected density of states was introduced into condensed matter physics by Friedel as a way of describing the electronic structure of surfaces. It is defined as a density in energy with the density for each energy being the probability that an electron (or other excitation) with that energy is in the given orbital of projection. For a finite system with normalized eigenstates $\{\Psi_\alpha\}$, then the projected density of states on an orbital u is:

$$g_u(E) = \sum_\alpha |u * \Psi_\alpha|^2 \delta(E - E_\alpha) \tag{4}$$

where $u^* \Psi_\alpha$ is the $\int u(r)^* \Psi_\alpha(r) dr^3$, overlap integral, whose magnitude squared is the probability and $\delta(E\text{-}E_\alpha)$ is a delta-distribution at $E=E_\alpha$.

It is an important property of the projected density of states that it remains finite as a system becomes (infinite) macroscopic because the number of states goes like the volume but the probability of each state goes like one over the volume. The total density of states increases like the volume, which of course, can be scaled by the volume to give an average density of states. The (total or) average density of states is given by the (unscaled or) scaled trace of the projected density of states,

$$\bar{g}(E) = \frac{1}{\Omega} \sum_\alpha g_\alpha(E) \tag{5}$$

where Ω is the volume and $\{g_\alpha(E)\}$ are the projected densities of states for a complete orthonormal set of orbitals. The units of $\bar{g}$ are states per unit volume per unit energy while those for g_u are just states per unit energy; there is a factor of states per unit volume between them.

f. Black Body Theorem

The significance of the projected density of states is that it is exponentially insensitive to changes in H far from the orbital of projection. This property is embodied in the quantum analogue of the electromagnetic black body theorem — namely that the density of modes inside a cavity is exponentially insensitive to the shape of the cavity. As a result the radiation from a black body cavity has a spectrum which is independent of its shape or the material from which it is made.

Since quantum mechanics is also a wave phenomena, the quantum analogue is obvious, but it is instructive to review the argument. Consider the eigenstates of H near the orbital u, in the case where they are real due to some boundary shown in Fig. 2. The projected density of states is large at E_α because Ψ_α has a maximum at u, but $g_u(E_\beta)$ is small because Ψ_β has a node at u. Suppose we now change the system at a distance L from u so that Ψ_α is shifted to the left by 1/4 of a wavelength. If L is large this is the same as shifting the peak in g_u from E_α to E_β where $E_\alpha - E_\beta$ is of order 1/L, as shown in Fig. 3. The same argument applies at all energies and so the conclusion is that the change in the boundary only changes g_u over energy ranges of order 1/L.

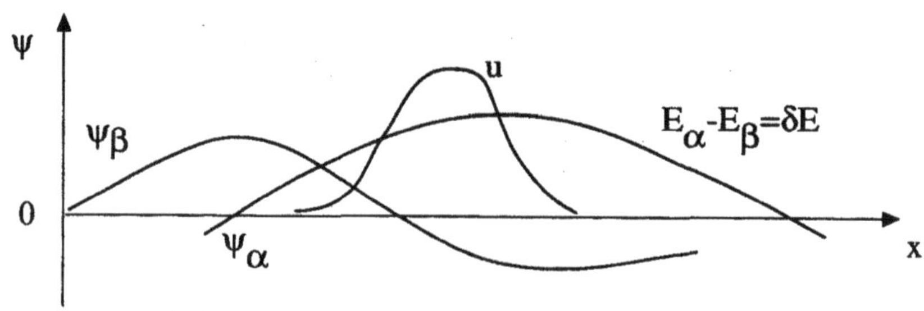

Fig. 2 Overlap of two eigenstates ψ_α and ψ_β with a localized state u.

Put another way, the solutions to the equations of motion near u depend only on H near u, but they are mixed by the boundary conditions at L. When this boundary changes, the solution at u can be changed from a maximum to a node, a large change, but when averaged over energy, this change only contributes to averages over ranges of energy less than δE.

If we now integrate the projected density of states over a smooth function, that is, one with an absolutely convergent power series in δE around E so that the coefficients decrease exponentially, the integral

$$\int_{\delta E} f(E) g_u(E) dE \qquad (6)$$

must converge exponentially for disturbances in H at increasing distances L.

This is the precise sense in which the projected density of states, the recursion method, and even Gaussian quadrature converge. It is convergence of integrals over smooth functions; it is convergence of a very general and useful kind; and it turns out it is just the convergence we need to calculate physical properties of systems of electrons. In physical terms it is just that the local properties of the electron are exponentially insensitive to their distant environment.

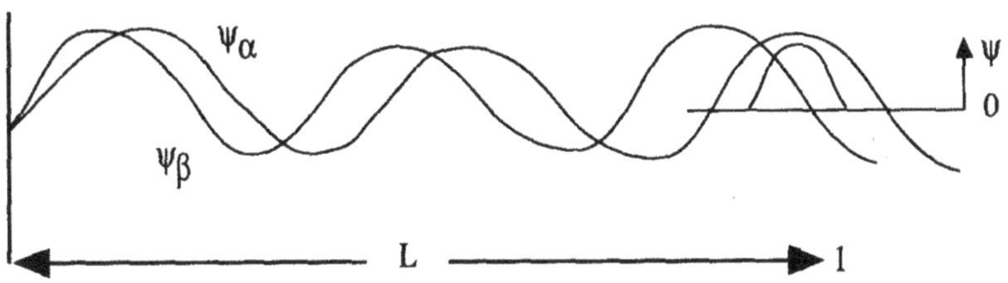

Fig. 3. Two waves whose energy difference changes the overlap with u from maximum to zero.

2. Quantum Path Counting and the Recursion Method

The purpose of today's lecture is to derive the recursion method from quantum path counting. This begins, however, with the resolvent (Greenian) operation and its asymptotic power series.

a. Resolvent Operator

The projected density of states is a mathematical quantity which is exponentially insensitive to disturbances distant from the orbital of projection. Defined as a sum of delta-distributions or as the limit of such a sum, it is difficult to see how to approximate it although it is clear from the black-body theorem that approximations should be organized in terms of increasingly distant environments.

The first step is to associate the projected density of states with some function whose approximations can be developed. The projected density of states is a weighted sum of delta-distributions for the finite system and the simplest function associated with the delta-distributions $\delta(E-E_\alpha)$ is the fraction $(E-E_\alpha)^{-1}$. Using this, we relate the function

$$R_u(E) = \sum_\alpha \frac{|(u * \Psi_\alpha)|^2}{E - E_\alpha} \tag{7}$$

to the projected density of states

$$g_u(E) = \sum_\alpha |u * \Psi_\alpha|^2 \delta(E - E_\alpha) \tag{8}$$

which is just the singular part of $R_u(E)/\pi$, as shown in Fig. 4.

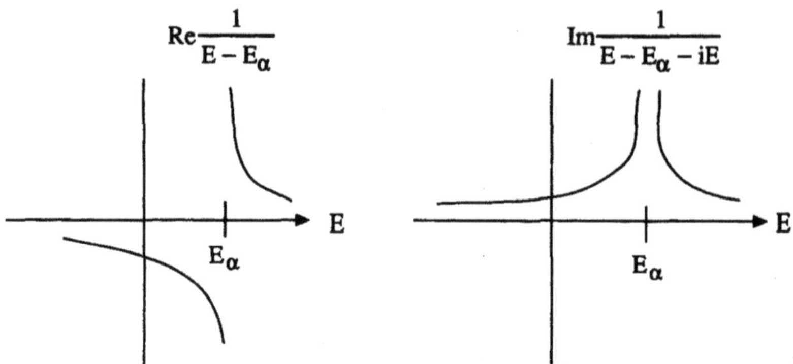

Fig. 4. Real and imaginary parts of $(E-E_\alpha)^{-1}$ near E_α.

This function $R_u(E)$ whose singular part is $\pi g_u(E)$ is called the diagonal element of the resolvent for u.

$R_u(E)$ can be extended to an operator on the states of the electrons by observing that it can be written as

$$R_u(E) = u^*\left(\sum_\alpha \frac{\Psi_\alpha \Psi_\alpha^*}{E-E_\alpha}\right)u \tag{9}$$

where u^* and Ψ_α^* are the bras of the kets u and Ψ_α. This may now be written as:

$$R_u(E) = u^*(E-H)^{-1}u, \tag{10}$$

where $(E\text{-}H)^{-1}$ is called the resolvent operator and has singularities at the stationary states and energies of the equations of motion.

b. Quantum Path Counting

Now that the projected density of states has been related to the singular part of the projected resolvent which is a function of E, we can expand it in powers of E,

$$R_u(E) = \sum_{n=0}^{\infty} \frac{(u^* H^n u)}{E^{n+1}} \tag{11}$$

Since $R_u(E)$ has singularities for values of E along the real energy axis, this series cannot be convergent. In the case where the spectrum (energies where R_uE is singular) is bounded and more generally provided, the moments

$$\mu_n = u^* H^n u \tag{12}$$

grow no faster than n!, it can be shown that the series is an asymptotic one which can be summed uniquely.

The moments have an interpretation as sums of quantum paths in the basis $\{\varphi_a\}$ for which H is defined. $u * H^0 u$ or $u * Iu$ is the sum of quantum paths of length zero from u to u and is defined to be unity. $u*Hu$ is the sum of quantum paths of length 1 from u to u; it is just the u-u diagonal elements of H. $u*H^2u$ is the sum of paths of length 2 which is

$$\mu_2 = \sum_{\alpha,\beta,\gamma} (u * \varphi_\alpha) t_{\alpha\beta} t_{\beta\gamma} (\varphi_\gamma * u) \tag{13}$$

and in general the sum of paths of length n from u to u is:

$$\mu_n = \sum_{\alpha_0,\alpha_1,\ldots,\alpha_n} (u * \varphi_{\alpha_0}) t_{\alpha_0\alpha_1}, \ldots t_{\alpha_{n-1}\alpha_n} (\varphi_{\alpha_n} * u) \tag{14}$$

The path counting may be better visualized by representing the equations of motion as a graph in terms of the sparse basis $\{\varphi_\alpha\}$ as shown in Fig. 5. We can obtain the moments for various projected densities of states by summing up the products of matrix elements for all the different paths beginning and ending on one of the φ_α's.

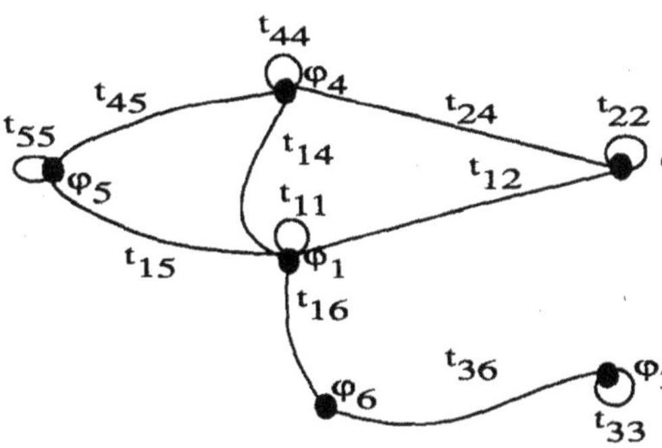

Fig. 5 A graphical representation of H for the purpose of path counting.

In terms of path counting, we may write the resolvent as

$$R_u(E) = \frac{1}{E} \sum_{n=0}^{\infty} \sum_{P_n(u,u)} \prod_{m=1}^{n} \left(P_{n,m}^{(u,u)} / E\right) \tag{15}$$

where $P_n(u,u)$ are paths of length n from u to u and $P_{n,m}^{(u,u)}$ is the mth matrix element of the path. This is the expression for the resolvent in terms of quantum rather than classical paths. For those who are interested, the propagation between u and v is:

$$k(t;u,v)=\sum_{n=0}^{\infty}\frac{1}{n!}\sum_{p_n(u,v)}\prod_{m=1}^{n}\left(-itP_{n,m}(u,v)\right) \tag{16}$$

c. Recursion as Renormalized Path Counting

Consider a very simple Hamiltonian for just one orbital u_0 with a matrix element a_0. The paths generated form a simple geometric series

$$\mu_0=1,\ \mu_1=a_0,\ \mu_2=a_0{}^2,\ \ldots,\ \mu_n=a_0{}^n$$

and so we may sum the series for the resolvent element,

$$R_0(E)=\frac{1}{E-a_0}. \tag{17}$$

Now suppose that the Hamiltonian is extended by adding basis states $\varphi_1 \ldots \varphi_2$ with matrix elements t_{0n} between u_0 and these other orbitals, but suppose there are no other matrix elements of H beyond these. The paths are now more complicated:

$$\mu_0=1,\ \mu_1=a_0,\ \mu_2=a_0^2+\sum_{n=1}^{7}t_{0n}^2,\ \mu_3=2a_0\sum t_{0n}^2+a_0^3,\ldots \tag{18}$$

This is a more complicated series, but with some thought, it can be seen to sum to

$$R_0(E)=\cfrac{1}{\cfrac{E}{E-a_0-\cfrac{\sum_{n=1}t_{0n}{}^2}{E}}} \tag{19}$$

which now counts into $R_0(E)$ all the paths between u_0, itself, and hops to $\varphi_1,\ldots\varphi_2$.

The next step is renormalization or transforming the Hamiltonian so that we make a new orbital

$$u_1=\frac{\left(\sum_{n=1}^{7}t_{0n}\varphi_n\right)}{b_1} \tag{20}$$

where b_1 is chosen to normalize u_1.

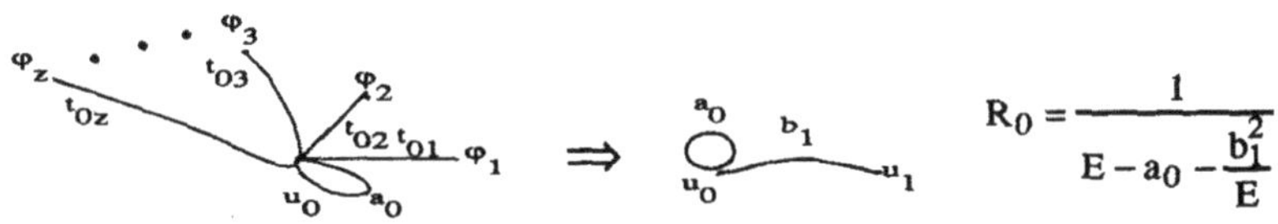

because $b_1{}^2 = \sum_{n=1}^{z} t_{0n}{}^2$. If we now extend the Hamiltonian so as to couple the $\varphi_1,\ldots\varphi_2$ to each other, themselves, and other orbitals, but do not add any matrix elements from u_0 to any orbitals, then any path which contributes to R_0 must pass through u_1, by definition. The paths which begin on u, and return to u, but never touch u_0 sum to the terms in a series for $R_1(E)$, the diagonal matrix element of the resolvent for H with the u_1 row and column excluded. Thus we may write $R_0(E)$ in terms of $R_1(E)$ as

$$R_0(E) = \frac{1}{E - a_0 - b_1{}^2 R_1(E)} \tag{21}$$

and we have performed a renormalization of R_0 by extracting u_0 and its matrix elements to leave the paths in $R_1(E)$.

The same process can now be carried out for $R_1(E)$ to get $R_2(E)$ and so on so that

$$u_0^*(E-H)^{-1}u_0 = \left(1/E - a_0 - b_1^2/E - a_1 - \ldots\right) \tag{22}$$

which is the continued fraction representation of the u_0-u_0 element of the resolvent.

d. Sparseness and Convergence

Let us return to what I said last time was the sense in which the projected density of states was a stable quantity, namely that for functions

$$f(E) = \sum_{n=u}^{\infty} \frac{1}{n!} f_n E^n \tag{23}$$

which are so smooth that their Taylor series converge in the whole E-plane,

$$\int f(E) g_u(E) dE \tag{24}$$

is exponentially insensitive to distant perturbations of H. In terms of the resolvent, and using Cauchy's theorem,

$$\int f(E)g_u(E)dE = \frac{1}{2\pi i}\oint_C f(E)R_u(E)dE \tag{25}$$

note: $\int E^n g_n(E)dE = \frac{1}{2\pi i}\oint E^n R_u(E) = \mu_n$ moments of $g_u(E)$ where C is some contour in the complex E-plane which encloses the spectrum of $R_u(E)$ as shown in Fig. 6. In terms of the moments.

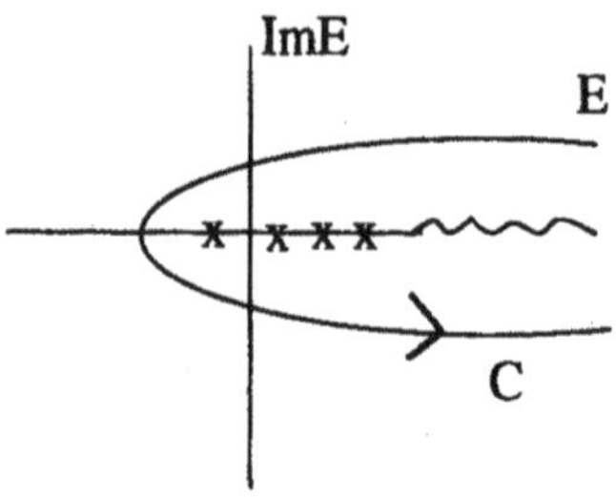

$$\int f(E)g_u(E)dE = \sum_{n=0}^{\infty} f_n \mu_v$$

Fig. 6 An example of a contour in the complex E-plane for integration.

The extreme case for f(E) to have no singularity is for f_n to vary like a constant to the power n, but if that is the case, then μ_n must not grow faster than n!, hence the bound on the growth of the moments.

The growth of the moments depends on the matrix elements of H which contribute to them through path counting. If a row or column of H has an infinite number of significant elements, then that clearly produces an infinite moment at some stage when the sum $\sum_\beta |t_{\alpha\beta}|^2$ comes into a path. On the other hand, if there are at most Z off diagonal matrix elements non-zero, and less than t in each row or column, then

$$\mu_{2n} < 2Z^{2n}t^{2n} \tag{26}$$

which satisfies the requirement. In fact the minimal requirements for a convergent path sum is somewhat weaker than strict sparseness.

We can also see that the dependence of $\int f(E)g_u(E)dE$ on a perturbation in the Hamiltonian is determined by what moment or what length paths the perturbation changes. If the perturbation changes no paths of length less than n, then the change of the integral is less than A^n where A depends on f(E) and $g_u(E)$, but is independent of n.

e. The Chain Model and Its Physical Interpretation

We can look at the renormalization of the path counting as a transformation of H into tridiagonal form,

Recursion: $Hu_n = a_n u_n + b_{n+1}u_{n+1} + b_n u_{n-1}$

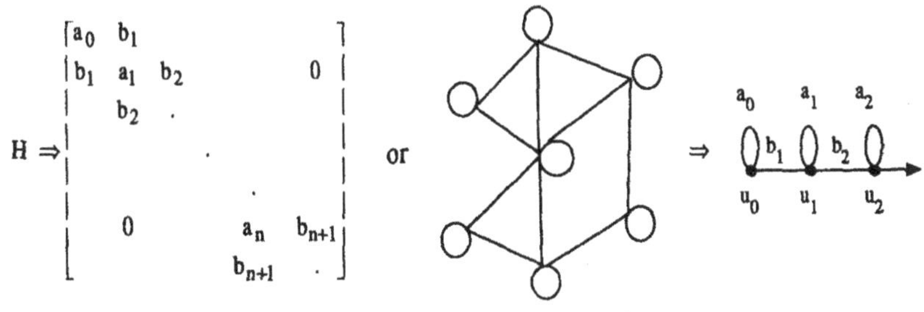

The transformed orbitals u_0, u_1, ... are a sequence of states or environments through which the system must pass in sequence, either forward or backward. The projected resolvent of $u_0{}^*(E-H)^{-1}u_0$ and hence the projected density of states is most influenced by the matrix elements of u_0, next by u_1, and so on. These are the environments to which physical quantities are exponentially insensitive.

A question which usually occurs to students of recursion is the H can be multidimensional, but the chain model 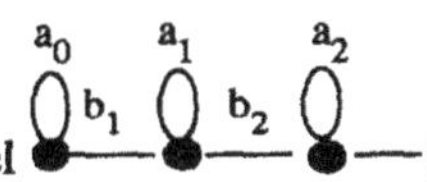 is one-dimensional. How can a multidimensional system be transformed exactly into a one-dimensional one? The answer is that u_0, u_1, u_2, ... do not span all the states of H if there is degeneracy present. The chain model only spans the states accessible from u_0, at most one out of each degenerate subspace of H. The accessible states are non-degenerate and therefore one-dimensional in an abstract sense. At least, they can be represented exactly by the chain model.

This emphasizes the importance of the choice of u_0 in applications of recursion. If the physical problem involves particular states, u_0 must couple strongly to them. H defines the model, but u_0 defines the problem. The question of sparseness can be restated as the requirement that the time-evolution of the system from u_0 be non-singular. This is sufficient for the moments to exist and grow slowly enough to converge physical quantities.

3. Orbital Peeling and the Calculation of Physical Quantities

So far we have established that integrals of smooth functions over the projected density of states converge exponentially in the lengths of paths summed, or in the number of moments included. The purpose of this lecture is to show how physical quantities can be calculated from the projected density of states, first at finite temperature which is relatively simple, but also at zero temperature which is more difficult.

a. Finite Temperature Expectation Values

In independent or adiabatic electron models, the expectation value of some physical quantity is the sum of the expectation values for each electronic state weighted by the occupation of that state in equilibrium.

A simple example of this kind of expectation is the occupation of an orbital u at temperature T and chemical potential μ,

$$N = \int f(T,\ \mu,\ E) g_u(E) dE \tag{27}$$

where f(T, μ, E) is the occupation probability for the state with energy E and $g_u(E)$ is the projected density of states for u.

At finite temperature, f(T, μ, E) is a smoothly varying function from 1 at low energies to zero at high energies, as shown below. It is least smooth near μ with its range of variation characterized by T as shown in Fig. 7.

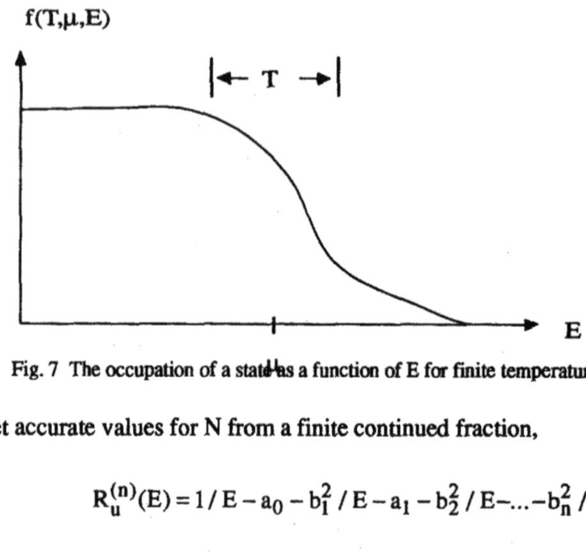

Fig. 7 The occupation of a state as a function of E for finite temperature.

In order to get accurate values for N from a finite continued fraction,

$$R_u^{(n)}(E) = 1/E - a_0 - b_1^2/E - a_1 - b_2^2/E - \ldots - b_n^2/E \tag{28}$$

we must include enough moments so that the variations in $g_u^{(n)}(E)$ produced by including higher moments, as shown in Fig. 8, are on energy scales less than T. In terms of $R_u(E)$ this means getting the poles of $R_u(E)$ closer than T to one another in the vicinity of μ. If the whole spectrum has width W, then the continued fraction must be calculated to order W/T levels to achieve the exponential convergence.

Other quantities which can be calculated are the contributions to the internal energy U from a projected density of states

$$U = \int E f(T,\ \mu,\ E) g_u(E) dE, \tag{29}$$

to which double counting terms must be added if appropriate. In general such expectation values can be written as:

$$\langle\theta\rangle = \int \theta(E) f(T,\mu E) g_u(E) dE \tag{30}$$

where the $\theta(E)$ usually varies smoothly with E, and the problems all come from the occupation probability which becomes sharper with lower T.

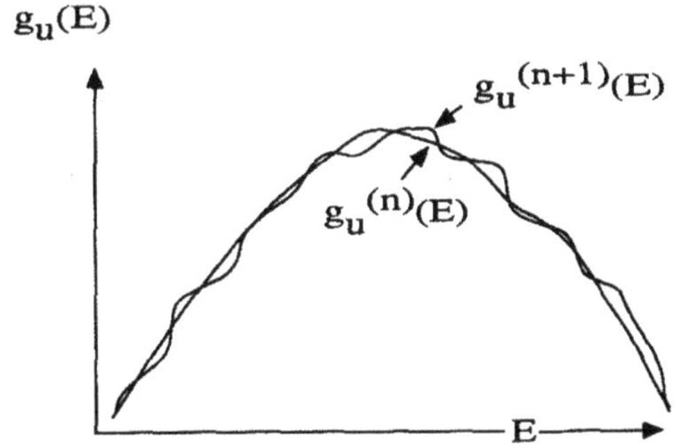

Fig. 8 Variation of approximate densities of states with n, the number of moments.

I want to return for a moment to the question of microscopic vs. macroscopic quantities. I have argued that using a microscopic theory for a macroscopic system we can only calculate microscopic quantities such as the occupations of orbitals, correlations between neighboring orbitals, energies contributed by orbitals, energy differences due to microscopic changes in the system, ... Quantities such as μ, T, E, ... the thermodynamic variables of the system depend on averages over the whole system which must either be calculated as the average of a number of microscopic quantities or a calculation involving a number of operations of the order of the volume, which converges as the inverse root of the volume and so is impractical, or try to find global solutions to the equations of motion, which are unstable.

Crystals are one exception to the above point because their average density of states can be calculated as the projected density of states of a single orbital u in a single unit cell. This is possible because all cells are equivalent and so have identical values of physical quantities. The orbital u, whose projected density of states is the average, must be constructed taking into account the internal cellular symmetries as well as the translation symmetry which is what makes the cellular density of states an average. Briefly, u must contain a component from each degenerate subspace weighted by the root of the degeneracy.

We now turn in the remainder of the lecture to the more difficult problems of ground state or T=0 properties of solids.

b. Integrated Projected Density of States

Let us first examine the convergence of the occupation of an orbital as $T \to 0$,

$$N_u(E_F) = \int f(E_F, E) g_u(E) dE \tag{31}$$

where $f(E_F,E)$ is unity below E_F and zero above, definitely not a smooth function of E. Despite this, we can still obtain useful results for $N_u(E_F)$ by bounding $f(E_F,E)$ above and below by polynomials of even order, as shown in Fig. 9.

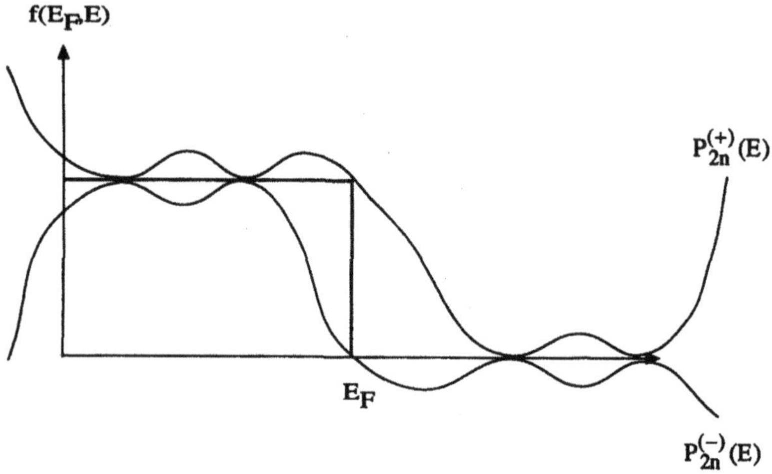

Fig. 9. Upper and lower bounding polynomials for the Fermi occupation function.

Integrals of $g_u(E)$ over the bounding polynomials depend only on the moment up to μ_{2n}, or on the continued fraction down to $b_n{}^2$,

$$R_u^{(2a)}(E) = 1/E - a_0 - b_1^2/E - a_1 - \ldots - b_n^2/E - a_n, \tag{32}$$

where I have also included a_n for a reason I shall now explain.

We can write lots of 2n order polynomials which bound $f(E_F,E)$ above and below, but the ones giving optimal bounds (least upper and greatest lower) are the ones which touch $f(E_F,E)$ at poles of $R_u^{(2n)}(E)$, when there is a pole at E_F. In order to put a pole at arbitrary values of E_F without changing the moments up to μ_{2n}, we need only vary a_n for which there is a simple relation with E_F.

Integrating $P_{2n}^{(\pm)}(E)$ over $g_u(E)$, we get bounds on $N_u(E_F)$

$$N_u^{(\pm)}(E_F) = \frac{1}{2\pi i}\oint_C P_{2n}^{(\pm)}(E) R_u^{(2n)}(E)dE = \begin{cases} \sum\limits_{E_m \le E_F} \omega_m = N_u^{(+)}(E) \\ \sum\limits_{E_m < E_F} \omega_m = N_u^{(-)}(E) \end{cases} \tag{33}$$

where $R_u^{(2n)}(E) = \sum_{m=0}^{n} \frac{\omega_m}{E - E_m}$. We see from this that the difference between the bounds on $N_u(E_F)$ is just the weight or residue of the pole at E_F. If E_F lies in a band-gap then ω_F decreases exponentially with the order of the approximation and we have recovered exponential convergence. If E_F falls within a band, then ω_F is of order 1/n for the $2n^{th}$ approximation and $N_u(E_F)$ converges very slowly as those calculating energies of metals know.

c. Energy As a Function of Occupation

There is a way to get around the slow convergence of ground state properties with respect to E_F by changing the independent variable to the occupation. Since the occupation itself is an integral over the projected density of states we have much more freedom to choose smooth functions than when we were forced into a sharp cutoff at E_F. Another way to look at this is as relating two quantities which are both sharp integrals over $g_u(E)$ and thereby canceling the error.

As an example consider the energy U as a function of the occupation N, defined implicitly by

$$U(N) = \int f(E_F, E) E g_u(E) dE$$

and

$$N = \int f(E_F, E) g_u(E) dE \ . \tag{34}$$

The source of the problem as we saw above is that we cannot define functions $f(E_F,E)$ to greater resolution than the spacing of poles in $R_u(E)$ and do the integrals accurately.

For a finite approximation to $R_u(E)$ we can only define integrands to the resolution of the poles, constraining $f(E_F,E)$ to be unity on the poles below E_F and zero on the ones above. We fix N by constraining the sum of the weights on the poles below E_F to add up to N by adjusting a_n as we did to get $N(E_F)$. For

$$R_u^{(2n)}(E) = \sum_{m=0}^{n} \frac{\omega_m}{E - E_m} = 1/E - a_0 - \ldots - b_n^2 / E - a_n \tag{35}$$

we adjust a_n so that for some F,

$$N = \sum_{m=0}^{F} \omega_m \tag{36}$$

which now leaves the Fermi level somewhere between E_F and E_{F+1}, but that doesn't matter because we aren't calculating it. The energy is now

$$U(N) = \sum_{m=0}^{F} E_m \omega_m \tag{37}$$

which converges exponentially in the number of moments or lengths of paths. In essence the trick is to approach $f(E_F,E)$ with a sequence of smooth functions constrained by the occupation.

d. Orbital Peeling — Microscopic Energy Differences

In structural relaxations it is important to calculate the change in the total energy due to a microscopic change in the system. What is difficult about this calculation for a solid is that the total energy is macroscopic while the energy difference is microscopic. One might think that the calculation is impossible because the bigger the solid, the higher the precision needed in the total energies in order to get the same precision in the energy difference.

This difficulty can be avoided by implicit cancellation in the macroscopic contributions to the total energies of two microscopic different systems. The microscopic change may be viewed as the change in a microscopic portion of H, the change in a few matrix elements of H. We break this change into a process of adding and removing (peeling) orbitals from the system. What we actually calculate is the difference in energy for H and H_0 which is the same as H but with the row and column corresponding to u_0 removed. The change in energy from H to H′ is a Hamiltonian with different elements from H in the row and column for u_0 is

$$\Delta U = \Delta U_0 - \Delta U_0' \quad , \tag{38}$$

the differences in energies for removing u_0 from H and H′.

We now calculate ΔU_0 as

$$\Delta U_0 = \sum_{occ} E_\alpha - \sum_{occ} E_\alpha^{(0)} \tag{39}$$

where E_α are the eigenvalues of H and $E_\alpha^{(0)}$ are those of H_0. We can write

$$\Delta U_0 = \frac{1}{2\pi i} \oint_{C_F} \left\{ \sum_\alpha \ell n \frac{1}{E - E_\alpha} - \sum_\rho \ell n \frac{1}{E - E_\rho^{(0)}} \right\} dE \tag{40}$$

or

$$\Delta U_0 = \frac{1}{2\pi i} \oint_{C_F} \ell n \frac{\det(E - H_0)}{\det(E - H)} dE. \tag{41}$$

But, from the formula for an element of the inverse of a matrix H,

$$\Delta U_0 = -\frac{1}{2\pi i} \oint_{C_F} \ell n \left[u_0 * (E - H)^{-1} u_0 \right] dE \tag{42}$$

Namely, the change in the total energy is just the integral of the logarithm of the projected resolvent, or the difference between the sum of the energies of the zeros of $R_0(E)$ (eigenvalues of H_0) below E_F and the sum of the energies of the poles of $R_0(E)$ (eigenvalues of H) below E_F. These differences can be expressed as the sum of differences of energies of neighboring zeros and poles (they interpolate) of $R_0(E)$.

The convergence of ΔU_0 is due to the property of $\ell n R_0(E)$ that it may be written as the difference between two projected densities of states $G_0(E)$ and $G_{00}(E)$ each with unit weights at its poles,

$$\Delta U_0 = \int f(E_F,\varepsilon)[G_0(\varepsilon) - G_{00}(\varepsilon)]dE \tag{43}$$

where, like U(N), $f(E_F,\varepsilon)$ can be written as a sequence of smooth functions provided the number of electrons is kept constant in going from H to H_0.

4. Ghosts, Paige's Theorem, and Dynamical Recursion

In the lectures so far we have treated arithmetic as exact and considered only the effects of finite numbers of moments on the accuracy of calculated quantities. The numerical applications of the recursion method are always in finite precision arithmetic and so to be thorough we must understand the relation between rounding errors and truncation errors.

In this last lecture I will discuss the remarkable property of recursion, namely that it gives answers to full machine precision, only weakly dependent on the number of operations. This involves the understanding of ghosts and Paige's theorem, and leads to the idea of dynamical recursion in which truly infinite systems may be approximately tridiagonalized.

a. Ghosts

There is a great barrier to thinking about rounding error in numerical methods because one tries to imagine what error ε in the answer results from a random error of δ in each arithmetic operation. As anyone who has tried knows, this is not the way. I will start with a description of the phenomena of rounding, namely ghosts, and go from there to a theory of errors.

If one takes a small non-degenerate, symmetric matrix H and a starting state u_0 which contains a component from each of the eigenvectors, and proceeds numerically with the tridiagonalization,

$$\begin{aligned}
&a_0 = u_0 * Hu_0\\
&b_1^2 = |(H - a_0)u_0|\\
&u_1 = (H - a_0)u_0 / b_1\\
&\quad\vdots\\
&a_n = u_n * Hu_n, \quad b_{n+1}^2 |(H - a_n)u_n = b_n u_{n-1}|\\
&u_{n+1} = ((H - a_n)u_n - b_n u_{n-1}) / b_{n+1}\\
&\quad\vdots
\end{aligned} \tag{44}$$

One notices that u_{10} is not really orthogonal to u_1 or u_2 and that if H was 100 by 100, b_{100} doesn't look any different from b_{99} or b_{101} even though in exact arithmetic b_{100} should be zero. This is the rapid loss of global orthogonality among the $\{u_n\}$ which takes place in recursion due to rounding error.

Does this mean that the method is useless? If one persists and plots the inner product of Ψ, a known eigenvector of H, with u_n against n, one sees the Lanczos phenomena, shown in Fig. 11, namely that every so often $\Psi^* u_n$ falls to a level of around $\sqrt{\delta}$ (δ is the rounding error) for two or more successive u_n and then starts to increase in magnitude. What this phenomena means is that between the arrows the recursion has converged Ψ and orthogonalized u to it to within the rounding error. Then, because the rounding is random, a new component of Ψ develops and the process of convergence and orthogonalization begins again.

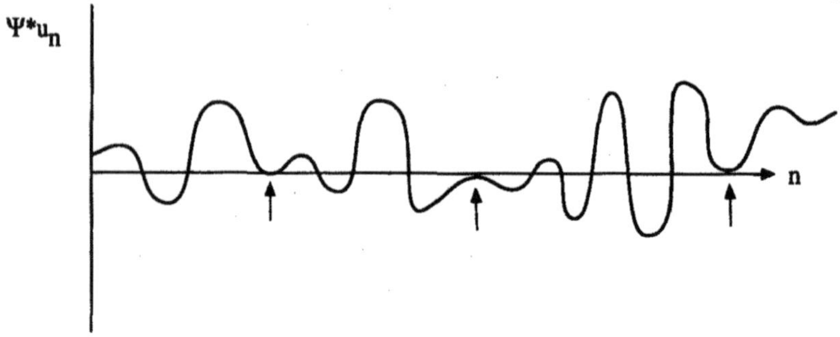

Fig. 10. A plot of overlap of an eigenvector with elements of the tridiagonal basis.

A consequence of this phenomena is that the eigenvalues obtained from the tridiagonalization are accurate to δ and the eigenvectors to $\sqrt{\delta}$. The loss of orthogonality observed is associated with the convergence of eigenvectors and their subsequent reconvergence. These extra copies of each eigenvector are called ghosts for obvious reasons.

b. Paige's Theorem

We can make a quantitative theory of errors in recursion paralleling Paige's theory of errors in the Lanczos method. The crucial idea is to take account of the error in the recursion explicitly,

$$Hu_n = a_n u_n + b_{n+1} u_{n+1} + b_n u_n + \Delta_n \tag{45}$$

where Δ_n is a vector which is the difference between Hu_n and $a_n u_n + b_{n+1} u_{n+1} + b_n u_{n-1}$, it is the vector of rounding or other errors. It is convenient to write the above equation in matrix form where H is an N by N matrix, J is an M by M tridiagonal matrix

$$J = \begin{bmatrix} a_0 & b_1 & 0 & & \\ b_1 & a_1 & b_2 & 0 & 0 \\ 0 & b_2 & a_2 & b_3 & 0 \\ & 0 & b_3 & \cdots & b_M \\ 0 & & 0 & b_M & a_M \end{bmatrix},$$

U is an N x M matrix of the column vectors $U = [u_0 u_1 ... u_{M-1}]$, and Δ is an N by M matrix of the Δ_n stuck together $\Delta=[\Delta_0\Delta_1...\Delta_{M-1}]$. Now the above three term recurrence may be written as

$$HU=UJ+\Delta \qquad \text{(Ron Te)} \tag{46}$$

an N x M matrix equation which accounts for the errors exactly and includes both H and the tridiagonal matrix J.

The resolvent we want to calculate is $(E\text{-}H)^{-1}$. The resolvent we actually calculate is $(E\text{-}J)^{-1}$ and they are related by subtracting the equation from

$$EU=UE \tag{47}$$

and multiplying by $(E\text{-}H)^{-1}$ on the left and $(E\text{-}J)^{-1}$ on the right to get

$$(E-H)^{-1}U=U(E-J)^{-1}+(E-H)^{-1}\Delta(E-J)^{-1} \tag{48}$$

Let ℓ_0 be the first column vector for the first column of J and u_0 be the column vector for the first column of H, assume we have constrained the numerics so that all the other u's are orthogonal to u_0, $U\ell_0 = u_0$ and $u_0 * U\ell_n = \delta_{0n}$.

$$u_0 * (E-H)^{-1} u_0 = \ell_0 * (E-J)^{-1} \ell_0 + \sum_\alpha u_0 * (E-H)^{-1} \varphi_\alpha \Delta_{\alpha n} \ell_n * (E-J)^{-1} \ell_0 \tag{49}$$

where the j_α are the basis for H and $\Delta_{\alpha n}$ is the component of φ_α in Δ_n.

This says that the errors in $\ell_0 * (E-J)^{-1} \ell_0$ are given by the products of resolvents and errors connecting u_0 to ℓ_0. For values of E off the real axis (those needed to do the contour integrals we use to evaluate physical quantities), it can be shown that $u_0 * (E-H)^{-1} \varphi_\alpha$ and $\ell_n (E-J)^{-1} \ell_0$ both decay exponentially with the length of H-paths from u_0 to φ_α and with the length of J-paths from ℓ_n to ℓ_0. Just as with other errors, those in the recursion have exponentially decreasing contributions, with their "distance" from u_0 and ℓ.

This result is a more precise form of Paige's theorem in that it gives an exact formula for the error in the resolvents. Because the errors contribute with decreasing exponentials, they are dominated by just a few matrix elements of Δ, hence the result that errors are of a order a single component of Δ, a single component of one of the u_n's not an accumulation of errors from all the components of Δ.

This error theory applies to truncation of J as well as errors in J. If we truncate J at level n, then Δ_n is large and its components near u_0 have the worst effects. A simple approximation is that the $\Delta_{\alpha n}$ are of order $b_1/\sqrt{N}$ and their effect is suppressed by at least the factor $\ell_n * (E-J)^{-1}\ell_0$, if the rest of the orthogonality is reasonably good, then there is a similar factor from H to give a total error of

$$u_0 * (E-H)^{-1} u_n \left(b_1/\sqrt{N}\right) \ell_n * (E-J)^{-1} \ell_0 \tag{50}$$

which is another, more precise, version of the black body theorem.

c. *Dynamic Recursion*

The remarkable property of recursion that errors do not accumulate suggests that one can take advantage of this to treat really enormous basis sets with controlled accuracy.

Suppose we have an infinite matrix H which is sparse in the sense that there are only a finite number of non-zero elements in each row and column. Suppose also, that we can store N components of each recursion vector u_n. If at each stage of the recursion we keep the N largest components of the new recursion vector we make an error in our resolvent which is proportional to the largest component we neglected. The idea is to discard deliberately the small components of u_{n+1} knowing that the errors do not accumulate.

In the worst case there would be N+1 components of u_{n+1} with equal magnitude. Discarding one of them produces an error of $b_{n+1}/\sqrt{N+1}$ in that component of Δ_n, which might not even turn out to be the biggest.

Implementation of a dynamic basis scheme requires indirect storage of the recursion vectors. The actual components must be accompanied by a list of which basis orbital each component corresponds to. The basis orbitals can be indexed by integers because H is sparse and each multiplication by H only increases the largest index, but always leaves it finite.

Dynamic recursion has actually been in use for years in the sense that densities of states always improve for a while after the recursion basis hits the boundary of the finite cluster used in static recursion. This is because the finiteness of a static cluster is only important when a substantial part of the weight of the recursion vector should be outside of it. This effect is explained by the same error theory as leads to dynamic recursion.

d. *Scaling of Recursion*

Error in a calculation goes as $A\exp\left\{-B\frac{M}{NZ}\right\}$ with a limit of $C/\sqrt{N}$ where M is the number of levels, N is the number of components stored for each u_n, Z is the number of non zero elements in each row or column of the Hamiltonian, and A, B, and C are geometrical constants.

e. More Recursion

There are many other ideas about recursion and its applications which I wish I had time to talk about. Instead I will list some papers which may be of interest.

(1) *Solid State Physics*;, **Vol. 35**, Ehrenreich, Seitz, and Turnbull, eds., Academic Press: New York, 1980 (general).
(2) Foulkes, M. and Haydock, R. *J. Phys. C* ***19***, (1986) 6573.
(3) Gibson, A., Haydock, R., and LaFemina, J. *J. Vac. Sci. Tech.*, A10 (1992) 2361.
(4) Blumberg, S. and Haydock, R. *Phys. Rev. B* **45** (1992)1550.
(5) Foulkes, W. M. C. and Haydock, R. *Phys. Rev. B* **39** (1989) 12520 (use with density functional).
(6) Godin, T. J. and Haydock, R. *Europhys. Lett.* ***14*** (1991)137 (quantum transport).
(7) Parlett, B. N. *The Symmetric Eigenvalue Problem*; Prentice Hall: Englewood Cliffs, NJ, 1980 (Paige's Theorem Lanczos Phenom.)
(8) Haydock, R. *Comp. Phys. Comm.* ***55*** (1989) 1 (dynamic recursion).

REAL SPACE ELECTRONIC STRUCTURE CALCULATION USING THE RECURSION METHOD

PATRICIO VARGAS C.
Physics Department
Universidad de Santiago de Chile
P.O. Box 307
Santiago-2
Chile

1. INTRODUCTION

The purpose of this lectures is to show, in a practical way, how to perform electronic structure calculations in the real space using the recursion method together with the Linear Muffin Tin Orbitals (LMTO) scheme for constructing the Hamiltonian.

I like to stress that this notes were elaborated within the spirit of the workshop and therefore they were intended for beginners in the field. More advanced topics can be found in other contributions of this workshop.

We start in Section 2 with a brief account of the recursion method, introducing a new way to terminate the continued fraction representation of the diagonal elements of the Green's function. In Section 3 we give a short introduction to the LMTO formalism in its tight-binding representation (TB-LMTO) which is particularly suitable for the calculation with recursion.

We give also some simple computer codes, which allow to perform electronic structure calculations based on TB-LMTO Hamiltonians and the recursion scheme. The source codes can be requested through internet in the adresses given in Section 3.9

2. THE RECURSION METHOD

2.1 Spectrum of an operator and green's function

The density of states $D(E)$, is related to the number of states $N(E)$ which energy less than E. This two quantities are coupled by the followings equation

$$N(E) = \int_{-\infty}^{E} dx\; D(x) \quad , \tag{1}$$

even more, if we know the eigenvalues $\{E_n\}$ of some Hamiltonian operator $\hat{H}$, then

$$N(E) = \sum_{n(\text{levels})} g_n \,\Theta(E - E_n) \quad . \tag{2}$$

The factor g_n is the degeneracy of level n, and $\Theta(x)$ is the Heaviside step-function

$$\Theta(x) = 1 \quad \text{if} \quad x > 0 \qquad \Theta(x) = 0 \quad \text{if} \quad x < 0 \quad .$$

From Eqs.(1) and (2) we have

$$D(E) = \frac{dN(E)}{dE} \, . \tag{3}$$

i.e.

$$D(E) = \sum_{n(\text{levels})} g_n \,\delta(E - E_n) \, ,$$

or in an equivalent form

$$D(E) = \sum_{\text{states}} \delta(E - E_n) \, . \tag{4}$$

where in the last summation we have suppressed the factor g_n, due to summation over states. Normally in the physical system under study, (which is represented by the Hamiltonian $\hat{H}$) we are dealing with some "basis" of wave functions, which is reazonably complete and orthonormal. Then $\hat{H}$ has a matricial representation and Eq.(4) can be written as:

$$D(E) = \mathbf{Tr}\, \delta(E - \hat{H}) \, . \tag{5}$$

This is a formal definition, because we normally do not know the spectrum of $\hat{H}$. However this definition is useful because it relates the DOS with the Green's operator $\hat{G}(E)$, which is defined as

$$\hat{G}(z) = \frac{1}{z - \hat{H}} \quad . \tag{6}$$

This operator can be calculated for all complex values z, provided we know $\hat{H}$ in some representation without an explicit knowledge of its eigenvalues. How we do that?.

Let's take the inverse of $(z - \hat{H})$. This can be done by calculating the Trace of $\hat{G}(z)$.

$$\textbf{Tr}\,\hat{G}(z) = \sum_s < s \mid \hat{G} \mid s > ,$$
$$= \sum_s G_{ss}(z) \ . \tag{7}$$

The trace is an invariant. In the representation which diagonalizes $\hat{H}$, we have

$$\textbf{Tr}\,\hat{G}(z) = \sum_s \frac{1}{z - E_s} \ . \tag{8}$$

Using Eq.(4) we can rewritte the last eq. as:

$$\textbf{Tr}\,\hat{G}(z) = \int_{-\infty}^{\infty} dx\, \frac{D(x)}{z - x} \ . \tag{9}$$

For a given energy E, the inversion of this relation can be achieved by taking the imaginary part of $\hat{G}(z)$ for $z = E - i\xi$.

$$\frac{1}{\pi}\textbf{Im Tr}\,\hat{G}(E - i\xi) = \frac{1}{\pi} \sum_s \frac{\xi}{(E - E_s)^2 + \xi^2} \ . \tag{10}$$

In the limit $\xi \to 0^+$ the Lorentz function, in the right hand side of the last equation, approximates a delta-function (in a distribution sense)

$$\left(\lim_{\xi \to 0+} \frac{1}{\pi} \frac{\xi}{x^2 + \xi^2} = \delta(x) \right) \ .$$

i.e. we have the following relation between the Total DOS and Green's operator

$$\lim_{\xi \to 0+} \frac{1}{\pi} \textbf{Im Tr}\,\hat{G}(E - i\xi) = \sum_s \delta(E - E_s) = D(E) \ . \tag{11}$$

Sometimes, we only need to calculate partial DOS, associated with some "orbital" α in our basis of functions. In that case we do not need to evaluate the trace, but just the diagonal element $G_{\alpha\alpha}(z)$ in that representation. The partial DOS associated with some particular orbital α, $D_\alpha(E)$, is then given by

$$D_\alpha(E) = \lim_{\xi \to 0+} \frac{1}{\pi} \textbf{Im}\, G_{\alpha\alpha}(E - i\xi) \ . \tag{12}$$

Now the point is to evaluate $G_{\alpha\alpha}(z)$ without solving the Schrödinger's equation ($\hat{H} \mid \psi >= E \mid \psi >$). We shall see that the Green's function $G_{\alpha\alpha}(z)$ is related with the moments μ_n of the DOS. Those μ_n are defined by

$$\mu_n = \int_{-\infty}^{\infty} d\epsilon\, \epsilon^n\, D_\alpha(\epsilon) \ ,$$

or

$$\mu_n = < \alpha \mid \hat{H}^n \mid \alpha > \ . \tag{13}$$

We can also write

$$\hat{H}^n = \underbrace{\hat{H}\ \dots\ \hat{H}}_{n-\text{times}} = \hat{H}\sum_\beta \mid \beta >< \beta \mid \hat{H}\sum_\gamma \mid \gamma >< \gamma \mid \hat{H}\cdots\sum_\epsilon \mid \epsilon >< \epsilon \mid \hat{H}$$

Then

$$\mu_n = \sum_{\beta\gamma\cdots\epsilon} < \alpha \mid \hat{H} \mid \beta >< \beta \mid \hat{H} \mid \gamma > \cdots < \epsilon \mid \hat{H} \mid \alpha > \ .$$

If $\hat{H}$ is given in a tight binding orthogonal representation, then the matrix elements $< \alpha \mid \hat{H} \mid \beta >$ couple only neighbor orbitals.

So, we see that the evaluation of the n-moment of the partial DOS associated with the α-orbital is related to "closed paths counting" in a network defined by the set of orbitals chosen (i.e. our tight-binding basis). We have to begin in the α-orbital, going through the neighbor orbitals β, γ and coming back to the starting point α.

Usually, the basis in real space is directly related to sites. That is the case of a tight-binding model for the electronic structure calculation of some materials, in this case, usually, the basis consist of an orthonormalized set of Wannier orbitals, each of them related to a particular atom. The moments calculation is then connected to closed paths counting in some particular lattice.

We shall see now, how the partial DOS (PDOS) can be constructed from its moments in a form of continued fraction. The starting point is the relation (9) for the diagonal element of the green's function.

$$G_{\alpha\alpha}(z) = \int_{-\infty}^{\infty} dx\, \frac{D_\alpha(x)}{z - x} \ . \tag{14}$$

If it convenient to rescale the energy range of non-zero $D_\alpha(x)$ to be in the interval $(-1, 1)$. i.e.

$$G_{\alpha\alpha}(E) = \int_{-1}^{1} dt\, \frac{D_\alpha(t)}{E - t} \ . \tag{15}$$

(we have made $t = (x - a)/2b$, where a is the band-center $((E_B + E_T)/2\ )$ and $2b$ is the band-width $(E_T - E_B)$, E_T = top of the band, E_B = bottom of the band).

By this way, $G_{\alpha\alpha}(E)$ has not imaginary part for $\mid E \mid > 1$. Expanding the integrand in a geometrical summation we have

$$G_{\alpha\alpha}(E) = \frac{1}{E}\int_{-1}^{1} dt\, D_\alpha(t)\left\{1 + \frac{t}{E} + \frac{t^2}{E^2} + \cdots + \frac{t^N}{E^N} + \frac{(t/E)^{N+1}}{1-(t/E)}\right\}$$
$$= \mu_0\left(\frac{1}{E}\right) + \cdots + \mu_N\left(\frac{1}{E}\right)^N + \left(\frac{1}{E}\right)^{N+1}\int_{-1}^{1} dt\,\frac{D_\alpha(t)\, t^{N+1}}{(E-t)} \qquad (16)$$

If we keep the last term in this summation this eq. is an identity. It is straightforward to demonstrate that this last equation (up to the order N of truncation) is equivalent to a continued fraction of the form[1] :

$$G_{\alpha\alpha}(E) = \cfrac{1}{E - a_1 - \cfrac{b_1^2}{E - a_2 - \cfrac{b_2^2}{E - a_3 - \cfrac{b_3^2}{E - a_4 \cdots}}}}\ . \qquad (17)$$

Where the $\{a_i, b_i\}$ coefficients are related to the moments $\{\mu_i\}$ through the following equations:

$$a_n = \frac{1}{\Omega_{n-1}}\left[\frac{D_{n-1}\Omega_n}{D_n} + \frac{D_n\Omega_{n-2}}{D_{n-1}}\right]\ ,$$

$$b_n^2 = \frac{D_n D_{n-2}}{D_{n-1}^2} \quad \text{for} \quad n = 1, 2, 3, \ldots \qquad (18)$$

and with $D_{-1} = \Omega_{-1} = 1$. D_n and Ω_n are Haenkel determinants.

$$D_n = \begin{vmatrix} \mu_0 & \mu_1 & \cdots & \mu_n \\ \mu_1 & \mu_2 & \cdots & \mu_{n+1} \\ \vdots & & & \\ \mu_n & \mu_{n+1} & \cdots & \mu_{2n} \end{vmatrix}$$

$$\Omega_n = \begin{vmatrix} \mu_1 & \mu_2 & \cdots & \mu_{n+1} \\ \mu_2 & \mu_3 & \cdots & \mu_{n+2} \\ \vdots & & & \\ \mu_{n+1} & \mu_{n+2} & \cdots & \mu_{2n+1} \end{vmatrix} \qquad (19)$$

Now we have the relation between the Green's function $G_{\alpha\alpha}(z)$ Eq.(14) and the moments of the partial DOS. (Eqs.(17),(18), (19)). Once we have calculated $2n$ moments of the partial DOS $D_\alpha(E)$. We can evaluate, using Eqs.(19) and (18), the first n-pairs of coefficients (a_i, b_i) , and from there a continued fraction (truncated at the n-level) expression for $G_{\alpha\alpha}(E)$. However the evaluation of a_n and b_n from Eqs.(18) and (19) is very unstable, and usually going beyond the level $n \sim 10$ is not possible. This is due to numerical instabilities. If $D_\alpha(E) \neq 0$ in $(-1,1)$ and $\mu_0 = 1$ (i.e. the partial DOS is normalized). Then $\mu_n = 0$ for odd n and

$$\mu_n = \int_{-1}^{1} d\epsilon\, \epsilon^n\, D_\alpha(\epsilon) \leq \frac{2\,\mathrm{MAX}}{n+1} \quad \text{for even } n\,,$$

where MAX is the maximun value of $D_\alpha(\epsilon)$ in $(-1,1)$. We see that $\mu_n \to 0$ if $n \to \infty$, i.e. D_n, Ω_n go to zero for larger n. Then, the evaluation of a_n and b_n using the eq.18 is very unstable.

Now we shall see, that the continued fraction representation of $G_{\alpha\alpha}(z)$ can be obtained via an alternative more stable recursive method.

2.2 The Recursion Method

The origin of the recursion method is due to C.Lanczos[2] and essentially it is a method of transforming a symmetrical matrix into a tridiagonal form. A modern version of the method, connecting with solid state physical problems is due to R.Haydock[3]. (See also his contribution in this Lectures)

The recursion method is a real-space method for the calculation of the diagonal elements of the Green's function which is associated with some Hamiltonian is some tight-binding basis $\{|\alpha>, |\beta>, |\gamma>, \ldots\}$.

For the calculation of the Partial DOS associated with some particular orbital-α. We need to evaluate

$$D_\alpha(E) = \frac{1}{\pi}\mathrm{Im}G_{\alpha\alpha}(E - i0^+)$$

We start, placing the "electron" in that orbital by defining a starting vector which is for instance:

$$|1> = (0, \underset{\alpha-\mathrm{times}}{\ldots} 1, 0, 0 \ldots)^\dagger$$

This is the first vector for the tridiagonalization procedure. From this starting vector we define a new subspace $\{|n>\}$. It turns out that in this new subspace the original Hamiltonian $\hat{H}$ has a tridiagonal form.

$$\hat{H} \rightarrow \begin{pmatrix} a_1 & b_1 & & \\ b_1 & a_2 & b_2 & \\ & b_2 & a_3 & b_3 \\ & & & \ddots \end{pmatrix} \tag{20}$$

The recursion relations are:

$$b_1 \mid 2 >= \hat{H} \mid 1 > -a_1 \mid 1 >$$

for the first component, and for the rest

$$b_n \mid n+1 >= \hat{H} \mid n > -a_n \mid n > -b_{n-1} \mid n-1 > \ , \tag{21}$$

valid for $n \ = \ 2, 3, \ldots$
a_n and b_{n-1} are the coefficients which make $\hat{H} \mid n >$ orthogonal to the preceding vector $\mid n >$ and $\mid n-1 >$ and b_n are the coefficients which makes $\mid n+1 >$ normalized to the unity. The $\{a_n b_n\}$ recursion coefficients are given by,

$$a_n =< n \mid \hat{H} \mid n > \ ,$$

$$b_n =< n+1 \mid \hat{H} \mid n > \ . \tag{22}$$

They describe, as we saw before, higher moments of the partial DOS and they represent the influence on the DOS of the atoms which are more and more distant from the atom to what the α-orbital belongs. The new vectors, defined by Eq.(21) span a orthonormal set in the Krylov subspace, i.e. in the following subspace.

$$\left\{ \mid 1 >, \hat{H} \mid 1 >, \hat{H}^2 \mid 1 >, \ldots, \hat{H}^n \mid 1 > \right\}$$

The basic recursion method can be viewed as a Gram-Schmidt orthogonalization procedure on the set of vectors $\{\mid 1 >, \hat{H} \mid 1 >, \hat{H}^2 \mid 1 >, \ldots, \hat{H}^{n-1} \mid 1 >\}$.

The new basis, is then generated by repetitive operations with $\hat{H}$ on the new created vectors. By this way (in a tight-binding sense) each operation connects more and more distant orbitals with the initial orbital α.

This recursive procedure is also very easy to implement numerically. We only need matrix multiplications. Also we need keep only two vectors $\mid n >$ and $\mid n-1 >$ to generate the $\mid n+1 >$ state. See Eq.(21).

Some care has to be taken with round off errors due to finite precision arithmetic on the computer. It can happen that after many recursions the new generated vectors are no more orthogonal to all preceding ones. In the practice, the Eq.(22) are replaced by

$$a_n = < n \mid (\hat{H} \mid n > - b_n \mid n-1 >) \quad ,$$

$$b_n = \mid \hat{H} \mid n > - a_n \mid n > - b_{n-1} \mid n-1 > \mid \tag{23}$$

Now, we see the expression for $G_{\alpha\alpha}(E)$ using the recursion method. By definition

$$\begin{aligned} G_{\alpha\alpha}(E) &= < \alpha \mid (E - \hat{H})^{-1} \mid \alpha > \\ &= < 1 \mid (E - \hat{H})^{-1} \mid 1 > \end{aligned} \tag{24}$$

because the starting vector $\mid 1 >$ is the α-orbital, i.e. one of the orbitals of the tight-binding basis. The matrix $E - \hat{H}$ is given by:

$$E - \hat{H} = \begin{pmatrix} E - a_1 & -b_1 & 0 & 0 \cdots \\ -b_1 & E - a_2 & -b_2 & 0 \cdots \\ 0 & -b_2 & E - a_3 & 0 \cdots \\ . & . & . & \cdots \end{pmatrix} \tag{25}$$

If D_o denotes the determinant of Eq.(25), and D_1 the determinant of Eq.(25) if we suppress first column and row, D_2 the determinant of Eq.(25) if we suppress the two first columns and rows, and so on. Then the leading element of $(E - \hat{H})^{-1}$, i.e. $< 1 \mid (E - \hat{H})^{-1} \mid 1 >$ is given by the "cofactor" divided by the determinant

$$< 1 \mid (E - H)^{-1} \mid 1 > = \frac{det D_1}{det D_o} = \frac{1}{det D_o / det D_1} \quad . \tag{26}$$

From the Cauchy expansion of a determinant we have $det D_o = (E - a_1) det D_1 - b_1^2 det D_2$ so that Eq.(26) can be written as

$$< 1 \mid (E - \hat{H})^{-1} \mid 1 > = \frac{1}{E - a_1 - \frac{b_1^2}{det D_1 / det D_2}} \quad .$$

Similarly we have

$$det D_n / det D_{n+1} = E - a_{n+1} - \frac{b_{n+1}^2}{det D_{n+1} / det D_{n+2}}$$

and the continued fraction representation for the Green's function Eq.(24) follows inmediately.

$$G_{\alpha\alpha}(E) = \cfrac{1}{E - a_1 - \cfrac{b_1^2}{E - a_2 - \cfrac{b_2^2}{E - a_3 - \cfrac{b_3^2}{\ddots}}}} \tag{27}$$

Thus, once the (a_n, b_n) coefficients have been determined, the continued fraction can be evaluated very fast by iteration for any number of E values required. By truncating the continued fraction at some level n, we only get n singularities, but in order to recover the infinity character of a DOS for an infinite system, and have continuous bands and van Hove singularities it is necessary to have a lot of recurrence.

In the practice this is done by terminating the continued fraction by a "Terminator" $T(z)$ after some level n. i.e.

$$G_{\alpha\alpha}(z) = \cfrac{1}{z - a_1 - \cfrac{b_1^2}{z - a_2 - \cfrac{b_2^2}{z - a_3 - \cfrac{b_3^2}{\ddots \cfrac{}{z - a_n - b_n^2\, T(z)}}}}} \tag{28}$$

Here $T(z)$ is a terminator applied at the nth-level. This introduction does not alter the first 2n-moments of the partial DOS. $T(z)$ emulates the effect of infinity coefficients and makes the spectrum a continuous one.

Now we will guess the form of this terminator. If after some level n we set all $a_i = a$ and $b_i = b$ we can sum the remainder of the continued fraction analytically, as when

$$T(z) = \cfrac{1}{z - a_{n+1} - \cfrac{b_{n+1}^2}{z - a_{n+2} - \cdots}} \tag{29}$$

is replaced by

$$T(z) = \frac{1}{z - a - b^2\, T(z)} \quad . \tag{30}$$

Then we can solve for

$$T(z) = \frac{(z-a) \pm \sqrt{(z-a)^2 - 4b^2}}{2b^2} \quad . \tag{31}$$

Let's us suposse that all recursion coefficients are equal to a and b. Then we get

$$D(E) = \frac{1}{\pi} \mathbf{Im}\, T(E - i0^+) = \begin{cases} \frac{\sqrt{4b^2 - (E-a)^2}}{2\pi\, b^2} & a - 2b \;<\; E \;<\; a + 2b \\ 0 & a - 2b \;>\; E \;>\; a + 2b \end{cases}$$

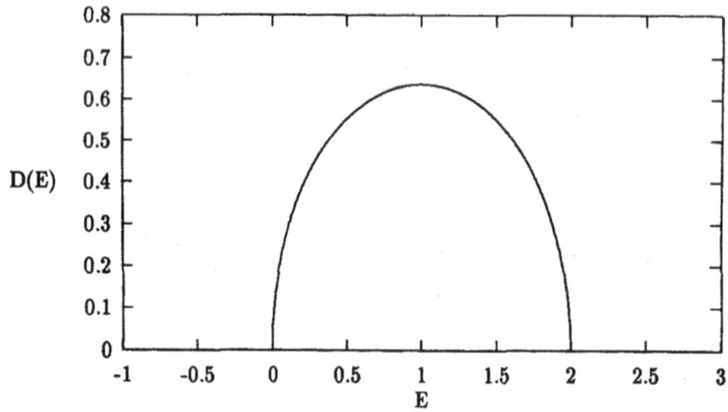

As we see in the figure, this "square-root" terminator produces a cut on the real axis of the complex z-plane between $a - 2b$ and $a + 2b$ with the character $| E - E_{edge} |^{1/2}$ near the edges.

The recursion coefficients (a_i, b_i), to whose continued fraction the $T(z)$ is appended, modulate the semi-elliptical spectrum of $T(z)$ with the particular local information extracted from $\hat{H}$.

One of the fastest ways to get the asymptotic values a, b from the know ones is due to Beer and Pettifor[4], that will be discussed in the next section.

Before we explain how to apply a terminator we summarize the main points of the recursion scheme.

1. Given an Hermitian Hamiltonian $\hat{H}$ in some particular "tight-binding basis" $\{| \alpha >, | \beta >, \ldots\}$ the total DOS is given by

$$D(E) = \mathbf{Tr}\ \delta(E - \hat{H}) = \mathbf{Tr}(\sum_i | i > \delta(E - E_i) < i |) \ ,$$

in the (unknown) representation which diagonalizes $\hat{H}$.

$$D(E) = \sum_\alpha < \alpha | \delta(E - \hat{H}) | \alpha > = \sum_{\alpha\, i} |< \alpha | i >|^2\, \delta(E - E_i)$$

$$= \sum_\alpha D_\alpha(E) \quad \text{(because the trace is an invariant)}$$

then the partial DOS associated with some particular "orbital" of the know basis is given formally by

$$D_\alpha(E) = \sum_{\substack{i \\ \text{exact eigenstates}}} |<\alpha \mid i>|^2 \, \delta(E - E_i)$$

2. The partial DOS $D_\alpha(E)$ is related to the Green's function

$G_{\alpha\alpha}(E) =< \alpha \mid \frac{1}{E-\hat{H}} \mid \alpha >$ by the following eq.:

$$G_{\alpha\alpha}(z) = \int_{-\infty}^{\infty} dx \frac{D_\alpha(x)}{(z-x)}$$

3. The partial DOS, $D_\alpha(E)$ is given by

$$D_\alpha(E) = \frac{1}{\pi}\mathbf{Im}\, G_{\alpha\alpha}(E - i0^+)$$

(inversion of the integral relation from above)

4. The moments of the Partial DOS, $D_\alpha(E)$ are defined as

$$\mu_n = \int_{-\infty}^{\infty} dx\, x^n D_\alpha(x) =< \alpha \mid \hat{H}^n \mid \alpha > \quad , \mu_o = 1$$

i.e. $D_\alpha(x)$ is normalized. This is related with closed paths counting.

5. $G_{\alpha\alpha}(z)$ has a continued fraction representation.

$$G_{\alpha\alpha}(z) = \cfrac{1}{z - a_1 - \cfrac{b_1^2}{z - a_2 - \cfrac{b_2^2}{z - a_3 - \cfrac{b_3^2}{z - a_4 \cdots}}}} \; .$$

where

$$a_n = \frac{1}{\Omega_{n-1}}\left[\frac{D_{n-1}\Omega_n}{D_n} + \frac{D_n\Omega_{n-2}}{D_{n-1}}\right]$$

$$b_n^2 = \frac{D_n D_{n-2}}{D_{n-1}^2} \quad , \quad D_{-1} = \Omega_{-1} = 1 \qquad n = 1, 2, 3, \ldots$$

6. The $\{a_i, b_i\}$ coefficients can be also obtained by the so-called RECURSION-METHOD, which is given by the following relations.

   Choose a starting vector $| 1 >=| \alpha >$.

   Then calculate b_1 and a_1 such as

$$b_1 \mid 2 >= \hat{H} \mid 1 > -a_1 \mid 1 >$$

$$< 1 \mid 2 >= 0 \quad \text{and} \quad < 2 \mid 2 >= 1$$

   then recur as follows.

$$b_n \mid n+1 >= \hat{H} \mid n > -a_n \mid n > -b_{n-1} \mid n-1 > \quad .$$

   $< n \mid m >= \delta_{nm}$ and $\hat{H}$ in this basis has a tridiagonal form

$$\hat{H} \rightarrow \begin{pmatrix} a_1 & b_1 & & \\ b_1 & a_2 & b_2 & \\ & b_2 & a_3 & b_3 \\ & & & \ddots \end{pmatrix}$$

7. The recursion coefficients are related with the moments of the partial DOS. The knowledge of $\{a_i, b_i\}$ for $i = 1, \ldots n$ implies that we know exactly the first $2n$-moments of the partial DOS.

8. The RECURSION-METHOD transforms the original Hamiltonian into a linear chain model

$$\hat{H} \rightarrow \sum_n | n > a_n < n | + \sum_n (| n > b_n < n+1 | + | n+1 > b_n < n |)$$

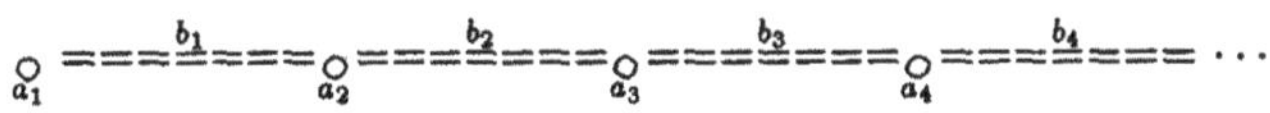

9. The evaluation of the partial DOS, $D_\alpha(E)$ is equivalent to evaluate the DOS in the "surface-atom" of the linear chain model.

10. The square-root terminator simulates an infinite medium which is appended at the "n-atom" of the linear chain transforming the problem in a semi-infinite linear chain model.

11. The inclusion of a terminator after the n-level of the recursion procedure does not alter the first 2n-moments of the partial DOS at the "surface atom".

12. The further apart we put the terminator the less influence has on the local DOS-shape of the first atom.

In the following section we shall see some general remarks on the terminator problem in the recursion method and give some examples.

2.3 Terminator

Chebyshev Polynomials

Let's write the continued fraction for $G_{\alpha\alpha}(z) = G_{11}(z)$ in the following form. (See Eq.(27))

$$G_{11}(z) = \frac{1}{z - a_1 - b_1^2 X_{22}(z)} \quad , \tag{32}$$

where $X_{22}(z)$ is defined by the relation

$$X_{ii}(z) = \frac{1}{z - a_i - b_i^2 X_{i+1\ i+1}(z)} \quad , \tag{33}$$

valid for $i = 2, 3, \ldots$. We note that X_{ii} does not represent some diagonal element of the Green's operator of Eq.(6).

Now, in order to relate the recursion coefficients $\{a_i, b_i\}$ with an expansion of complex exponentials, we recall some properties of the Chebyshev polynomials

$$\int_{-1}^{1} \frac{\sqrt{1-x^2}}{y-x} U_n(x)\, dx = \pi T_{n+1}(y) \ , \tag{34}$$

where

$$U_n(x) = (-)^n U_n(-x) = \frac{\sin(n+1)\phi}{\sin\phi} \ ,$$

and

$$T_n(x) = (-)^n T_n(-x) = \cos(n\phi). \tag{35}$$

Here $x = \cos\phi$.

The partial DOS which arises when all recursion coefficients are equal ($a_i = a$, $b_i = b$ $\forall_i$) is proportional to $\sqrt{1-x^2}$. (See Eq. (31) and below, with $x = \frac{E-a}{2b}$). This function is exactly the weight function of the Chebyshev polynomial. (Eq. (34)).

Then, it is a good idea to write the real DOS as: (defined in $(-1, 1)$ with a suitable change of variables).

$$D(x) = \sqrt{1-x^2}\, f(x) \ . \tag{36}$$

Here the DOS defined by Eq. (31) (the square root) is "perturbed " by the $f(x)$ function to get the correct real shape. This is shown in the following scheme.

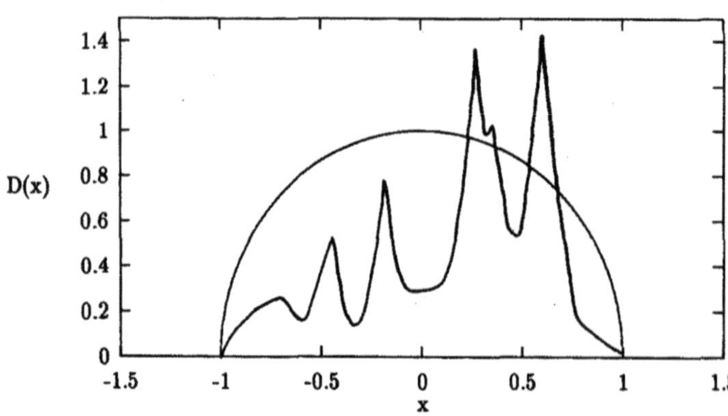

FIG. 1. Density of States scheme of Equation (36), showing the shape of a "real " DOS (normalized in (-1,1) interval), as well as the shape of the weight function of the Chebyshev polynomials. It is assumed that the ratio of this two quantities, $f(x)$ has a finite Chebyshev expansion

Moreover, we make the following Ansatz, $f(x)$, defined in $(-1,1)$, admits an expansion in Chebyshev polynomials, i.e.

$$f(x) = \sum_{n=0}^{\infty} A_n U_n(x) \quad . \tag{37}$$

Now, we evaluate $G_{\alpha\alpha}(z) = G_{11}(E - i\eta)$ given in Eq.(15) (the Hilbert transformation) to get.

$$G_{11}(E - i\eta) = \int_{-1}^{1} \frac{(E-t)D(t)}{(E-t)^2 + \eta^2}\, dt \; - i \int_{-1}^{1} \frac{\eta D(t)}{(E-t)^2 + \eta^2}\, dt \quad . \tag{38}$$

Here $t = (x - a)/2b$ and $E = (z - a)/2b$, where $a = a_\infty$ and $b = b_\infty$ are the asymptotic limits.

When $\eta \to 0^+$, the first integral is the principal part and the second term is proportional to a delta-function.

Then

$$G_{11}(E) = \fint_{-1}^{1} \frac{D(t)}{(E-t)}\, dt \; - i\pi\, D(E) \quad , \tag{39}$$

from which the standard relation of Eq.(11) follows.

(i.e. $D(E) = +\frac{1}{\pi}\mathrm{Im}\, G_{11}(E - i0^+)$)

By inserting in Eq.(39) the definition of Eq.(36) we get

$$G_{11}(E) = \fint_{-1}^{1} \frac{\sqrt{1-t^2}}{(E-t)} f(t)\, dt - i\pi\sqrt{1-E^2}\, f(E) \quad . \tag{40}$$

Now, we use the Eq.(37) and the integral relation of Eq.(34) to obtain:

$$G_{11}(E) = \sum_{n=0}^{\infty} A_n \left[\pi T_{n+1}(E) - i\pi\sqrt{1-E^2}\, U_n(E)\right] \tag{41}$$

Using $E = \cos\phi$ and $\sin\phi = \sqrt{1-E^2}$ (Note, we have E defined in $(-1,1)$), we arrive to the following relation

$$G_{11}(E) = \pi \sum_{n=0}^{\infty} A_n \exp\left(-i(n+1)\phi\right) \quad , \tag{42}$$

or in more compact form:

$$G_{11}(E) = \pi \sum_{n=0}^{\infty} A_n^{(1)} \xi^{n+1} \quad , \tag{43}$$

Here $\xi = e^{-i\phi}$. The index in A_n is introduced to relate then with G_{11}. We proceed now to calculate the $A_n^{(1)}$ coefficients.

We rewritte Eq.(32) in the following form ($E = (z-a)/(2b)$)

$$b\, G_{11} = \frac{1}{2E - \bar{a}_1 - \bar{b}_1^2\, b\, X_{22}} \quad , \tag{44}$$

Where X_{22} is defined in Eq.(33) and $\bar{a}_i = \frac{a_i - a}{b}$ and $\bar{b}_i = \frac{b_i}{b}$.

We note that the imaginary part of G_{11} vanishes outside the interval $(-1,1)$. It is easy to see that X_{22} also vanishes outside $(-1,1)$ (their imaginary part), then the imaginary part of X_{ii} also vanishes for points outside $(-1,1)$.

Therefore X_{ii} also admits a Chebyshev expansion like.

$$X_{ii}(E) = \pi \sum_{n=0}^{\infty} A_n^{(i)} \exp\left(-i(n+1)\phi\right) \quad , \tag{45}$$

or better

$$X_{ii}(E) = \pi \sum_{n=0}^{\infty} A_n^{(i)}\, \xi^{n+1} \quad .$$

Moreover we have

$$X_{ii}(E) = \frac{1}{2E - \bar{a}_i - \bar{b}_i^2\, b\, X_{i+1}(E)} \quad . \tag{46}$$

By inserting the relation (45) in Eq.(46) we get

$$(2E - \bar{a}_i) \sum_{n=0}^{\infty} A_n^{(i)} \xi^{n+1} \; - \; b\, \bar{b}_i^2 \pi \sum_{n=0}^{\infty} A_n^{(i)} \xi^{n+1} \sum_{m=0}^{\infty} A_m^{(i+1)} \xi^{m+1} \; = \; \frac{1}{\pi b} \quad . \tag{47}$$

This equation must be valid for all ξ, then all the coefficients of the polynomial in ξ must vanish, besides the free term.

Then:

$$A_0^{(i)} = \frac{1}{\pi b} \quad ,$$

$$A_1^{(i)} = \frac{\bar{a}_i}{\pi b} \quad , \tag{48}$$

$$A_{n+2}^{(i)} = -A_n^{(i)} + \bar{a}_i \, A_{n+1}^{(i)} + b \, \bar{b}_i^2 \, \pi \sum_{l=0}^{n} A_l^{(i)} \, A_{n-l}^{(i+1)} \ .$$

This recurrence relations allow us to calculate $A_n^{(1)}$ from the N pairs of $(\bar{a}_i, \bar{b}_i)$. We use them to eliminate successively the vectors $A_n^{(k)}$ starting from $A_o^{(2N-1)}$. It can be easily seen that the calculation scheme proceeds in this case as indicated below

$$\begin{array}{ccccccccc}
A_o^{(1)} & & A_1^{(1)} & & A_2^{(1)} & \cdots & A_{2N-2}^{(1)} & & A_{2N-1}^{(1)} \\
\uparrow & \nearrow & \uparrow & \nearrow & \uparrow & & \uparrow & \nearrow & \\
A_o^{(2)} & & A_1^{(2)} & & A_2^{(2)} & \cdots & A_{2N-2}^{(2)} & & \\
\vdots & & & & & & & & \\
A_o^{(2N-2)} & & A_1^{(2N-2)} & & & & & & \\
\uparrow & \nearrow & & & & & & & \\
A_o & & & & & & & &
\end{array}$$

Once we get the $2N-1$ Chebyshev coefficients. $A_n^{(1)}$ from the first N pairs of recursion coefficients we "extrapolate" them using the following Ansatz. We suppose that after some level N the remainder coefficients are <u>linearly</u> related to the preceding ones:

$$A_n^{(1)} = \sum_{i=1}^{N} c_i \, A_{n-i}^{(1)} \quad . \tag{49}$$

This equation holds for $n > N$. The above conjecture is completely equivalent to assume that the Chebyshev coefficients can be written as: (we suppress the superscript (1)).

$$A_n = \sum_{i=1}^{N} d_i \, z_i^n \quad , \tag{50}$$

where the z_i values are the complex roots of the following polynomial.

$$z^N - \sum_{j=1}^{N} c_j \, z^{N-j} = 0 \ . \tag{51}$$

When the asymptotic limits (a, b) are correct, $\lim_{n\to\infty} A_n = 0$ (because expression 42 converges).

This implies in Eq.(50) that $| z_i |< 1 \quad 1 \leq i \leq N$.

This criterion allow us to demand that all the roots of the polynomial of Eq.(51) lie inside the unitary circle in the z-plane.

This is not always possible, sometimes the calculated A_i coefficients are wrong because the a and b asymptotic values are inaccurate. This fact guides us to find numerically the correct bandwidth and bandcenter.

The employed numerical recipe is as follows: from Eq.(48) and an initial guess for a and b calculate $2N$ coefficients A_n. Use Eq.(49) to calculate the N coefficients c_i. Now use these c-coefficients to find the complex roots of the polynomial in Eq.(51). If all $| z_i |\leq 1$ then stop and use Eq.(49) to extrapolate A_n for $n > 2N$. If not, then change the asymptotic limits a and b and start again.

After the extrapolation procedure is successfully ended we proceed to calculate the partial DOS by calculating the following series:

$$D(E) = \sum_{n=0}^{\infty} A_n \sin\left[(n+1)\cos^{-1}\left(\frac{E-a}{2b}\right)\right] . \tag{52}$$

In practice the upper limit in the summation is of the order of 100.

It is very useful to start with a good choice of a and b. That is best done using Beer and Pettifor procedure[4]

Their idea is to find the band limits $(a - 2b, a + 2b)$, from a finite number of recursion coefficients, in such a way that the minimum width $(4b)$ does not let any delta functions outside the band limits. In a mathematical sense one has to impose simultaneously a divergence of G_{11} in the band limits.

Let us write again the continued fraction representation for $G_{11}(E)$;

$$G_{11}(E) = \cfrac{1}{E - a_1 - \cfrac{b_1^2}{E-a_2-\cfrac{b_2^2}{z-a_3-\cfrac{b_3^2}{\ddots\, \overline{E-a_n-b_n^2\, T(z)}}}}} \tag{53}$$

for constant coefficients $T(E)$ is

$$T(E) = \frac{1}{2}\left(-(E-a) \pm \sqrt{(E-a)^2 - 4b^2}\right)$$

In the band limits $T(a \pm 2b) = \pm b$, therefore, by substituting this value in the Eq. (53) we get

$$G_{11}(a \pm 2b) = \cfrac{1/2}{\pm b - (a_1 - a)/2 - \cfrac{b_1^2/2}{\pm b - (a_2 - a)/2 - \cdots \cfrac{b_{n-1}^2/2}{\pm b - (a_n - a)/2}}} \tag{54}$$

This continued fraction corresponds to the leading element (i.e. element 1,1) of the inverse of the following matrix.

$$\begin{pmatrix} \pm b - (a_1 - a)/2 & b_1/2 & 0 & \cdots \\ b_1/2 & \pm b - (a_2 - a)/2 & & \\ \vdots & \vdots & & \\ & & & b_{n-1}/\sqrt{2} \\ 0 & 0 & b_{n-1}/\sqrt{2} & \pm b - (a_n - a) \end{pmatrix} \tag{55}$$

Then the values a and b are obtained from the eigenvalues of this matrix in such a way that $| b_{\max} |=| b_{\min} |= b$.

Normally one starts choosing $a = a_n$ or $a = \frac{1}{n}\sum_{i=1}^{n} a_i$ then get the maximum and minimum eigenvalue of the matrix $b_{\max}$ and $b_{\min}$. If $| b_{\max} |\neq| b_{\min} |$ change a to:

$$a_{new} = a_{old} + b_{\max} + b_{\min} \quad ,$$

repeat until $b_{\max} + b_{\min} = 0$, i.e. $b =| b_{\max} |=| b_{\min} |$.

2.4 Examples

Now we examine some analytical examples.

Let us take the following analytical form.

$$D(E) = \begin{cases} \frac{3}{2}\sqrt{E} & 0 \leq E < 1 \\ 0 & 0 > E > 1 \end{cases} \tag{56}$$

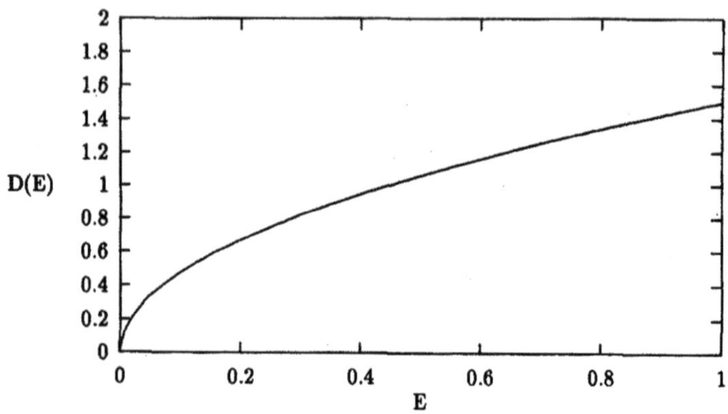

FIG. 2. A DOS with a square root shape
from Eq.(56) we calculate all moments.

$$\mu_n = \int_0^1 d\epsilon \epsilon^n D(\epsilon) = \frac{3}{2n+3} \ .$$

Then, by evaluating the Haenkel Determinants we obtain the following set of coefficients.

i	a_i	b_i
1	0.60000	1.00000
2	0.51111	0.26186
3	0.50427	0.25324
4	0.50226	0.25151
5	0.50140	0.25088
6	0.50095	0.25057
7	0.50069	0.25040
8	0.50052	0.25030

Now we use Beer-Pettifor method[4] and our Chebyshev extrapolation procedure and evaluate again the DOS. The results are shown in the next 2 figures.

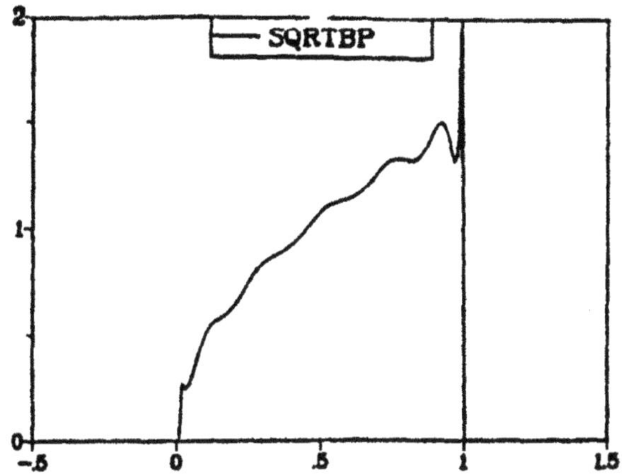

FIG. 3. Reconstruction of D(E) (Eq.56) using Beer-Pettifor method[4] to find a and b and the square-root terminator

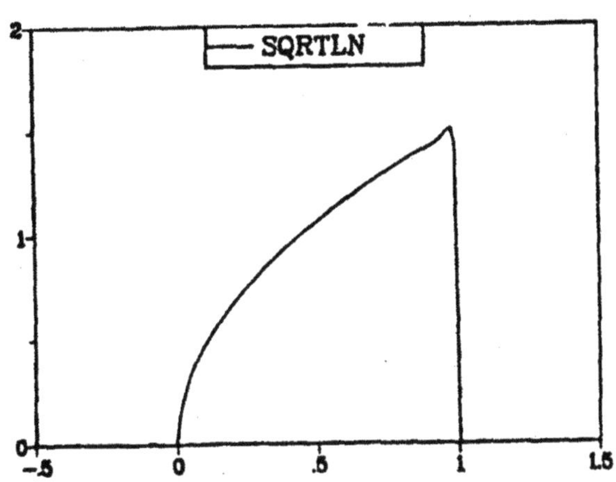

FIG. 4. Reconstruction of D(E) (Eq.56) using the Chebyshev method and Eq.(52) with 50 terms

A lot of "academic" examples can be done using the relation between the Haenkel determinants and the recursion coefficients. In the following flux diagram, the necessary steps to do a test calculations are indicated.

Give analytical $D(E)$

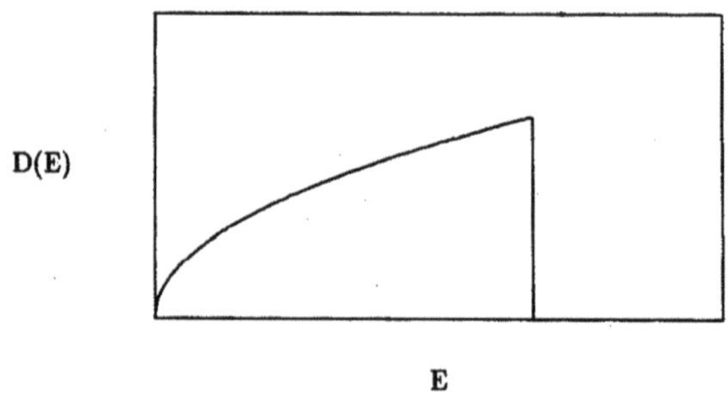

Evaluate the moments of $D(E)$

$$\mu_n = \int_{-\infty}^{\infty} d\epsilon \, \epsilon^n \, D_\alpha(\epsilon) \quad n = 1, 2, \ldots, 2N+1$$

Evaluate the recursion coefficients (N pairs) (Eqs.(18) and (19))

$$a_n = \frac{1}{\Omega_{n-1}} \left[\frac{D_{n-1}\Omega_n}{D_n} + \frac{D_n \Omega_{n-2}}{D_{n-1}} \right]$$

$$b_n^2 = \frac{D_n D_{n-2}}{D_{n-1}^2} \quad \text{for} \quad n = 1, 2, 3, \ldots, N$$

$$D_n = \begin{vmatrix} \mu_0 & \mu_1 & \cdots & \mu_n \\ \mu_1 & \mu_2 & \cdots & \mu_{n+1} \\ \vdots & & & \\ \mu_n & \mu_{n+1} & \cdots & \mu_{2n} \end{vmatrix}$$

$$\Omega_n = \begin{vmatrix} \mu_1 & \mu_2 & \cdots & \mu_{n+1} \\ \mu_2 & \mu_3 & \cdots & \mu_{n+2} \\ \vdots & & & \\ \mu_{n+1} & \mu_{n+2} & \cdots & \mu_{2n+1} \end{vmatrix}$$

Build up the Green's function (Eq.(28))

$$G_{\alpha\alpha}(z) = \cfrac{1}{z - a_1 - \cfrac{b_1^2}{z - a_2 - \cdots \cfrac{}{z - a_n - b_n^2\, T(z)}}}$$

Put some terminator $T(z)$ and evaluate the DOS $D(E)$ (Eqs.(12) and (52))

$$D(E) = \frac{1}{\pi}\mathbf{Im}\, G_{\alpha\alpha}(E - i0^+)$$

or

$$D(E) = \sum_{n=0}^{\infty} A_n \sin\left[(n+1)\cos^{-1}(\frac{E-a}{2b})\right]$$

compare with the analytical expression

A new example (very ill conditionated) is to have a large gap, the case in which the recursion coefficients are $a_n = (-1)^{n+1}$ and $b_n = 1$ for all n.

This case gives a DOS which extends in $(-\sqrt{5/2}, \sqrt{5/2})$, with a gap between $(-1, 1)$ and an infinity discontinuity in -1. The analytic result for this case is given by:

$$D(E) = -\frac{1}{\pi}\mathbf{Im}\, \frac{\sqrt{E^4 - 6E + 5}}{2(E+1)} \tag{57}$$

We see that the extrapolation procedure reduces all the spurious oscillations, but it can not avoid a small region with negative $D(E)$ in the gap, close to the van Hove singularity.

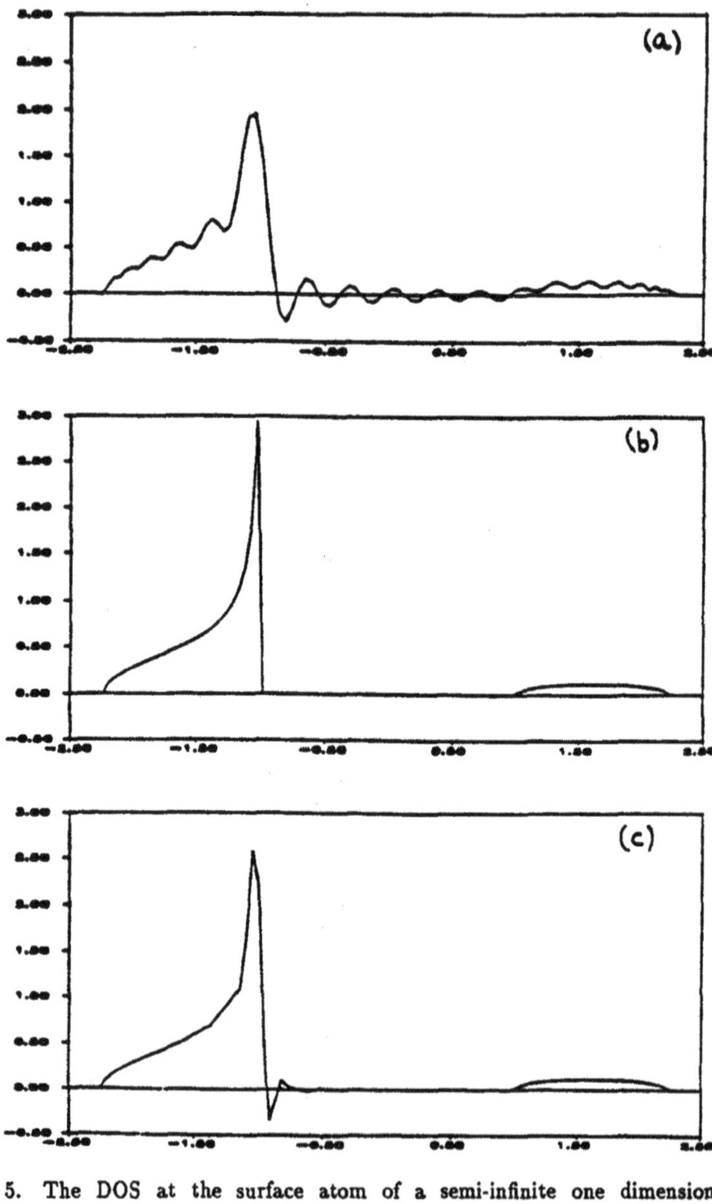

FIG. 5. The DOS at the surface atom of a semi-infinite one dimensional alloy ABABAB.... The exact result is shown in (b), and the one obtained with the extrapolated coefficients in (c). For comparison we show in (a) the result obtained using only 40 exact A_n coefficients with no extrapolation. From Vargas et.al.[5]

Another trivial and very ill behaved example is the case of a tight binding model for an infinite linear chain of atoms all having the same on site energies $a_n = 0$ and the same hopping integrals $b_n = 1$. The partial DOS has the following analytic form:

$$D(E) = \frac{1}{2\pi} \frac{1}{\sqrt{1-(E/2)^2}} \quad , \tag{58}$$

In this case the function $f(E)$ defined in Eq.(36) is

$$f(E) = \frac{1}{2\pi} \frac{1}{1-(E/2)^2}$$

and we can see that this function does not admit a finite expansion in Chebyshev polynomials as formally stated in Eq.(37).

The results obtained with the present extrapolation method are given in the next figure. As in the previous example a very small region of negative $D(E)$ is found outside the band close to the singularity.

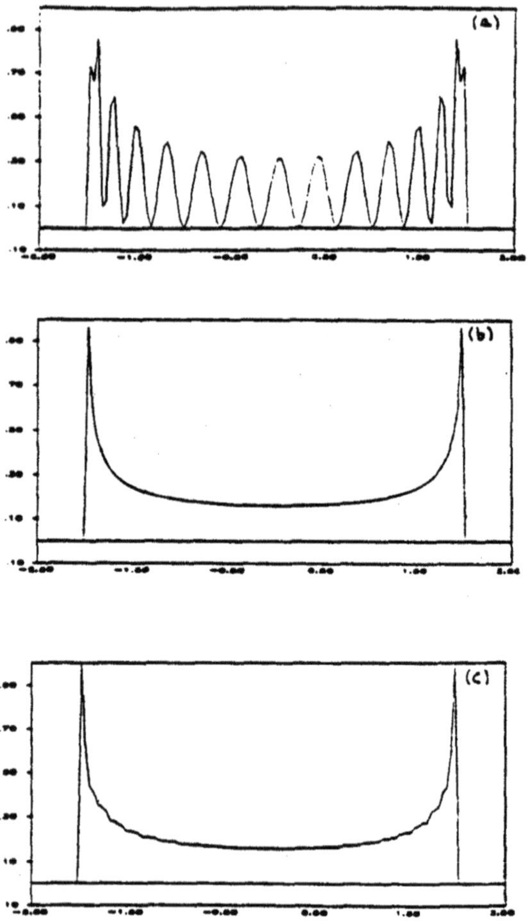

FIG. 6. DOS of an ordered infinite one-dimensional chain. The exact result is shown in (b), and the one obtained with the extrapolated coefficients in (c). For comparison we show in (a) the result obtained using only 15 exact A_n coefficients with no extrapolation. From Vargas et.al[5]

From the above mentioned examples, we can see that the extrapolation algorithm suppresses artificial oscillations, improving greatly the shape of the DOS. One can also see that when a gap is present this procedure leads to negative values. This fact is evident from Eq.(52) when an infinite sum is truncated after same level, the reconstructed function always oscillates near the discontinuity. This is the same problem of any truncated Fourier series representation of a function that has a discontinuity. To avoid the negative values in the gap one should return to the continued fraction representation by recalculating the extrapolated a_n and b_n values, as in this representation $D(E)$ is positive defined.

The representation of the Partial DOS is not limited to the Chebyshev polynomials, one could in principle use any family of orthogonal polynomials.

The election of the best family depends strongly on the singularities and behaviour of the DOS at the band edges. If the weight function of these polynomials has already the correct behavior at the band edges the rest of the function $f(E)$ admits a rapidly convergent expansion in these polynomials. However that family of polynomials must satisfy the relation

$$\mathrm{Re}G_{11}(E - i0^+) = -\frac{1}{\pi} \fint_{-\infty}^{\infty} dx \, \frac{\mathrm{Im}G_{11}(x - i0^+)}{x - E} \quad . \tag{59}$$

Certainly the Chebyshev polynomials employed here will be the best choice for 3-Dimensional problems in which the partial DOS vanishes as $\sqrt{E}$ at the band edges, and will be worse for low dimensional cases with singularities at the band edges.

For a review of a recursion method and all kind of terminators see Heine et. al.[6,7] and references therein.

3 LMTO TIGHT-BINDING + RECURSION

3.1 Introduction

In order to calculate the electronic structure of topologically disordered solids and liquids it is customary to solve the one-electron Schrödinger equation with an iterative, real-space scheme and to obtain local quantities such as orbital-projected densities of states. The local point of view has been extensively discussed by Heine, Bullet, Haydock and Kelly[6] who advocate the recursion method as the most efficient real-space scheme.

In this notes we explain how first principles electronic structure calculations with the recursion method may conveniently be carried out using a basis of linear muffin-tin orbitals (LMTO's) in a tight-binding (TB) representation.

As is well known, the LMTO set is a minimal basis, and it is complete for the muffin-tin or atomic sphere potential used for its reconstruction and over an energy range of about 1.5 Rydberg. For a complete review of TB-LMTO formalism we advice the reader to the works of Andersen and Jepsen[8] and Andersen et.al.[9,10,11]

In the present notes we only state and explain the results needed in order to apply the recursion method with LMTO's.

3.2 Calculating Structure Constants

In the present we use LMTO's which have no kinetic energy, so they have envelope functions which are solutions of the Laplace equation.

A conventional solid-state LMTO, $\chi^{o}_{RL}(\underline{r}-\underline{R})$, thus has an envelope function proportional to the $2^l - pole$ field $|\,\underline{r}-\underline{R}\,|^{-l-1}\, Y_L(\underline{r}-\underline{R})$.

If $\underline{R}$ takes the values of all the sites in the solid and if L takes the 4(sp), 9(spd), or the 16(spdf), angular-momentum values, we obtain the set of conventional LMTO's $\{|\,\chi^{o}>\}$. This set of functions may be linearly transformed into an equivalent set of "screened" LMTO's, $|\,\chi^{\alpha}>$.

The envelope of the screened LMTO, $\chi^{\alpha}_{RL}(\underline{r}-\underline{R})$ has $2^l - pole$ character near $\underline{R}$ but, in addition, it has screening poles at the neighboring sites. Each set of screened LMTO's is characterized by a set α of screening constants, α_{RL}, and these may, for instance, be determined so as to give the LMTO set short range, or so as to make it nearly orthonormal. For the conventional set, the screening constants are zero. The following two, site independent sets of screening constants, *sp*-screening and *spd*-screening have been found numerically to give short-ranged envelope function for all, reasonably homogeneous, three dimensional structures.

$$\underline{\alpha} = \{\alpha_s, \alpha_p, \alpha_{l>1}\} \tag{60}$$

$$= \{0.28723, 0.02582, 0.0\} \equiv \underline{\alpha}_1$$

(sp-screening)

$$\underline{\alpha} = \{\alpha_s, \alpha_p, \alpha_d, \alpha_{l>2}\} \tag{61}$$

$$= \{0.34850, 0.05303, 0.01071, 0.0\} \equiv \underline{\alpha}_2$$

(sp-screening).

The corresponding LMTO sets are referred to as "tight-binding" sets. (We note however that this is not the only way to have a tight-binding representation, it is only the simplest possible way).

The envelope of $\chi^{\alpha}_{RL}(\underline{r} - \underline{R})$ may be expanded about the site $\underline{R}'$ in a spherical-harmonic series and the expansion coefficients $S^{\alpha}_{R'L,RL}$ are the so-called "screened structure constants". For a given $\underline{\alpha}$, they form a Hermitian matrix which will turn out to be the structural factor of the two-center Hamiltonian for the LMTO set.

The conventional structure constants have the following off-site elements. (conventional means $\alpha = 0$).

$$
\begin{aligned}
S^{o}_{ss\sigma}(d/w) = -2(w/d),\ \ S^{o}_{sp\sigma}(d/w) = (2\sqrt{3})(w/d)^2, \\
S^{o}_{pp\{\sigma,\pi\}}(d/w) = 6\{2,-1\}(w/d)^3,\ \ S^{o}_{sd\sigma}(d/w) = -(2\sqrt{5})(w/d)^3, \\
S^{o}_{pd\{\sigma,\pi\}}(d/w) = (6\sqrt{5})\{-\sqrt{3},1\}(w/d)^4, \\
S^{o}_{dd\{\sigma,\pi,\delta\}}(d/w) = 10\{-6,4,-1\}(w/d)^5 \quad ,\ldots .
\end{aligned}
\tag{62}
$$

Where the z-axis is chosen along $\underline{R} - \underline{R}'$ and $d = \mid \underline{R} - \underline{R}' \mid$ is the interatomic distance. A length w is introduced to make the structure constants dimensionless (w = wigner seits radius). For a general formula of the conventional structure constants see Andersen[12,13] or the program we give to perform simple calculations (see section examples).

The on-site elements of the conventional structure constants vanish per definition. The screened structure constants are given in terms of the conventional ones and the screening by the Dyson equation.

$$
S^{\alpha}_{RL,R'L'} = S^{o}_{RL,R'L'} + \sum_{R''L''} S^{\alpha}_{RL,R''L''}\alpha_{R''L''}S^{o}_{R''L'',R'L'}, \tag{63}
$$

or, in matrix notation,

$$
S^{\alpha} = S^{o}(1 - \alpha S^{o})^{-1} = \alpha^{-1}[(\alpha^{-1} - S^{o})^{-1} - \alpha]\alpha^{-1}, \tag{64}
$$

where α is considered a diagonal matrix with elements α_{RL}. If is given by Eq.(60) or Eq.(61), and if the length ω entering in the definition of S^{o} is taken to be the average Wigner Seitz radius

$$
\frac{4\pi}{3}w^3 = V_{site} \tag{65}
$$

where V_{site} is the volume per site, then the screened structure constants are so localized that they effectively vanish when the interatomic distance d exceeds the radius of a cluster containing about 20 atoms for *spd*-screening and about 50 atoms for *sp*-screening (convince yourself trying the programs). These structure constants are named the tight-binding structure constants S^{α}.

The choice (65) for w is indeed not crucial; for instance, the structure constants with $\underline{\alpha} \equiv \underline{\alpha}_1$ or $\underline{\alpha}_2$ are well localized, even for the extremely inhomogeneous structure consisting of a surface of a semi-infinite solid when w is taken to be the WS radius of the bulk.

For an amorphous system one could thus calculate the tight-binding structure constants by inverting, for each site, the Hermitian, positive-definite matrix $\alpha^{-1} - S^0$ in Eq. (64) for the cluster centered at that site. The dimension of the square matrix to be inverted is approximately 20×9 for *spd*-screening and 50×4 for *sp*-screening, that is, about 200.

The off-site elements of the tight-binding structure constants, which will turn out to be structural factors of the "hopping integrals", follow the universal interpolation formula, see Sob et.al.[14]

$$S^{\alpha}_{ll'm}(d/w) = A^{\alpha}_{ll'm} \exp(-\lambda^{\alpha}_{ll'm} d/w), \tag{66}$$

rather closely, provided that the distribution of sites within a distance of order $10w/\lambda$ is reasonably homogeneous and that w is taken to be the local average Wigner Seitz radius. The parameters A and λ are given in the following Table.

TABLE I. Parameters of the Interpolation Formula $S^{\alpha}_{ll'm} = Ae^{-\lambda d/w}$ for the off-site elements of the TB structure constants. We have used *sp*-screening (set I) and *spd*-screening (set II). The screening constants $\underline{\alpha}_I$ and $\underline{\alpha}_{II}$ are given by Eqs.(60) and (61), respectively. w is the average Wigner-Seitz radius given by (65).

	spd-screening		*sp*-screening	
	A_{II}	λ_{II}	A_I	λ_I
$ss\sigma$	-184.7	3.293	-43.57	2.559
$sp\sigma$	371.7	3.301	81.77	2.503
$pp\sigma$	791.0	3.331	182.7	2.529
$sd\sigma$	-575.0	3.440	-84.51	2.444
$pd\sigma$	-1422.0	3.535	-288.6	2.671
$dd\sigma$	-3685.0	3.905	-1018.0	3.199
$pp\pi$	-359.9	3.935	-241.5	3.589
$pd\pi$	837.0	3.965	487.6	3.558
$dd\pi$	1997.0	3.998	1272.0	3.657
$dd\delta$	-844.0	4.708	-821.0	4.494

The on-site elements of the tight binding structure constants do not vanish. They depend sensitively on the local environment and have <u>no</u> approximation in terms of a simple, structure-independent interpolation formula. These "crystal-field" elements are, however, given explicitly in terms of the off-site elements by the Dyson equation (63). Since S^0 has no on-site elements, only the off-site elements of S^α appear on the right-hand side of Eq.(63).

A simple scheme to evaluate the on-site elements of a tight-binding structure matrix constants is thus obtained by using Eq.(63), together with Eq.(62) for S^o and the approximation (66) for the tight-binding hopping integrals, S^α.

3.3 The two-center Hamiltonian

Given an envelope function, the LMTO is now constructed by continuous and differentiable augmentation inside certain muffin-tin or atomic spheres with functions, $\varphi_{RL}(\underline{r}-\underline{R})$ and $\dot{\varphi}^\alpha_{R'L'}(\underline{r}-\underline{R}')$, to be specified later. We may thus write.

$$\chi^\alpha_{RL}(\mathbf{r}-\mathbf{R}) = \kappa^\alpha_{RL}(\mathbf{r}-\mathbf{R}) + \varphi_{RL}(\mathbf{r}-\mathbf{R}) + \sum_{R'L'} \dot{\varphi}^\alpha_{R'L'}(\mathbf{r}-\mathbf{R}')h^\alpha_{R'L',RL} \quad , \tag{67}$$

where $\kappa^\alpha_{RL}(\underline{r}-\underline{R})$ is the envelope function truncated inside all spheres and the φ and $\dot{\varphi}$ functions are truncated outside their respective spheres. In shorthand matrix notation we express the $\alpha-set$ of LMTO's as:

$$\mid \underline{\chi}^\alpha > = \mid \underline{\kappa}^\alpha > + \mid \underline{\varphi} > + \mid \underline{\dot{\varphi}}^\alpha > h^\alpha \ , \tag{68}$$

where, for instance, $< \underline{r} \mid \underline{\varphi} >$ is a row vector with elements $< \underline{r} \mid \varphi(RL) > \equiv \varphi_{RL}(\underline{r}-\underline{R})$, and $< \underline{\varphi} \mid \underline{r} >$ is a column vector with elements $< \varphi_{RL} \mid \underline{r} > = \varphi^*_{RL}(\underline{r}-\underline{R})$. The Eq.(68) says that the "head" of the envelope function is given the proper atomic wiggles through smooth augmentation with a φ function, plus a $\dot{\varphi}$ function times the on-site element of h^α, and that the "tail" of the envelope function is orthogonalized to the core states of the neighboring atoms through smooth augmentation with $\dot{\varphi}$ functions. The latter are constructed in such a way that, when multiplied with the proper coefficients h^α, they smoothly match into the envelope function. This means that the $\dot{\varphi}$ functions by construction must have the proper radial logarithmic derivative at the sphere boundary, and that the coefficients h^α are determined by the condition of continuity of the orbital. They are explicitly given by

$$\begin{aligned} H^\alpha_{RL,R'L'} &\equiv \epsilon_{\nu R\ell}\delta_{RR'}\delta_{LL'} + h^\alpha_{RL,R'L'} \\ &= C^\alpha_{R\ell}\,\delta_{RR'}\,\delta_{LL'} + \sqrt{d^\alpha_{R\ell}}\ S^\alpha_{RL,R'L'}\ \sqrt{d^\alpha_{R'\ell'}}, \end{aligned} \tag{69}$$

in terms of the structure constants and certain potential parameters, $\underline{C}^\alpha$, $\underline{\epsilon}_\nu$ and $\underline{d}^\alpha$ which depend on the values of φ and $\dot\varphi$, and which will be given below. The Hermitian matrix defined by Eq.(69) turns out to be the "first order Hamiltonian" in the atomic-sphere approximation (ASA), it is seen to have the two center form.

3.4 Potential Parameters

The augmentation functions φ and $\dot\varphi$ for a given atomic sphere radius S_R are obtained from the spherical average, $v_R(\underline{r}-\underline{R})$, of the potential $V(\underline{r})$ by solution of the corresponding radial Schrödinger equation at some chosen energies $\epsilon_{\nu R\ell}$ and at the neighboring ones $\epsilon_{\nu R\ell}+d\epsilon$. Specifically, if $\varphi_{RL}(\epsilon,r)$ is the solution of the l-th radial Schrödinger equation, normalized to unity in the sphere,

$$\int_0^{s_R} \varphi_{R\ell}(\epsilon,r)^2 r^2 dr \equiv \langle \varphi_{RL}(\epsilon)^2 \rangle = 1 \quad , \tag{70}$$

then the φ-function is defined by

$$\varphi_{RL}(r) \equiv \varphi_{RL}(\epsilon_{\nu R\ell}) \quad ,$$

and the energy-derivative function,

$$\partial\varphi_{R\ell}(\epsilon,r)/\partial\epsilon\,|_{\epsilon_{\nu R\ell}} \equiv \dot\varphi_{R\ell}(r) \quad ,$$

is seen to be orthogonal to φ in the sphere. The augmentation-function $\dot\varphi$ for the $\underline{\alpha}$-set is now defined as

$$\dot\varphi^\alpha(r) \equiv \dot\varphi(r) + \varphi(r)o^\alpha = \partial[\{1+(\epsilon-\epsilon_\nu)o^\alpha\}\varphi(\epsilon,r)]/\partial\epsilon\,|_{\epsilon_\nu} \quad , \tag{71}$$

with the constant o^α which will be specified latter. This constant gives $\dot\varphi$ the same radial logarithmic derivative at the sphere as the envelope tails, i.e.

$$\begin{aligned} D\{\dot\varphi^\alpha(s)\} &\equiv \partial\ln|\dot\varphi^\alpha(r)|/\partial\ln r\,|_s \\ &= D\{(s/w)^{-\ell-1} - (s/w)^\ell\alpha/[2(2\ell+1)]\} \equiv D^\alpha(s) \quad , \end{aligned} \tag{72}$$

The definition of $D^\alpha(s)$ is seen to be equivalent to (see for instance the Varena notes[9], pag. 81, Eqs. 49-58)

$$\alpha = \frac{(s/w)^{2\ell+1}}{2(2\ell+1)} \frac{D^\alpha(s)-\ell}{D^\alpha(s)+\ell+1} \quad . \tag{73}$$

The position - and width - potential parameters C^{α}_{RL} and d^{α}_{RL}, which enter in the definition (69) of the two-center, first-order Hamiltonian, may be expressed in terms of the value and the radial logarithmic derivative of φ at the sphere boundary as follows:

$$C^{\alpha} = \epsilon_{\nu} + s\varphi^2(s)\frac{[D\{\varphi(s)\} + \ell + 1][D^{\alpha}(s) - D\{\varphi(s)\}]}{D^{\alpha}(s) + \ell + 1} ,$$

and

$$(d^{\alpha})^{1/2} = (s/2)^{1/2}(s/w)^{\ell+1/2}\varphi(s)\frac{D^{\alpha}(s) - D\{\varphi(s)\}}{D^{\alpha}(s) + \ell + 1} . \tag{74}$$

It is the existence of a Wronskian relation between φ and $\dot{\varphi}$ which make it possible to evaluate the two-center, first-order Hamiltonian without calculation of $\dot{\varphi}$. This means, for instance, that a rough estimate of this Hamiltonian may be obtained merely from atomic orbitals, chosing the $\epsilon_{\nu R\ell}$'s as the atomic energies and the $\varphi_{R\ell}(r)$'s as the atomic orbitals renormalized to the spheres. This estimate corresponds to taking the potentials in the spheres as truncated atomic potentials. More realistic estimates of the position parameters, the C's, can be obtained by including the overlap of the neighboring atomic potentials as a perturbation.

The potential parameter needed to specify $\dot{\varphi}$ in Eq.(71) is the so-called "*overlap parameter*" :

$$\begin{aligned} o^{\alpha} &= < \dot{\varphi}^{\alpha}|\varphi > \\ &= \left(s\varphi(s)^2 \frac{[D\{\varphi(s)\} - D^{\alpha}(s)][D\{\dot{\varphi}(s)\} - D\{\varphi(s)\}]}{D\{\dot{\varphi}(s)\} - D^{\alpha}(s)} \right)^{-1} , \end{aligned} \tag{75}$$

which depends on the potential parameter $\varphi(s)$, $D\{\varphi(s)\}$ and $D\{\dot{\varphi}(s)\}$.

As an example, we give in the following Table the potential parameters for Fe-bcc whose band structure we shall consider latter. The potential parameters depend on the scale constant w through the definition (73) of $D^{\alpha}(s)$. The width parameter d^{α}, according to (74), also contains the $(s/w)^{2l+1}$.

TABLE II. Self-consistent potential parameters for Fe. The *spd*-screening parameters $\underline{\alpha}_{II}$ Eq.(61) were used. s_{Fe}=w=2.662 a.u.

Rydbergs	c	d	o^{-1}	ϵ_{ν}	$p^{-1/2}$
Fe-s	-0.306	0.159	-2.12	-0.474	5.1
Fe-p	0.330	0.0671	-1.73	-0.325	6.6
Fe-d	-0.189	0.0154	1.78	-0.263	0.80

3.5 Overlap matrices

For the LMTO basis of Eq. (68) one finds the following expression for the overlap matrix :

$$\mathcal{O} \equiv <\underline{\chi}|\underline{\chi}> = (1 + ho)(1 + oh) + hph + <\underline{\kappa}|\underline{\kappa}> \quad , \tag{76}$$

Here we drop the superscripts α when the equation also holds for a general α. We used matrix notation with 1 being the unit matrix and with the potential parameters o and p being diagonal matrices with elements o_{Rl} and

$$p_{Rl} \equiv \int_0^{s_R} \dot{\varphi}_{Rl}(r)^2 r^2 dr \equiv <(\dot{\varphi}_{RL})^2> \quad ,$$

respectively. The potential parameters p are the small parameters of a linear method and $\epsilon_{\nu Rl} \pm \frac{1}{2}(p_{Rl})^{-1/2}$ are the "energy windows" inside which such a method can be expected to yield useful results. Fortunately the widths of these windows are usually several Rydbergs. The last term in Eq. (76) is the integral over the product of two envelope functions in the interstitial region. This integral can be calculated analytically for the conventional, unscreened envelopes, and the transformation to a tight-binding representation then requires left and right multiplications with the tight-binding structure constants.

For the Hamiltonian matrix one finds.

$$\begin{aligned} \mathcal{H} &\equiv <\underline{\chi}| - \nabla^2 + V(r)|\underline{\chi}> \\ &= h(1 + oh) + <\underline{\chi}|\epsilon_\nu|\underline{\chi}> + <\underline{\chi}|V_{NMT}(\mathbf{r})|\underline{\chi}> \quad . \end{aligned} \tag{77}$$

Here,

$$<\underline{\chi}|\epsilon_\nu|\underline{\chi}> \equiv (1 + ho)\epsilon_\nu(1 + oh) + h\epsilon_\nu ph \quad ,$$

can be lumped together with the overlap matrix, provided that the matrix ϵ_ν commutes with all other matrices, i.e. if all $\epsilon_{\nu Rl}$'s have been chosen identical. $V_{NMT}(\underline{r})$ is the difference between the full potential and the potential used to construct the LMTO set, that is,

$$V_{NMT}(\mathbf{r}) \equiv V(\mathbf{r}) - \sum_R v_R(|\mathbf{r} - \mathbf{R}|) \quad ,$$

is the full potential in the interstitial region and the non spherical part inside the spheres.

In the ASA approximation the interstitial region and the nonspherical part of the potential are supposed to vanish so that, in each of the expressions (76) and (77) for the overlap and Hamiltonian matrices, the last term is neglected. The ASA is a reasonable approximation provided that there are very few electrons in the interstitial region or that the touching muffin-tin spheres can be substituted by "space-filling" atomic spheres i.e. atomic Wigner Seitz spheres, whose overlap does not exceed 20 %. i.e. if

$$s_R + s_{R'} \leq 1.20|\mathbf{R} + \mathbf{R}'|$$

This substitution is possible for closely packed materials, as well as for those open structures which, through the addition of interstitial (empty) spheres, become closely packed.

3.6 Standard potential parameters

The position parameter C^γ (74) of the nearly orthonormal representation (i.e. in the basis $\mid \underline{\chi}^\gamma >=\mid \underline{\chi} > (1 + oh)^{-1}$) is the second-order estimate of the energy corresponding to the boundary condition $D(\varphi(\epsilon, s)) = -l - 1$. Similarly, the width parameter d^γ (74) is the first order estimate of $\frac{1}{2}s(\frac{s}{\omega})^{2l+1}\varphi(\epsilon, s)^2 \approx \frac{1}{2}\omega\varphi(\epsilon, \omega)^2$. This means that C^γ and d^γ are more insensitive to the (somewhat arbitrary) choice of ϵ_ν than are C^α and d^α which are merely first - and zeroth - order estimates, respectively. For tabulations, one therefore uses C^γ and d^γ or, in the "standard" notation $C \equiv C^\gamma$ and $\Delta \equiv d^\gamma$. Given tabulated values of C, Δ and γ, one can obtain C^α, d^α and o^α, for any choice of α and viceversa, using the expressions.

$$\frac{c^\alpha - \epsilon_\nu}{C - \epsilon_\nu} = \left(\frac{d^\alpha}{\Delta}\right)^{\frac{1}{2}} = \frac{\alpha - \gamma}{o^\alpha \, \Delta} = 1 + \frac{\alpha - \gamma}{\Delta}(C - \epsilon_\nu)\,. \tag{78}$$

For nearly all elemental, closed-packed metals, the four potential parameters, C, Δ, γ and p have been obtained from self consistent LMTO-ASA band structure calculations and are tabulated in the Varena Lectures at the equilibrium Wigner Seitz radii ($s \equiv \omega$). In that reference, the self consistently calculated atomic volume derivaties , *e.g.* $dC(s \equiv w)/d\ln s, \;|_{s\equiv\omega}$, the pressure dependencies have been tabulated as well. Recently, (see Kumm[15]) the equilibrium potential parameters for almost all elements of the periodic system have been recalculated and tabulated, providing a unique tool to start new calculations.

These atomic-sphere potentials thus correspond to atomic spheres which are kept neutral. Alternatively, from the values of C, Δ, γ, p and the tabulated value of the potential at the sphere one can, using Eqs. (205 - (207) of the Varena Lectures[9], find the atomic volume derivatives, denoted $\frac{\delta}{\delta \ln s}$, of the potential parameters corresponding to frozen atomic-sphere potentials.

Although computed for crystalline phases, the potential parameters tables in the References may be used to estimate the potential parameters for non-crystalline phases. This is much simpler than performing charge-self-consistent recursion-method calculation as we shall describe later.

If the noncrystalline phase is monoatomic, the tabulated parameters should merely be extrapolated to the proper density (i.e. $s_x \equiv w_x \rightarrow s_{non-x} \equiv w_{non-x}$) using the tabulated atomic- volume derivatives $\frac{d}{d \ln s}$, which hold for radius deviation of about $\pm$ 5 % .

In case the noncrystalline structure is so open that there are spheres which overlap by more than 20 % one should, after the extrapolation to the proper density (w_{non-x}), choose a new $s_{non-x} (< w_{non-x})$ in such a way that the overlap is now 20 % or less. The corresponding extrapolation parameters to the smaller radius (s_{non-x}) should be performed using the frozen potential derivatives ($\frac{\delta}{\delta \ln s}$). Subsequently, Δ and γ should be multiplied by the factor $(\frac{s_{non-x}}{w_{non-x}})^{2l+1}$ in order to take into account that w no longer equals s, but the average Wigner Seitz radius given by (65). The parameters of the two center tight-binding Hamiltonian may finally be calculated using Eq. 78 with $\alpha = \alpha_{II}$.

Due to change transfer, it is harder to estimate the potential parameters of an intermetallic alloy but, if it is possible to fill space with overlapping spheres whose sizes do not deviate grossly from those of the constituent metals, that is, if Vegard's law holds approximately, then one can usually assume that the spheres are sufficiently neutral that the above mentioned construction works. Otherwise, a charge self consistent calculation for the noncrystalline material seems to be needed.

3.7 Test calculations

In order to demonstrate the quality of recursion calculation we first considered a crystal and compared the PDOS obtained by a standard band structure calculation with those obtained by recursion calculation for a finite cluster, consisting of 1000 atoms coordinated further to seconds nearest neighbors, and the terminator described in the first part of this lectures.

We did not perform self-consistent calculations but used the self-consistent potential parameters listed in Table III of the Varena Lectures[9]. With the help of Eq.(78), we then transformed these standard potential parameters into the position and width parameters C^α and d^α, needed in the two-center tight-binding Hamiltonian

$$H = C^\alpha + \sqrt{d^\alpha} S^\alpha \sqrt{d^\alpha} \ ,$$

in Eq.(69).

These parameters are listed in the table II, together with those additional parameters that would be needed to construct the third order Hamiltonian.

The structure matrix S^α was generated as described before, by inversion of the $(15 \times 9) \times (15 \times 9) = 135 \times 135$ matrix $\alpha^{-1} - S^o$ obtained when including the 15-site cluster which contains the central site and its two first-neighbor shells (bcc structure).

The results for the PDOS of the most central atom as well as the set of recursion coefficients for the p_x, p_y, p_z- orbitals are shown in the following figures.

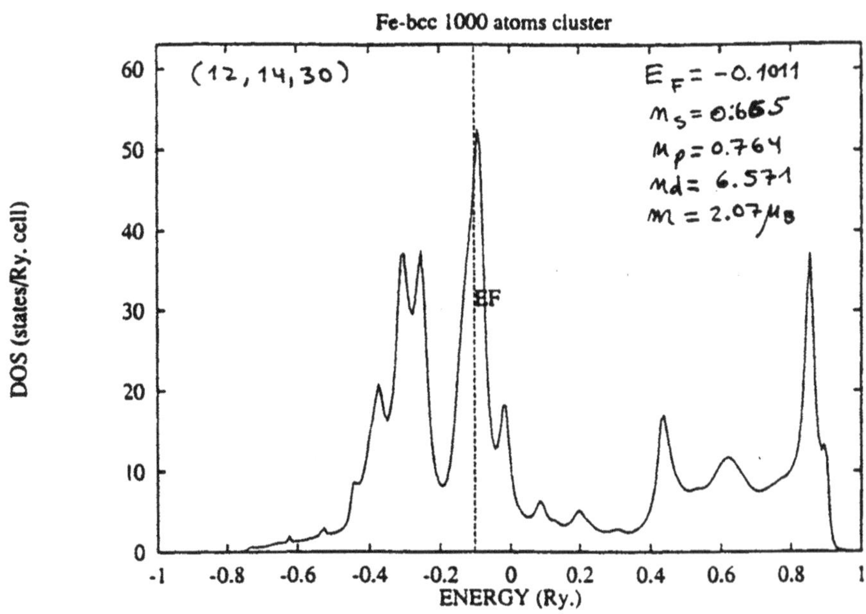

FIG. 7. The total DOS on the most central atom of Fe in a cluster of 1000 atoms with a bcc-structure first order Hamiltonian was used, α_{II}-screening and the potential parameters for the equilibrium volumen (i.e. a = 2.86 Åor s = 2.662 a.u.).

For the calculation 12, 14 and 30 levels of recursion coefficients were used for computing the partial s, p and d density of states respectively.

Fermi level and partial occupation numbers are in close agreement with a standard k-space band structure calculation using a better Hamiltonian, including f-electrons.

(E_F = -0.125 Ry, n_s = 0.633, n_p = 0.751, n_d = 6.528, n_f= 0.088)

The magnetization $m = 2.07\mu_B$ is obtained using the Stoner construction from a paramagnetic DOS curve. (See sec. 6.2 in Varena Lectures[9]).

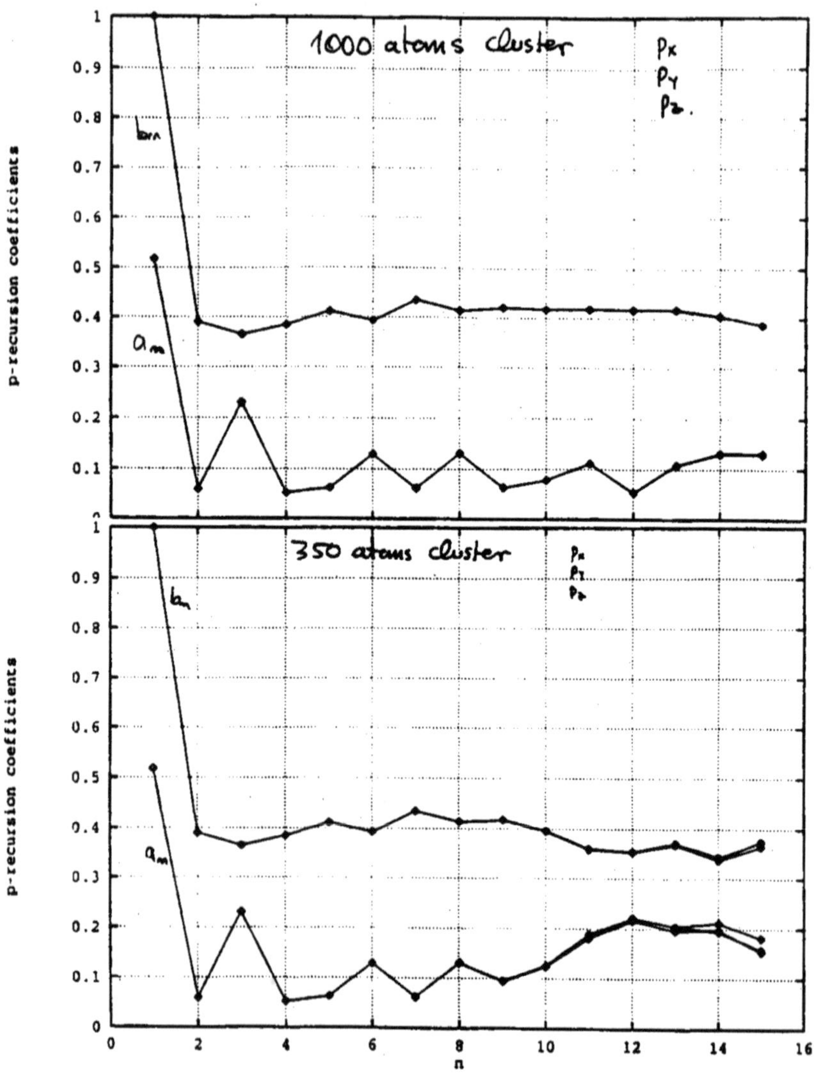

FIG. 8. Recursion coefficients for the *p*-orbitals of the most central Fe atom in two bc

clusters of different sizes.

In the small cluster, after the 10th level the p-orbitals loose the cubic symmetry due to finite size effects.

3.8 Amorphous systems

In the next diagram we show the self consistent scheme employed for amorphous systems. (The scheme we used the work of Nowak et.al.[17])

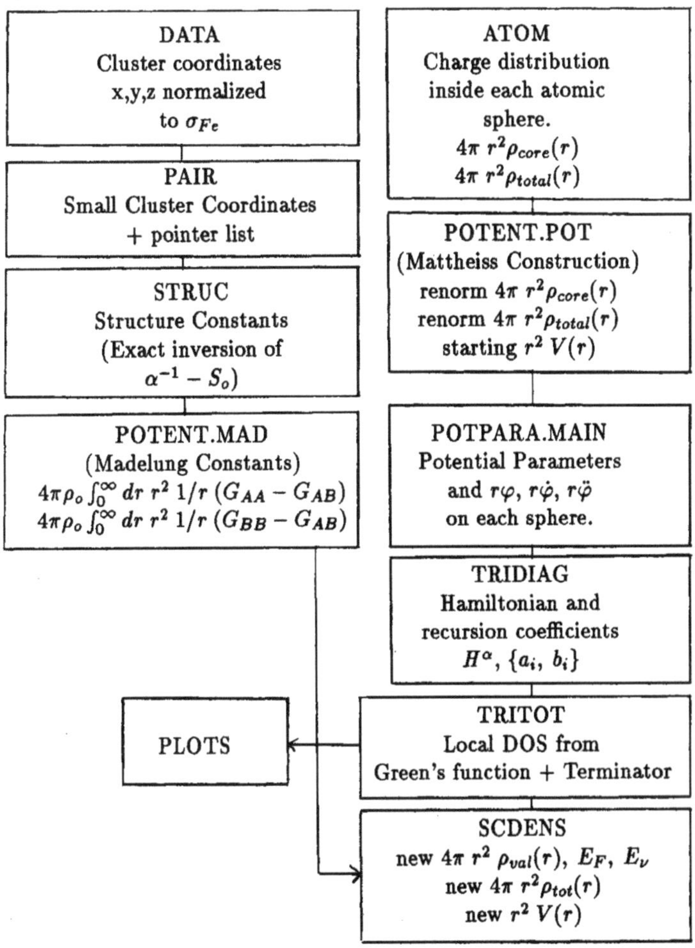

Now, we show the results of a self consistent calculation for amorphous $Fe_{80}B_{20}$.

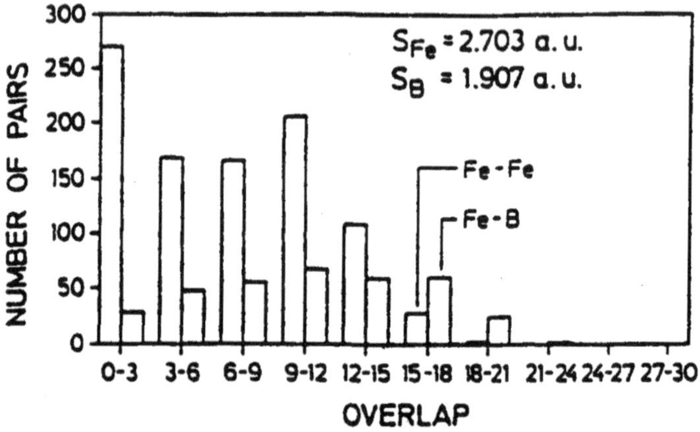

FIG. 9. Histogram showing the sphere overlap (in percent) of Fe-B and Fe-Fe pairs in amorphous $Fe_{80}B_{20}$. The overlap is defined as $(s_1 + s_2 - d_{12})/d_{12}$. Very few pairs have an overlap exceeding 20 %.

TABLE III. Self consistent potential parameters for $Fe_{80}B_{20}$. The *spd*-screening parameters $\underline{\alpha}_{II}$ given in Eq.(61) were used.

Rydbergs	c	d	o^{-1}	ϵ_ν	$p^{-1/2}$
Fe-s	-0.415	0.153	-1.68	-0.584	5.1
Fe-p	0.172	0.0655	-1.42	-0.419	6.7
Fe-d	-0.250	0.0188	1.66	-0.315	0.77
B-s	-0.842	0.206	4.50	-0.763	6.8
B-p	0.275	0.098	5.24	-0.455	5.6

s_{Fe}=2.703 a.u., s_B=1.907 a.u. and w=2.581 a.u.

The structural model for amorphous $Fe_{80}B_{20}$ was a relaxed (Morse Potential) random packing of 1500 atoms (hard spheres), see Fujiwara et.al.[16]

The potential parameters were calculated selfconsistently for the average Fe and the average B atom, and using the first-order Hamiltonian. The ϵ_ν parameters were, as usual, chosen at the centers of gravity of the occupied PDOS. The charge density was averaged over all Fe and over all B spheres, and the potential was assumed to be the same in all Fe and in all B spheres. After selfconsistency was achieved, the final PDOS calculation were carried on to second order, using the Hamiltonian $H^\alpha - h^\alpha\, o^\alpha\, h^\alpha$.

Selfconsistency on each atomic site can also be performed, see for instance Peduto et al.[18]

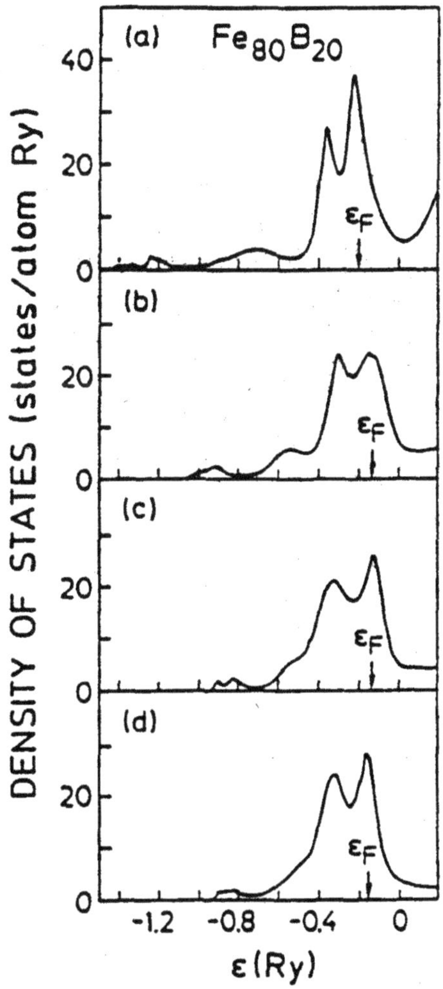

FIG. 10. Average DOS per atom in amorphous $Fe_{80}B_{20}$, calculated with different approximations. (a)-(c): First-order Hamiltonian. (d): Second-order Hamiltonian. The structure constants were obtained from: (a) the interpolation formula for the off-site elements and the average of the bcc- and fcc-values for the on-site elements, (b) the interpolation formula for the off-site elements and using the Dyson equation (63) for the on site elements, (c) and

(d) exact inversion. From Nowak et.al.[17]

3.9 Do it yourself

Now, we give an extremely simple procedure to perform partial DOS calculation on simple systems.

It is not a selfconsistent procedure, and it was written for training purposes, but if you use correct (properly scaled) potential parameters, you get 'ab initio' results for the PDOS and occupations on bulk atoms.

The programs need the following sequence:

$$\boxed{\text{GENLATT}} \Longrightarrow \boxed{\text{PAIR}} \Longrightarrow \boxed{\text{STRUC}} \Longrightarrow \boxed{\text{NHAMIL}} \Longrightarrow \boxed{\text{NTRIDIAG}} \cdots$$

$$\cdots \Longrightarrow \boxed{\text{RECU}} \Longrightarrow \boxed{\text{GNUDOS}}$$

Now a brief explanation of the INPUT-OUTPUT procedures:

- program GENLATT builds clusters with the most common structures i.e. *sc*, *bcc*, *fcc*, *hcp* and *diamond* structure. Also, *diamond + interstitial spheres* structure is included (i.e. a topological *bcc* lattice)
    - inputs: name of the output file type of the cluster to build
    - outputs: ascii-file containing the cluster coordinates ordered by distances from the central atom.
- program PAIR constructs small clusters around each of IQ atoms each cluster contains IAT-atoms. Builds a pointer list on each small cluster: NUM(IQ,IAT), then K=NUM(I,J) means that the J-atom of the I-cluster is the K-atom in the original big cluster.
    - inputs: name of file output and file with the input coordinates (i.e. the output from genlatt).
    - outputs: file with small clusters and the pointer list.
    - parameters: IBIG total number of atoms in the big cluster. IQ number of small clusters. IAT number of atoms in each small cluster.

- program STRUC. This program calculates the screened structure constant matrix S^α by inverting IQ-times small matrices and keeping every time the first row. The total S^α matrix is then constructed using the pointer list as an array of dimensions (IQ×9)×(IAT×9)
  - inputs: file with the small clusters coordinates and pointer list, ascii file with the average wigner Seitz radius in units of the lattice parameter used in the original cluster (this makes the S^α matrix scale independent).
    Name of the output file.
  - outputs:file containing the S^α matrix and pointer list (binary file). file STRUC.OUT with some elements of the unscreened and screened structure matrix for testing.
  - parameters: IQ and IAT, as before.
- program NHAMIL. This program calculates the first order hamiltonian
  $H^\alpha = C^\alpha + (d^\alpha)^{1/2}\, S^\alpha\, (d^\alpha)^{1/2}$
  - inputs: file with the S^α matrix and pointer list, ascii file with the screened potential parameters, $c^\alpha, d^\alpha, E_\nu$ for each type of atoms, LMAX=lmax+1 value for each type of atom. You can get this parameters for any atom using the tabulated values in the varena notes, and the program varena.f (explained at the end of this file). The last input is the name of the output file.
  - outputs:binary file with H^α and pointer list. file HAMIL.OUT for testing. (ascii file)
  - parameters: IQ and IAT as before.
- program NTRIDIAG. This program performs the recursion procedure for a particular atom (NATOM) of the IQ possible choices. It calculates all 9 sets of recursion coefficients (s, p, d) for that particular atom.
  - inputs: binary file with the Hamiltonian and pointer list (i.e. the output of NHAMIL program) ascii file with the parameters NATOM,NS,NP,ND which are: atom selected, number of recursion levels to be calculated for each particular ℓ-orbital. Usual numbers for NS, NP and ND are 7,9,15 respectively for the most central atom. This is for clusters with IQ approx. 400 atoms The reason for that are the finite size effects on the recursion coefficients. s and p orbitals are in general less localized than d-orbitals.
  - outputs: ascii file with the sets of recursion coefficients.

– parameters: IQ and IAT as before.

- program RECU. This program performs calculation of the partial DOS, E_{Fermi} and partial occupations, for the chosen atoms. Uses the continued fraction representation of the green's function plus a terminator.
  - inputs: file with the sets of recursion coefficients (i.e. the output of NTRIDIAG program) average number of valence electrons for that particular atom. Energy window for the PDOS calculations. Name of the output file. Type of terminator to be used.
  - outputs: E_{Fermi}, partial occupations, n_s, n_p and n_d. All the PDOS each in 250 points in the energy window, and in the following order:

1	2	3	4	5	6	7	8	9	10	11	12	13	14	15
s	px	py	pz	dxy	dyz	dzx	d3z3	dx-y	stot	ptot	dtot	totdos	dt2g	deg

where dt2g = dxy+dyz+dzx and deg = dx-y + d3z2

- program GNU is a driver for the public domain gnuplot package. It generates a suitable file for PDOS plots.
  - inputs: output file of the RECU program label of the PDOS to be plotted (i.e. a number from 1 to 15) Fermi level (to be marked in the Plot) type of output (i.e. file or screen)
  - outputs: Plot of the chosen PDOS.

An additional small program VARENA.F is also written for this package. The program reads "unscrenned potential parameters " (for a given element) exactly in the way they are given in the Varena Notes[9] or in the new Tables of Kumm[15]. The program then converts the parameters to the most screened representation.

The programs (in FORTRAN source code) are easy to read and consequently can rapidly be altered to fit the customer own desires. They can be requested in the following (internet) addresses: jepsen@radix1.mpi-stuttgart.mpg.de (Ove Jepsen) or pvargas@lauca.usach.cl (Patricio Vargas)

ACKNOWLEDGMENTS

Finally the author acknowledge the always warm hospitality of the Andersen's group in Stuttgart (Max Planck Institute) and also to V. Kumar and the ICTP for the invitation to participate in the workshop during August of 1992.

[1]H.S. Wall, in *Analytical Theory of Continued Fractions* (Toronto, Van Nostrand, 1948).

[2]C. Lanczos *J. Res. Natl. Bur. Stand.* Sec. B**45** (1950) 255

[3]R. Haydock in *Solid State Physics* **35** *129*, (Eds. H.Ehrenreich F.Seitz and D.Turnbull Springer-Verlag) (1980).

[4]N. Beer and D. G. Pettifor in *The Electronic Structure of Complex Systems*, edited by P. Phariseau and W.M. Temmerman (Plenum, New-York, 1984).

[5]P. Vargas, M. Weissmann, J.E. Ure and N. Majlis, Sol. State Commun. **74**, 703 (1990).

[6]Solid State Physics, **35**, Eds. H.Ehrenreich F.Seitz and D.Turnbull Springer-Verlag) (1980).

[7]The Recursion Method and Its Applications, Springer Series in Solid State Science 58, Eds. D.G.Pettitor and D.L.Weaire. Springer Verlag, 1984.

[8]O. K. Andersen and O. Jepsen, Phys. Rev. Lett. **53**, 2571 (1984).

[9]O. K. Andersen, O. Jepsen and D. Glötzel, in *Highlights of Condensed-Matter Theory,* edited by F. Bassani, F. Fumi and M. P. Tosi (North-Holland, New York, 1985).

[10]O. K. Andersen, Z. Pawlowska and O. Jepsen, Phys. Rev. B**34**, 5253 (1986).

[11]O. K. Andersen, O. Jepsen and M. Sob, in *Electronic Band Structure and its Applications,* ed. M. Yussouff (Springer Lecture Notes, 1987).

[12]O.K. Andersen, Solid State Comm. **13**, 133 (1973).

[13]O.K. Andersen, Phys. Rev. B. **12**, 3060 (1975).

[14]M. Sob, O. Jepsen and O. K. Andersen, Z. Phys. Chem (Neue Folge), **157**, 515 (1988).

[15]Andreas Kumm, Ph.D. Thesis, unpublished (1992) and O.K. Andersen this lectures.

[16]T. Fujiwara, H. S. Chen and Y. Waseda, Z. Naturforsch., **37a**, 611 (1982);

[17]H.J. Nowak, O.K. Andersen, T.Fujiwara, O. Jepsen and P. Vargas, Phys. Rev. B, **44**, 3577 (1991).

[18]P.R. Peduto, S. Frota-Pessoa and M. Methfessel Phys. Rev. B. **44**, 13 2893 (1991)

THE AUGMENTED SPACE AND ELECTRONIC STRUCTURE OF RANDOM BINARY ALLOYS

ABHIJIT MOOKERJEE
S.N.Bose National Centre for Basic Sciences
DB 17, Sector 1, Salt Lake City,
Calcutta 700064, India

ABSTRACT

We discuss the application of the Augmented Space methodology to directly obtain configuration averaged physical properties of random binary alloys. To do this we bring together the ideas of the tight-binding LMTO, the recursion method and the ideas of the augmented space and illustrate how to obtain configuration averaged density of states of random binary alloys.

1. Introduction.

Almost a decade ago Andersen and Jepsen[1] introduced the tight-binding, linearized muffin-tin orbital (TB-LMTO) method. This method allows us to calculate, self-consistently and from first principles, an effective hamiltonian from which the electronic structure can be obtained with considerable accuracy and speed. The hamiltonian is sparse in real space. This allows the method to be ideally suited for describing substitutionally or structurally disordered systems. Moreover, the recursion method introduced in the early seventies by Haydock *et.al.* [2] allows us to work directly in real space and is certainly is the ideal technique to use in conjunction with the TB-LMTO when we are dealing with disordered systems. The hamiltonian of such systems does not possess lattice translational symmetry and we cannot use the Bloch theorem and k-space methods to reduce the problem to inversion and diagonalization of manageably small matrices. Both these methodologies have been described in some detail in this volume [3,4].

In this article I wish to concentrate on the idea of configuration averaging. Configuration averaging is of central interest in the study of disordered systems. In an earlier article [5] the reason why we wish to configuration average and what are the properties which should be averaged have been discussed at length. Here, I wish to illustrate that the augmented space method (ASM) for configuration averaging [6] is probably one of the computationally most efficient and accurate methods for dealing with configuration averaging in conjunction with the TB-LMTO and Recursion Methods. It allows us to take into account correlated scattering from clusters of reasonably large sizes. The recent *termination procedures*[7] for the Recursion Method ensures that the correct *herglotz* analyticity properties are maintained when we extrapolate from a necessarily finite computer calculation to the infinite system. So far, two different approaches have been used for configuration averaging : the mean field methods which include the coherent potential approximation (CPA) and methods which attempt to generalize this to include cluster effects, and the brute force computational methods which generate large number of different random configurations for the hamiltonian and then directly calculate physical properties for a given configuration and average this over the different configurations generated. Both the methods have their own drawbacks. Perhaps the most satisfactory and widely used method is the CPA [8]. It maintains the *herglotz* properties throughout the spectrum and for all degrees of disorder. However, the nested self-consistent loops involved in the calculations are computationally extremely expensive and involve delicate analytic continuations from off the real energy axis onto the real spectrum. This has been described elsewhere in this volume [9,10]. Moreover, the CPA is a single site, mean field approximation. Local chemical ordering or local lattice relaxations cannot be satisfactorily dealt with within the single site approximation. Attempts to go beyond the single site approximations have their limitations. The cluster embedding method [11] is not self-consistent and may yield unphysical structures in the spectrum because of the extra scattering at the boundary of the embedded cluster. Analyticity preserving generalizations of the CPA have been proposed [12-14]. However, they are usually mathematically cumbersome and computationally expensive. Applications to realistic alloy systems have been made so far within the empirical tight-binding method [15] only. Until the proposed KKR-CCPA and LMTO-CCPA have been carried out on realistic alloy systems, the final comment on the usefulness of these generalizations cannot be made. In contrast, we wish to illustrate that the marriage between the ASF, TB-LMTO and Recursion Methods should provide one of the most tractable and fast techniques for studying the electronic structure of, for example, random binary alloys.

2. The Augmented Space Method.

2.1 An Example.

Let us begin with a very simple example in order to illustrate the basics of the methodology. We first note that the probability density of a random variable x has the following properties :

$$p(x) \geq 0. \tag{1a}$$

$$\int_{-\infty}^{\infty} p(x)dx \quad = \quad 1 \quad < \quad \infty. \tag{1b}$$

In addition, we may impose the following condition :

$$\int_{-\infty}^{\infty} x^n p(x)dx \quad < \quad \infty. \tag{1c}$$

This is not a very serious restriction, since almost all physically important probability distributions obey this property. One of the most common exception is the Lorenzian distribution. This can, however, be treated by complex integration methods. It should be noted that if this condition is violated, then the methodology described hereafter fails and the continued fraction expansion described below diverges.

The interesting thing to note from the above is that the probability density p(x) has properties very similar to the density of states of a hamiltonian. The question we wish to ask is the following : Is it possible to find an operator **M** in a certain Hilbert Space Φ such that the probability density can be expressed as the imaginary part of a matrix element of its resolvent ?

$$p(x) \quad = \quad -\frac{1}{\pi}\Im m\langle 1|(x^+ I - M)^{-1}|1\rangle. \tag{2}$$

x^+ means $x+\imath\delta$ with $\delta \to 0$

The problem is exactly the inverse of the Recursion Method. There one is given a representation of the Hamiltonian operator. One finds a tri-diagonal representation and then expresses the density of states as the imaginary part of a continued fraction involving the tridiagonal elements. Inversely, if we can find a convergent continued fraction expansion of the probability density, then the tridiagonal matrix formed out of the continued fraction coefficients is a representation of the operator **M** we are seeking. A simple example will illustrate this :

Let us take a bimodal distribution :

$$p(x) = c\delta(x-1) + (1-c)\delta(x). \tag{3a}$$

We may expand the right hand side of (3a) as a continued fraction :

$$\begin{aligned} p(x) &= -\frac{1}{\pi}\Im m\left(\frac{c}{x^+ - 1} + \frac{1-c}{x^+}\right) \\ &= -\frac{1}{\pi}\Im m\left(\cfrac{1}{x^+ - c - \cfrac{c(1-c)}{x^+ - (1-c)}}\right) \\ &= -\frac{1}{\pi}\Im m\langle 1|(x^+ I - M)^{-1}|1\rangle. \end{aligned} \tag{3b}$$

where,

$$M = \begin{pmatrix} c & \sqrt{c(1-c)} \\ \sqrt{c(1-c)} & 1-c \end{pmatrix} \tag{3c}$$

Since the distribution is bimodal, the rank of the matrix **M** is 2. This matrix is the representation of an operator **M** in a *configuration space* Φ also of rank 2. It is easy to check that the eigenvalues of the matrix are 0 and 1, i.e. the two values taken by the random variable. The tri-diagonal representation shown in Eq. (3c) is in a basis such that the basis vectors can be expressed in terms of the two eigenvectors as :

$$
\begin{aligned}
|1\rangle &= \sqrt{c}|\mu = 0\rangle + \sqrt{1-c}|\mu = 1\rangle \\
|2\rangle &= \sqrt{1-c}|\mu = 0\rangle - \sqrt{c}|\mu = 1\rangle .
\end{aligned} \tag{4}
$$

If we define the projection operators $\mathbf{P}_1 = |1\rangle\langle 1|$ and $\mathbf{P}_2 = |2\rangle\langle 2|$, and the transfer operator $\mathbf{T}_{12}$ as $|1\rangle\langle 2|$,then we may express the operator **M** as follows :

$$
M = cP_1 + (1-c)P_2 + \sqrt{c(1-c)}\,[T_{12} + T_{21}] \tag{5}
$$

It is now an easy task to directly check that the bimodal probability density is :

$$
p(x) = -\frac{1}{\pi}\Im m \Re_{11}(x, M)
$$

where $\Re$(x,M) is the resolvent of the operator **M**.

Note that the procedure sketched above is quite general. If instead of the bimodal distribution we had any other distribution obeying the restrictions (1a)-(1c), we proceed in a very similar way. First we obtain a convergent continued fraction expansion of the type :

$$
p(x) = \cfrac{1}{x - a_1 - \cfrac{b_1^2}{x - a_2 - \cfrac{b_2^2}{x - a_3 - \ldots}}} \tag{6}
$$

For the standard distributions the continued fraction coefficients are known. For example, for the Gaussian distribution $a_n = a, b_n = b\sqrt{n}$, while for the semicircular distribution $a_n = a, b_n = b$. The required representation of the operator **M** is then a tridiagonal matrix with a_n down the diagonal and b_n down the off-diagonal positions. In these cases, of course, since the distributions and continuous, the rank of the configuration space is ∞, however, the basis is countable and both inner products and norms may be suitably defined.

2.2 The Augmented Space Theorem.

Let us now consider the average of a function f(x) of the random variable x. We may proceed as follows :

$$\langle f(x)\rangle_{av} = \int_{-\infty}^{\infty} f(x) \quad p(x) \quad dx$$
$$= -\frac{1}{\pi}\Im m \int_{-\infty}^{\infty} f(x)\langle 1|(xI - M)^{-1}|1\rangle dx \tag{7a}$$

Now, using the completeness of the eigenvectors of the operator **M** on Φ we obtain from the above :

$$\langle f(x)\rangle_{av} \quad = \quad \frac{1}{\pi}\Im m\langle 1|\sum_{\mu}\sum_{\mu'}\int_{-\infty}^{\infty} f(x)|\mu\rangle\langle\mu|(xI - M)^{-1}|\mu'\rangle\langle\mu'|1\rangle dx$$

But (x**I** - **M**)$^{-1}$ is diagonal in the eigenbasis of **M** and is equal to $\frac{1}{(x-\mu)}\ \delta_{\mu\mu'}$, so that

$$\langle f(x)\rangle_{av} = \langle 1|\sum_{\mu}\int_{-\infty}^{\infty} f(x)|\mu\rangle\left[-\frac{1}{\pi}\Im m(x-\mu)^{-1}\right]\rangle\mu|1\rangle dx$$
$$= \langle 1|\sum_{\mu}\int_{-\infty}^{\infty} f(x)|\mu\rangle\delta(x-\mu)\langle\mu|1\rangle dx$$
$$= \langle 1|\left[\sum_{\mu}|\mu\rangle f(\mu)\rangle\mu|\right]|1\rangle \tag{7b}$$

But, by definition the operator function $\tilde{f}(\mathbf{M})$ of the operator **M** is defined as the operator

$$\sum_{\mu}|\mu\rangle f(\mu)\langle\mu| \tag{7c}$$

So that, we have the central result :

$$\langle f(x)\rangle_{av} = \langle 1|\tilde{f}(M)|1\rangle. \tag{8}$$

This is the **Augmented Space Theorem**. We have reduced the problem of configuration averaging to one of obtaining *one special* matrix element of an operator in the configuration space Φ constructed according to the above given prescription from the probability distribution of the random variable. In general, in complex problems the calculation of matrix elements is a much easier task than carrying out the configuration averaging by a brute force integration method. Moreover, the result in Eq. (8) is *exact*. Once a representation of the required operator is obtained, it is much easier to maintain analytic properties when we use approximations, than in the original problem.

For the very simple bimodal distribution, since the matrix $\mathbf{M}$ is idempotent $\mathbf{M}^n = \mathbf{M}$, we may easily check the following averages :

(1) $f(x) = x \Rightarrow \tilde{f}(\mathbf{M}) = \mathbf{M}$. So that $\langle x\rangle_{av} = M_{11} = c$.

(2) $f(x) = x^2 \Rightarrow \tilde{f}(\mathbf{M}) = \mathbf{M}^2 = \mathbf{M}$. So that $\langle x^2\rangle_{av} = M_{11} = c$.

(3) $f(x) = \ln(1+x) = \sum_{n=1}^{\infty}(-1)^{n-1}\frac{x^n}{n} \Rightarrow \tilde{f}(\mathbf{M}) = \sum_{n=1}^{\infty}(-1)^{n-1}\frac{M^n}{n}$.

So that $\langle \ln(1+x)\rangle_{av} \;=\; c\sum_{n-1}^{\infty}\frac{(-1)^{n-1}}{n} \;=\; c\ln 2$

2.3 The averaged Green function.

Let us illustrate the generalization of the above theorem, by considering a simple, tight-binding hamiltonian with one orbital per site and diagonal disorder with bimodal, independent distributions. Note that this is for illustrative purposes only. Generalizations to more than one orbital per site[15], off- diagonal disorder[16] and correlated distributions[17] have been discussed earlier. The hamiltonian is an operator on a hilbert space $\mathcal{H}$ which is spanned by the tight-binding basis $\{|i\rangle\}$:

$$H = \sum_i \varepsilon_i P_i + \sum_i \sum_{j\neq i} t_{ij}\,(T_{ij} + T_{ji}) \tag{9a}$$

where

$$\begin{aligned}\varepsilon_i &= \varepsilon_A x_i + \varepsilon_B(1 - x_i)\\ &= \varepsilon_B + \delta\varepsilon x_i\end{aligned} \tag{9b}$$

Here x_i is a random variable which takes the value 1 if the site labelled i is occupied by an A type of atom and 0 if not. We may rewrite Eq.(9a) as :

$$H = H_B + \delta\varepsilon \sum_i x_i P_i \tag{9c}$$

$\mathbf{H}_B$ is the hamiltonian for the constituent B alone. The variables $\{x_i\}$ are assumed to be independently and identically distributed. This is a case of homogeneous disorder. There are situations when this breaks down. For example , when we are considering an alloy surface. It is known that alloy surfaces do show segrigation or enrichment of one constituent, although deep in the bulk one has homogeneous randomness[18]. In such situations it is not difficult to generalize to the situation when the probability distribution is, for example, dependent on which layer below the surface the site is situated. The probability density of each x_i is given by Eq. (3a).

For *each* variable x_i we now introduce a configuration space Φ_i of rank 2, spanned by a basis $\{|1_i\rangle, |2_i\rangle\}$. Associated with the probability distribution of *each* variable is an operator $M^{(i)}$ constructed according to the prescription described earlier.

The full configuration space is then the direct product $\Psi = \prod^{\otimes} \Phi_i$. It is of rank 2^N if the rank of H is N. At this point we should like to introduce a change of notation. This is because subsequently we should like to use some of the methodologies common to statistical mechanics of Ising systems. Instead of naming the basis vectors of each Φ_i as $|1_i\rangle$ and $|2_i\rangle$, we shall call them $|\uparrow_i\rangle$ and $|\downarrow_i\rangle$. We should carefully note that this nomenclature is only to draw an analogy between methodologies. There are no *spins* associated with these bases at all.

The **Augmented Space Theorem** then tells us that :

$$\langle G_{ij}(z,\{x_i\})\rangle_{av} = \langle \uparrow_1 \ldots \uparrow_i \ldots |\tilde{G}_{ij}(z,\{M^{(i)}\})| \uparrow_1 \ldots \uparrow_i \ldots\rangle. \tag{10}$$

Here,

$$\tilde{G}_{ij}(z,\{M^{(i)}\}) \quad = \quad \langle i|(z\tilde{I} - \tilde{H})^{-1}|j\rangle. \tag{11}$$

and the effective hamiltonian in the full **Augmented Space** $\Upsilon = H \otimes \Phi$ is given by

$$\tilde{H} = H_B \otimes I + \delta\varepsilon \sum_i P_i \otimes M^i. \tag{12}$$

We first note that the expression in Eq.(11) is *exact*. As we mentioned earlier, we also see that the problem of computing the configuration average has been reduced to calculating a particular matrix element of the resolvent of an effective hamiltonian $\tilde{H}$ in the Augmented Space Υ. Before we discuss how to apply the recursion method in calculating this matrix element, we have to first discuss how one represents the basis in this abstract configuration space Φ in a mathematically simple manner.

A general member of the basis in Φ can be written as follows :

$$|\uparrow_1\uparrow_2 \ldots \downarrow_{i_1} \ldots \downarrow_{i_2} \ldots\ldots \downarrow_{i_C} \ldots\rangle$$

The *number* of sites in which we have a $\downarrow$ state, in this case C, is called the *cardinality* of the basis member. The *sequence* of sites $\{i_1, i_2 \ldots i_C\}$ at which we have these $\downarrow$ states is called the *cardinality sequence*[19]. It is denoted by σ. Note that the cardinality sequence completely describes the basis member. I have emphasized that the cardinality set is a sequence. The order is very important. For example, the two basis members $|\uparrow_1\downarrow_2\downarrow_3\rangle$ and $|\downarrow_1\uparrow_2\downarrow_3\rangle$ may have the same cardinality, but the order in the cardinality sequence being different are actually different members of the basis. Eq.(11) may be rewritten as :

$$\langle G_{ij}(z)\rangle_{av} = \langle i\emptyset|\tilde{G}(z)|j\emptyset\rangle. \tag{13}$$

We define the inner product of each component of the basis in terms of :

$$\langle\uparrow_i|\uparrow_i\rangle = 1$$

$$\langle\downarrow_i|\downarrow_i\rangle = 1$$

$$\langle\uparrow_i|\downarrow_i\rangle = 0$$

and the total inner product as the simple product of the inner product of each component. This definition satisfies the properties required for a well defined inner product. In terms of this definition it is easy to see that :

$$\langle\sigma|\sigma\prime\rangle = \delta_{\sigma\sigma\prime} \tag{14}$$

3. Calculations in Augmented Space.

3.1 Path Counting.

The idea of path counting to illustrate calculations on the Recursion Technique has been introduced by Haydock[20] and Mookerjee[21]. To illustrate this for the calculations in Augmented Space, let us continue with the example of the previous section. Let us first see the effect of the Augmented Space Hamiltonian $\tilde{H}$ on a general member of the basis in the augmented space :

$$\begin{aligned}\tilde{H}|i\sigma\rangle &= H_B|i\sigma\rangle + \delta\varepsilon\sum_k P_k\otimes M^k|i\sigma\rangle\\ &= \varepsilon_B|i\sigma\rangle + \sum_{j\neq i} t_{ij}|j\sigma\rangle + \delta\varepsilon\left[\hat{c}|i\sigma\rangle + \sqrt{c(1-c)}|i\sigma\prime\rangle\right]\end{aligned} \tag{15a}$$

Here

$$\hat{c} = \begin{cases} c, & \text{if } i\notin\sigma;\\ 1-c & \text{if } i\in\sigma.\end{cases}$$

$$\sigma\prime = \begin{cases} \sigma-\{i\} & \text{if } i\in\sigma;\\ \sigma+\{i\} & \text{if } i\notin\sigma.\end{cases}$$

So that,

$$\tilde{H}|i\sigma\rangle = \hat{\varepsilon}|i\sigma\rangle + \sum_{j\neq i} t_{ij}|j\sigma\rangle + \delta\varepsilon\sqrt{c(1-c)}|i\sigma'\rangle. \qquad (15b)$$

$$\hat{\varepsilon} = \begin{cases} c\varepsilon_A + (1-c)\varepsilon_B & \text{if i} \notin \sigma; \\ (1-c)\varepsilon_A + c\varepsilon_B & \text{if i} \in \sigma. \end{cases}$$

Now we interpret Eq.(15) as follows : Every member of the basis in the Augmented Space is represented by a *vertex*. So that the countable basis is represented by a *lattice* in the augmented space. The operation of $\tilde{H}$ on any member of the basis is represented by a *path*. The path is of length zero for the first term in Eq.(15) and of length one, which we shall call a *link*, for the subsequent terms. The vertices $|j\sigma\rangle$ and $|i\sigma'\rangle$ are called the *neighbours* of $|i\sigma\rangle$. Figure 1 illustrates the path equivalent of Eq.(15).

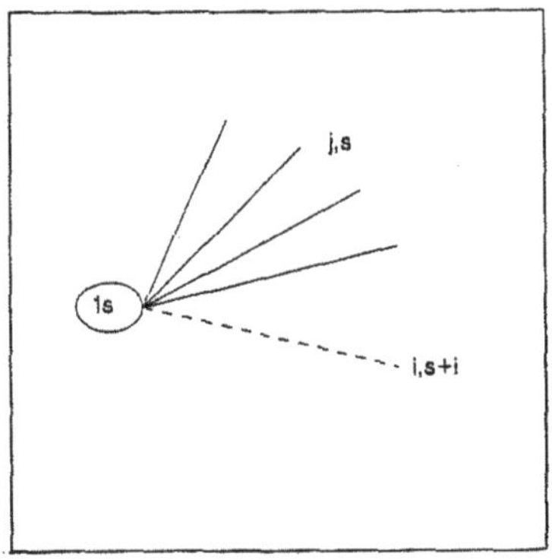

Fig.1 This figure shows the path version of the Eq.(15) showing the action of $\tilde{H}$ on a specific element of the basis in augmented space.

The calculation of the Green function can now proceed by introducing the path counting equivalent of the Feenberg Perturbation expression[22]. This has been discussed in an earlier publication[5]. We shall discuss only the essential points here. The expansion follows directly if we consider the first term on the right-hand side of Eq.(15) as the *unperturbed* hamiltonian, and the subsequent terms as the *perturbation*. Then,

$$\begin{aligned} \tilde{G}(z) &= (z\tilde{I} - \tilde{H})^{-1} \\ &= (z\tilde{I} - \tilde{H}_0 - \tilde{H}')^{-1} \\ &= g_0 + g_0\tilde{H}'g_0 + g_0\tilde{H}'g_0\tilde{H}'g_0 + \dots \end{aligned} \qquad (16)$$

where, $g_0 = (z\tilde{I} - \tilde{H}_0)^{-1}$, because $\tilde{H}_0$ is totally diagonal we have

$$\langle i\sigma|g_0|j\sigma'\rangle = \frac{1}{(z-\varepsilon_B)}\delta_{ij}\delta_{\sigma\sigma'}$$

If we now introduce the following rules :

(1) Contribution of a vertex v, $\kappa(v) = \langle v|g_0|v\rangle$

(2) Contribution of a link ℓ joining the two vertices v and v' , $\kappa(\ell) = \langle v|\tilde{H}'|v'\rangle$

(3) A Path of length n, P_n is defined to be a sequence of vertices and links : $\{v_1, \ell_{12}v_2 \ldots v_n\ell_{n,n+1}v_{n+1}\}$

(4) Contribution of a Path $\kappa(P_n) = \prod_{i=1}^{n+1}\kappa(v_i)\prod_{i=1}^{n}\kappa(\ell_{i,i+1})$

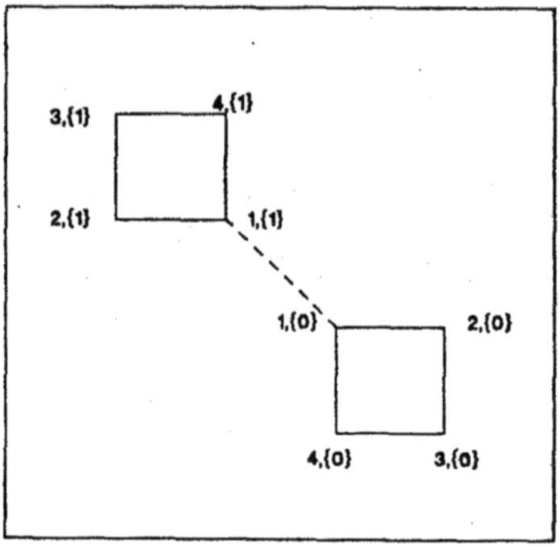

Fig.2 The initial part of the graph in Augmented Space around the vertex labelled by (1,∅). Solid lines refer to links in real space, while the dashed lines refer to links in configuration space.

Eq.(16) may now be written in terms of path counting as follows :

$$\tilde{G}_{ij}(z) \quad = \quad \sum_{n=0}^{\infty}\sum_{C_n}\kappa(P_n) \tag{17}$$

where P_n is a path of length joining the vertices v_i and v_j and C_n is the set of all such paths on the basis spanning the space.

The renormalization procedure of Feenberg then modifies the above in terms of self- avoiding paths. The effect of this is to renormalize the contribution of the vertices. The contribution of a vertex in a path now depends on where this vertex lies in the sequence of vertices which define the path. The expression for the Green function remains exactly the same as Eq.(17) only P_n now refers only to ***self-avoiding paths*** and the rule (1) is replaced by :

(1a) If the path P_n involves the sequence of vertices { $v_1, v_2 \ldots v_n$ } with $v_1 = v_i$ and $v_n = v_j$, then the contribution of a vertex v_s in the sequence is $\kappa(v_s) = \tilde{G}_{ss}^{(1,2\ldots s-1)}$. The superscripted variables indicate that the Green function is calculated in a space spanned by the original basis from which $\{|1\rangle, |2\rangle \ldots |s-1\rangle\}$ have been removed.

This forms the basis of the path counting method described by Haydock. Let us illustrate the calculation of the Green function in Augmented Space using the path counting ideas. The following figure illustrates the initial part of the graph starting from a vertex labelled by $(1,\emptyset)$.

We note that the paths which involve both real space and configuration space links renormalize the vertex and link contributions in real space alone. Mookerjee[22] has shown that this renormalization leads to a Dyson equation and a self-energy $\Sigma(z)$ which is given by an expression exactly similar to Eq.(17) with the contributions of the initial and final vertices removed. We then note that :

$$\tilde{G}_{1\emptyset,1\emptyset}(z) = \frac{1}{z-\tilde{\epsilon}\sum t_{1j}^2 G^{(1\emptyset)}_{2\emptyset,2\emptyset} - W^2\Gamma - R(z)} \tag{18}$$

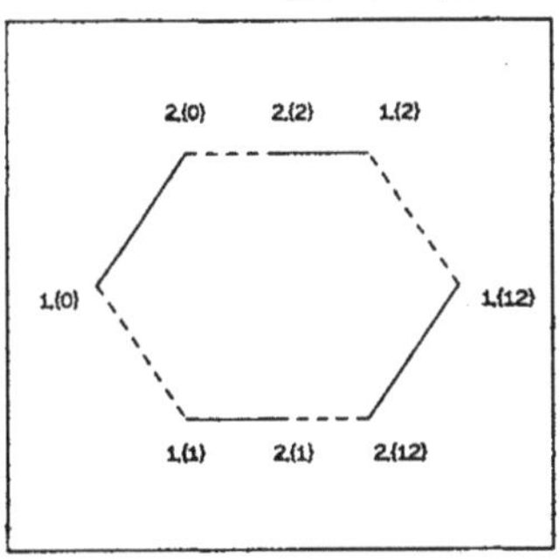

Fig.3 The shortest closed, polygonal paths which involve both real and configuration space links.

$W = \sqrt{c(1-c)}\delta\epsilon$. The $W^2\Gamma$ contribution comes from the path $P_2 = \{ v_{1\emptyset}\ v_{1,\{1\}}\ v_{1,\emptyset}\}$, while R(z) comes from contributions of longer polygonal paths starting from the vertex $v_{1\emptyset}$ and back, and which involve both real and configuration space links. The shortest of these is of length eight and is shown in the following figure. We note that this path involves two sites and their four configurations. Kumar *et.al.*[12] have argued that contributions of these paths are related to correlated scatterings from two sites. The single site coherent potential approximation (CPA) then involves ignoring contributions of such paths in the calculation of all Green function elements, including that in the second term of the denominator in Eq.(18). If we do this we note that the self energy is totally diagonal and :

$$\langle G_{11}(z)\rangle_{av} \quad = \quad \tilde{G}_{1\emptyset,1\emptyset}(z) \quad = \quad g_{11}(z-\Sigma)$$

where

$$\Sigma(z) = \bar{\epsilon} + W^2 G^{(1\emptyset)}_{1\{1\},1\{1\}}$$

Bishop and Mookerjee[21] have shown that this leads to the familiar CPA equation. The generalizations should then involve in retaining in our calculations the contributions of closed, polygonal paths. However, one has to systematically include infinite classes of paths, otherwise the crucial herglotz analytic properties of the approximated Green function are violated. This had been one of the principal difficulties with proposed generalizations. We shall discuss two different procedures : first, the method using the partition theorem (also referred to as downfolding) and secondly the direct usage of the Recursion Method and its herglotzicity preserving terminating procedures.

3.2 Generalizations of the CPA.

Let us continue with our tight-binding example of § 2. We shall first choose a cluster C in real space and partition the Augmented space Υ into two subspaces : (1) Υ_1 spanned by $\{|i\sigma\rangle\}$ where i∈C, and $\sigma \in$ all configurations of C. If the size of this cluster is n, then the rank of this subspace is n× 2^n. (2) $\Upsilon_2 = \Upsilon\backslash\Upsilon_1$.

Then,

$$\tilde{H} = \begin{pmatrix} \tilde{H}_1 & \tilde{H}_{12} \\ \tilde{H}_{21} & \tilde{H}_2 \end{pmatrix}$$

The mean field apprximation then involves in projecting the hamiltonian in the subspace Υ_2 onto an effective hamiltonian which has no configuration space links at all. That is,

$$\tilde{H}_2 \simeq H'_2 = \left((\bar{\epsilon}+\Sigma_0)\sum_{i\notin C} P_i + \sum_{i\notin C}\sum_{j\notin C}(t_{ij}+\Sigma_{ij})T_{ij} \right) \otimes P_\emptyset$$

$$\tilde{H}_{12} \simeq H'_{12} = \sum_{i\in C}\sum_{j\notin C}(t_{ij}+\Sigma_{ij})$$

The Partition theorem now allows us to obtain the Green function elements only in the subspace Υ_1. In the LMTO language, we may downfold into the much smaller subspace Υ_1. If $\tilde{R}(z) = (z\tilde{I} - \tilde{H})$ then

$$\tilde{R}^{-P_1} = (\tilde{R}_1 - \tilde{R}'_{12}\tilde{R}'^{-P_2}_2\tilde{R}'_{21})^{-1} = \hat{R}^{-1}$$

R^{-P_1} denotes the inverse of the operator R in the subspace labelled by 1. Let us call $\tilde{R}'_{12}\tilde{R}'^{-P_2}_2\tilde{R}_{21} = \tilde{R}^M(\Sigma_0, \Sigma_{ij})$.

We shall again partition Υ_1 into the space $\Upsilon_\emptyset$ spanned by $\{|i\emptyset\rangle\}$ with $i \in C$ and its complement . The matrix elements of the resolvent $\hat{R}^{-1}$ in this subspace gives us the configuration averaged Green functions according to the Augmented Space Theorem.

$$\hat{R} = \begin{pmatrix} \hat{R}_1 & \hat{R}_{12} \\ \hat{R}_{21} & \hat{R}_2 \end{pmatrix}$$

$$\langle G(z)\rangle_{av} = \left(\hat{R}_1 - \hat{R}_{12}\hat{R}_2^{-P_2}\hat{R}_{21}\right)^{-P_1} \tag{20a}$$

Now, by definition of the self-energy

$$\langle G(z)\rangle_{av} = \left[zI - \sum_i \left(\bar{\varepsilon} + \tilde{R}^M_{ii} + \Sigma_0\right) P_i \sum_i\sum_j \left(t_{ij} + \tilde{R}^M_{ij} + \Sigma_{ij}\right) T_{ij}\right]^{-1} \tag{20b}$$

But from Eq.(20a) we obtain,

$$\langle G(z)\rangle_{av} = \left[zI - \sum_i \left(\bar{\varepsilon} + \tilde{R}^M_{ii} + W\Gamma_{ii}W\right) P_i \sum_i\sum_j \left(t_{ij} + \tilde{R}^M_{ij} + W\Gamma_{ij}W\right) T_{ij}\right]^{-1} \tag{20c}$$

where, $W \;=\; \sqrt{c(1-c)}\delta\varepsilon$ and, $\Gamma_{ij} \;=\; \langle i,\{i\}|\hat{R}_2^{-P_2}|j,\{j\}\rangle$

Comparing the Eqs.(20a)-(20c) we obtain the CCPA self-consistent equations :

$$\Sigma_0 = W\langle i\{i\}|\hat{R}_2^{-P_2}|i\{i\}\rangle W$$

$$\Sigma_{ij} = W\langle i\{i\}|\hat{R}_2^{-P_2}|j\{j\}\rangle W \tag{21}$$

It has been rigorously shown that this generalization always leads to herglotz Green function[24]. We note that the above CCPA is used to calculate the Green function $G_{ii}(z)$ where the vertex i is chosen to be the centre of the cluster C chosen by us. The CCPA then includes the effects on $G_{ii}(z)$ of the correlated scattering of sites within this cluster. Since the choice of the cluster centre is immeterial, the averaged Green function is also translationally invariant. This is easily seen from the fact that in Augmented Space the hamiltonian is invariant under the translation $\mathbf{T} : \mathbf{T}\ |i;\{i_1,i_2\ldots i_C\}\rangle \quad = \quad |i+\chi;\{i_1+\chi,i_2+\chi\ldots i_C+\chi\}\rangle$.

3.3 Recursion on Augmented Space.

The generalization of the CPA is often rather cumbersome for practical computational purposes. Eq.(21) involves NL coupled equations, where N is the size of the cluster and L the number of orbitals per site. For Cu on a fcc lattice, even the nearest neighbour cluster CPA would involve 117×117 equations. We may reduce these considerably by using local point-group symmetries. However, we would like to propose an alternative approach which is both fast, accurate and whose approximations are controllable. This is the use of the Recursion Method directly on the Augmented Space.

The LMTO first order hamiltonian in the most tight-binding representation may be set up immediately for a random binary alloy :

$$H = \sum_i C_i P_i + \sum_i \sum_j \Delta_i^{-1/2} S_{ij} \Delta_j^{-1/2} T_{ij}$$

with,

$$C_i = C_{i,L}\delta_{LL'} = C_A x_i + C_B(1-x_i)$$

$$\Delta_i^{-1/2} = \Delta_{iL}^{-1/2}\delta_{LL'} = \Delta_A^{-1/2} x_i + \Delta_B^{-1/2}(1-x_i)$$

$$S_{ij} = S_{iL,jL'} \tag{22}$$

where j can either be = i, or its nearest or next-nearest neighbours on the lattice. A liitle algebra transforms this hamiltonian to :

$$H = H_B + \sum_i \delta C \ \ x_i P_i + \sum_i \sum_j [\Delta^{-1/2} x_i S_{ij} \Delta_B^{-1/2} + \ldots$$
$$+ \Delta_B^{-1/2} S_{ij} \Delta^{-1/2} \ \ x_j + \Delta^{-1/2} \ \ x_i S_{ij} \Delta^{-1/2} \ \ x_j] T_{ij}$$

$$\delta C = C_B - C_A \qquad \Delta^{-1/2} = \Delta_B^{-1/2} - \Delta_A^{-1/2}$$

We may now convert this into the Augmented Space hamiltonian according to the prescriptions discussed earlier,

$$\tilde{H} = H_B \otimes \tilde{I} + \sum_i \delta C \;\; P_i \otimes M^{(i)} + \sum_i \sum_j \left[\Delta^{-1/2} S_{ij} \Delta_B^{-1/2} T_{ij} \otimes M^{(i)} + \ldots \right.$$
$$\left. + \Delta_B^{-1/2} S_{ij} \Delta^{-1/2} T_{ij} \otimes M^{(j)} + \Delta^{-1/2} S_{ij} \Delta^{-1/2} T_{ij} \otimes M^{(i)} \otimes M^{(j)} \right.$$

or,

$$\tilde{H} = \tilde{H}_B + \tilde{H}_1 + \tilde{H}_{2a} + \tilde{H}_{2b} + \tilde{H}_3$$

We have already seen how $\tilde{H}_1$ acts on a member of the augmented space basis. For illustrative purposes, let us explicitly write down how the three new extra bits of the Augmented Space hamiltonian acts on a member of the Augmented Space basis. These new bits arise because of off-diagonal disorder in the TB-LMTO hamiltonian.

$$\tilde{H}_{2a} |i\sigma\rangle = \sum_j c \Xi_{ij} |j;\sigma\rangle + \sum_j \sqrt{c(1-c)} \Xi_{ij} |j; \sigma + \{j\}\rangle$$

with $i \notin \sigma$ and $\Xi_{ij} = \Delta^{-1/2} S_{ij} \Delta_B^{-1/2}$

Similarly,

$$\tilde{H}_3 |i\sigma\rangle = \sum_{j \notin \sigma} c^2 \Xi_{ij} |j;\sigma\rangle + \sum_{j \in \sigma} c(1-c) \Xi_{ij} |j;\sigma\rangle + \sum_j c\sqrt{c(1-c)} \Xi_{ij} |j;\sigma \pm \{j\}\rangle + \ldots$$
$$+ \sum_{j \in \sigma} c\sqrt{c(1-c)} \Xi_{ij} |j;\sigma \pm \{i\}\rangle + \sum_{j \notin \sigma} (1-c)\sqrt{c(1-c)} \Xi_{ij} |j;\sigma \pm \{i\}\rangle + \ldots$$
$$+ \sum_j c(1-c) \Xi_{ij} |j;\sigma \pm \{i\} \pm \{j\}\rangle$$

where $i \notin \sigma$ and $\Xi_{ij} = \Delta^{-1/2} S_{ij} \Delta^{-1/2}$ Combining the above equations it is easy to note that we may divide the hamiltonian into four distinct types :

(1) $H^{(0)}$ which acts only on the real space, leaving the configuration space part unchanged. A little algebra shows us that this is just $H_{VCA} \otimes I$.

(2) $H^{(1)}$ which acts only on configuration space. This is of the type : $P_i \otimes M^{(i)}$.

(3) $H^{(3)}$ which acts both on the real and configuration space, but changes the configuration only at one site. This is of the type : $T_{ij} \otimes M^{(i)}$ or $T_{ij} \otimes M^{(j)}$.

(4) $H^{(4)}$ which acts on both the real and configuration space and changes the configuration at two sites. This is of the type : $T_{ij} \otimes M^{(i)} \otimes M^{(j)}$.

The configuration is stored as a binary string of 0 and 1 values. These strings are stored in bits of words. On a 32 bit machine, for example, a string of 31 0*s* and 1*s* representing the configuration of 31 sites can be stored. Note that 1 bit is always used up to denote the sign. A recursion calculation always begins by a calculation on a finite part of the lattice. If this involves, say, 3100 sites then only 100 words W(1),... W(100) are sufficient to store the configurations of this quite large system. This is a very efficient way of storing the configurations. Such a storage and manipulation has been extensively used in numerical Monte Carlo work on the Statistical Mechanics of Ising systems. We shall borrow from the expertise in that field.

Note that the configuration of the ith site is stored in the jth word where j = $\frac{i}{N}$+1 , when we are working with a N+1 bit machine. Moreover, it is stored at the kth bit of this word, where k = i - (j-1)× N. If we wish to find out the actual configuration at this site, we can use the command ic(i) = IBITS(W(j),k). This will give an integer 0 or 1 corresponding to the two different configurations of the site i.

The inner product of two configurations stored in the two arrays W1 and W2 can be obtained from :

$$(W1, W2) = \prod_i IBITS(W1(j), k) \times IBITS(W2(j), k)$$

It is easy to check that this inner product is 0 unless the two configurations are identical. Using these operations we may now begin with the required basis member in Augmented Space : $|i\emptyset\rangle = |1\}$. Operations by the Hamiltonian involve operations of the type :

(i,W)→ (i,W′) where W′(j) = ICHNG(W(j),k) where ICHNG is either IBSET or IBCLR according to whether ic(i) is 0 or 1.

(i,W) → (N_i,W′) where N_i is a near neighbour of i and W′(j) = ICHNG(W(j),k).

(i,W) → (N_i,W′) where N_i is a near neighbour of i and W′(j') = ICHNG(W(j'),k'). j' and k' correspond to N_i.

(i,W) → (N_i,W′) where N_i is a near neighbour of i and W′(j) = ICHNG(W(j),k); W′ (j') = ICHNG(W(j'),k').

Given the lattice in real space we first create a neighbour map in real space. Then starting from $|1\}$ = (1,W) where W(m)=0 ∀m, and the above opeartions we create the neighbour map in Augmented Space. Two bases are called neighbours if the are reached from each other by one operation by the hamiltonian. Unlike the case of pure materials, the contributions in the recursion calculations from different kinds of neighbours are not the same. This has to be recorded by a pointer.

The recursion coefficients may then be generated from the recursive calculation :

$$|n\} = \tilde{H}|n-1\} - \alpha_{n-1}|n-1\} - \beta_{n-1}^2|n-2\}$$

$$\alpha_n = \{n|\tilde{H}|n\}$$
$$\beta_n^2 = \{n|n\}\backslash\{n-1|n-1\}$$
$$|0\} = 0$$

Once the coefficients are generated upto a finite number of steps, say, p, we have :

$$\langle G_{ii}(z)\rangle_{av} = \cfrac{1}{z-\alpha_1-\cfrac{\beta_1^2}{z-\alpha_2-\cfrac{\beta_2^2}{\cdots\cfrac{\cdots}{z-\alpha_p-T(z)}}}}$$

The terminator T(z) is calculated directly by processing the coefficients $\{\alpha_n, \beta_n\}$ n=1,2 ... p. This is described in detail by Haydock[4]. The terminator possesses herglotz analytic properties and gives the band edges, band widths and band weights accurately. The procedure is fast, accurate and stable. We propose that in conjunction with the TB-LMTO and Recursion packages, the Augmented Space approach should replace cumbersome and numerically much more costly mean field approaches and their generalizations.

4. Acknowledgements.

The author would like to thank Prof. Abdus Salam and the International Centre for Theoretical Physics, Trieste, Italy for providing the entire financial backing with which the Workshop, at which this work was presented, was conducted at Trieste. Useful discussions with R. Haydock, J. Kudrnovský, N-X Chen and R. Prasad and computational help from C.M.M.Nex, O. Jepsen, P. Vargas and G.P.Das are gratefully acknowledged.

5. References

1. O. K. Andersen, O. Jepsen and M. Sob in *Electronic Band Structure and its Applications*, ed. M.Yussouff (Springer Verlag Notes, 1987)

2. R. Haydock, V. Heine and M.J. Kelly, *J. Phys. C : Solid State* **5** (1972) 2845.

3. O.K. Andersen et al in *this volume*

4. R. Haydock in *this volume*

5. A. Mookerjee in *Electronic band Structure and its Applications*, ed. M. Yussouff (Spring Verlag Notes, 1987).

6. A. Mookerjee, *J. Phys. C : Solid State* **6** (1973) L205 ; 1340.

7. R. Haydock and C.M.M.. Nex ; C.M.M. Nex, *J. Phys. A: Math Phys* . **11** (1978) 653.

8. P. Soven, *Phys. Rev.* **156** (1967) 809. ; D.W. Taylor, *Phys. Rev.* **156** (1967) 1017.

9. J. Kudrnovsky et al in *this volume*

10. R. Prasad in *this volume*

11. A. Gonis and G.M. Stocks, *Phys Rev* . **B25** (1982) 659.

12. V. Kumar, A. Mookerjee and V.K. Srivastava, *J. Phys. C : Solid State* **15** (1982) 1939.

13. A. Mookerjee, *J. Phys. F: Metal Phys* . **17** (1987) 1511.

14. L.J. Grey and T. Kaplan, *Phys. Rev* . **B15** (1977) 3260 ; *ibid* **B21** (1980) 4230 ; *ib* **B29** (1984) 3684 ; T. Kaplan, P.L. Leath, L.J. Gray and H.W. Diehl, *ibid* **B21** (1980) 423(

15. A. Mookerjee, V.K. Srivastava and V. Choudhry, *J. Phys. C : Solid State* **16** (198 4555.; P.K. Thakur, A. Mookerjee and V.A. Singh, *J. Phys. F : Metal Phys* . **17** (1987) 1523 S.S. Rajput, S.S.A. Razee, A. Mookerjee and R. Prasad, *J. Phys. Condens. Matter* **2** (199 2653.

16. A. Mookerjee, *J. Phys. C : Solid State* **8** (1975) 2943.

17. T. Kaplan and L.J. Grey, *J. Phys. C : Solid State* **C9** (1976) L303.

18. D. Kumar, A. Mookerjee and V. Kumar, *J. Phys.* **F6**, 725 (1976).

19. D. Paquet and P. Leroux-Hugon, *Phys. Rev.* **B29** (1984) 593.

20. R. Haydock in *Ph.D Thesis* (University of Cambridge, U.K., 1972).

21. A.R. Bishop and A. Mookerjee, *J. Phys. C : Solid State* **7** (1974) 2165.

22. R. Haydock and A. Mookerjee, *J. Phys. C : Solid State* **7** (1974) 3001.

23. A. Mookerjee, *J. Phys. C : Solid State* **8** (1974) 1524 ; *ibid* **8** (1974) 1688.

24. S.S.A. Razee, A. Mookerjee and R. Prasad, *J. Phys. Condens. Matter* **3** (1991) 3301.

KKR APPROACH TO RANDOM ALLOYS

R. PRASAD
Physics Department, Indian Institute of Technology
Kanpur 208016, India

ABSTRACT

The methods of electronic structure calculations for random alloys in the Korringa-Kohn-Rostoker (KKR) framework will be reviewed. The KKR coherent-potential approximation (KKR-CPA) has emerged as the most powerful tool to study the electronic structure of random alloys. In this method the alloy is replaced by an ordered effective medium which is determined by a self-consistent condition. Within the density functional formalism, the KKR-CPA provides a first-principles parameter-free method of calculating the electronic structure of random alloys. Some of the recent applications of the method will be discussed. The KKR-CPA is a single-site approximation and might fail when correlated scattering from different sites is important e.g. in presence of short-range order or clustering. To include the correlated scattering, a cluster generalization of the KKR-CPA, called KKR cluster CPA (KKR-CCPA) has been recently proposed. I shall also review this method and its applications.

1. Introduction

Random alloys lack the translational symmetry, therefore , usual methods of the band theory developed for pure ordered solids are not applicable to these systems. During the last two decades, however, there has been a great progress in understanding the electronic structure of random alloys. This has been made possible by the application of the coherent-potential approximation (CPA) to muffin-tin Hamiltonians[1-4]. The muffin-tin model, in which atoms are represented by spherically symmetric and non-overlapping potentials, has already been very successful in understanding the electronic structure of pure ordered solids. Thus it was a natural starting point for the study of random metallic alloys. An application of the CPA to muffin-tin Hamiltonian was first considered by Soven[5] and later by several workers[6]. In the CPA, a random alloy is replaced by an ordered effective medium which is determined by requiring that the average scattering from a single atom embedded in the effective medium be zero. Thus the CPA is a mean-field approximation, which restores periodicity and hence some of the simplicity of the ordered solids. This formulation reduces to the Korringa-Kohn-Rostoker (KKR) formulation in the perfect limit and is called KKR-CPA. Within the density functional theory , it provides a first-principles parameter-free theory for random alloys.

The KKR-CPA is a single site approximation, which neglects the correlated scattering between two or more sites. Therefore, it might fail whenever a single-site effective description becomes inadequate, e.g., in presence of strong local environmental effects, such as short-range ordering or clustering. There is some experimental evidence on Cu-Pd and Cu-Pt systems to show that the CPA is not adequate for these systems[7]. These local environmental effects can be investigated accurately only through a cluster or multisite approximation. Recently, a self-consistent cluster generalization of the KKR-CPA has been proposed by combining augmented space technique with the KKR Green's function formalism[8-10]. This formulation, which is called KKR cluster CPA (KKR-CCPA), guarantees positive and single-valued density of states at all energies. This can be made fully self-consistent within the density functional formulation and like KKR-CPA, is a first-principles parameter-free theory of the electronic structure of random alloys. The method has been recently generalized to disordered alloys with short-range order[11]. It has been also developed in the tight-binding LMTO framework [12].

In this article I shall briefly discuss the KKR-CPA and KKR-CCPA formulations with some applications. I shall focus on substitutional random binary alloys which are the simplest disordered systems. Applications of the KKR-CPA to magnetic alloys and phase stability are discussed by Staunton *et al* in this volume. The outline of the article is as follows. In Sec. 2 the KKR-CPA and its applications are discussed. Some selected results are shown to highlight the effects of alloying on average density of states, energy bands, Fermi surface and surface states. In Sec. 3 the KKR-CCPA and its application to one-dimensinal muffin-tin alloy are discussed.

2. KKR-CPA and Its Applications

2.1 Formulation of the KKR-CPA

Consider a substtutional random binary alloy A_xB_{1-x} of A and B atoms where x is the atomic concentartion of the A type atoms in the alloy. Let us consider a particular configuration of this alloy described by the one electron Hamiltonian

$$H = H_0 + V, \tag{1}$$

where H_0 is the free electron Hamiltonian and

$$V(\mathbf{r}) = \sum_i v_i(\mathbf{r}_i), \tag{2}$$

where $\mathbf{r}_i = \mathbf{r} - \mathbf{R}_i$, $\mathbf{R}_i$ being the position vector of the ith site. $v_i(\mathbf{r}_i)$ is the potential of A(B) type centred on the site i and is assumed to be of muffin-tin form i.e. spherically symmetric inside a sphere of radius r_m, called muffin-tin radius and constant in the region between the sphere and the cell boundary. r_m is so chosen that the muffin-tin spheres do not overlap.

In the developement of alloy theory , the one electron Green's function has played a central role. For the particular configuration it is defined as

$$G(E) = (E + i0^+ - H)^{-1}. \tag{3}$$

Let ψ_n and E_n be the eigenfunctions and eigenvalues of H. The Green's function in **r**-representation then can be expressed as

$$G(\mathbf{r},\mathbf{r}';E) = \sum_i \frac{\psi_n(\mathbf{r})\psi_n^*(\mathbf{r}')}{(E - E_n + i0^+)}. \tag{4}$$

The density of states and the charge density can be written as

$$\rho(E) = \frac{1}{\pi N} \int ImG(\mathbf{r},\mathbf{r};E)d^3r, \tag{5}$$

$$\rho(\mathbf{r}) = \frac{1}{\pi} \int_{-\infty}^{E_F} ImG(\mathbf{r},\mathbf{r};E)dE, \tag{6}$$

where E_F is the Fermi energy. Note that expressions are derived for a particular configuration of the alloy. What we are interested in are averages of these quantities over all the configurations of the alloy consistent with its definition. The configurational average of the density of states is given by

$$\langle\rho(E)\rangle = \frac{1}{\pi N} \int Im\langle G(\mathbf{r},\mathbf{r};E)\rangle d^3r, \tag{7}$$

where $\langle G\rangle$ denotes the average Green's function. Similarly, the average charge density can be expressed in terms of the average Green's function. Having expresed the relevant expressions in terms of the configurational average of the Green's function, we now seek for a method to evaluate this quantity. For this purpose we start with the Dyson's equation

$$G = G_0 + G_0VG, \tag{8}$$

where

$$G_0 = (E - H_0)^{-1} \tag{9}$$

is the free electron Green's function. Eq. (8) then can be written as

$$G = G_0 + G_0VG_0 + G_0VG_0VG_0 + \cdots = G_0 + G_0TG_0 \tag{10}$$

where the scattering operator T(E) is given by

$$\begin{aligned} T &= V + VG_0V + \cdots \\ &= \sum_i v_i + \sum_i v_iG_0\sum_j v_j + \cdots \\ &= \sum_i t_i + \sum_i t_iG_0\sum_{j\neq i} t_j + \cdots \end{aligned} \tag{11}$$

Here t_i are the atomic t-matrices given by

$$t_i = v_i + v_i G_0 v_i + \cdots = v_i(1 - G_0 v_i)^{-1} \tag{12}$$

Now we introduce the decomposition[13]

$$T = \sum_{i,j} T_{ij} \tag{13}$$

where

$$T_{ij} = t_i \delta_{ij} + t_i G_0 \sum_{k \neq j} T_{kj} \tag{14}$$

T_{ij} are called path operators and describe all possible scattering processes beginning at i and ending at j. Note that Eq. (14) is exact and represents infinite coupled operator equations for T_{ij}. These equations are formally solvable only for the muffin-tin model. Because of the spherical symmetry of the potential, the use of angular momentum representation is particularly convenient and simplifies the algebra; non-overlapping of the potentials reduces Eq. (14) to a set of algebraic equations involving only their energy shell-matrix elements. The matrix-elements of t_i and T_{ij} in the coordinate representation can be expanded as

$$t_i(\mathbf{r}_i, \mathbf{r}_i') = \sum_L Y_L(\mathbf{r}_i) t_i^l(r_i, r_i) Y_L(\mathbf{r}_i') \tag{15}$$

$$T_{ij}(\mathbf{r}_i, \mathbf{r}_j') = \sum_{LL'} Y_L(\mathbf{r}_i) T_{ij}^{LL'}(r_i, r_j) Y_L'(\mathbf{r}_j') \tag{16}$$

where L is a composite index for (lm) and Y_L are real spherical harmonics. Similarly, the Fourier transform of $t_i(\mathbf{r}_i, \mathbf{r}_i)$ can be written as

$$t_i(\mathbf{k}, \mathbf{k}') = \int\int e^{-i\mathbf{k}\cdot\mathbf{r}_i} t_i(\mathbf{r}_i, \mathbf{r}_i') e^{i\mathbf{k}'\cdot\mathbf{r}_i'} d^3 r_i d^3 r_i' \tag{17}$$

The $k = k' = \kappa = \sqrt{E}$ matrix element of t_i^l, generally called as energy shell matrix elements, are easily evaluated in terms of phase-shifts $\delta_i^l(\kappa)$ of potential $v_i(r)$

$$t_i^l(\kappa, \kappa) = t_i(\kappa) = -\frac{1}{\kappa} e^{i\delta_i^l} \sin \delta_i^l \tag{18}$$

It turns out that most of the properties of our interest can be expressed in terms of only energy shell matrix elements of t_i and T_{ij}. For energy shell matrix elements, Eq. (14) reduces to the following system of linear algebraic equations

$$T_{ij}^{LL'}(\kappa) = t_i^l(\kappa) \delta_{ij} \delta_{LL'} + t_i^l \sum_{kL_1} B_{ik}^{LL_1}(\kappa) T_{kj}^{L_1 L'}(\kappa) \tag{19}$$

where $B_{ij}^{LL'}(\kappa)$ are related to the Fourier transform of the usual KKR structure constants, $B(\mathbf{k}, E)$ and depend only on the lattice structure:

$$B_{ij}^{LL'}(\kappa) = -(4\pi i \kappa) \sum_{L_1} i^{l-l'-l_1} C_{LL'L_1} Y_{L1}(\mathbf{R}_{ij}) h_{l_1}^{+}(\kappa R_{ij})(1 - \delta_{ij}) \tag{20}$$

where

$$C_{LL'L_1} = \int Y_L(\mathbf{r})Y_{L'}(\mathbf{r})Y_{L_1}(\mathbf{r})d\Omega \tag{21}$$

and h_l^+ is the outgoing Hankel function. The solution of Eq. (19) is

$$T_{ij}^{LL'}(\kappa) = [(t^{-1}(\kappa) - B)^{-1}]_{ij}^{LL'} \tag{22}$$

Having obtained $T_{ij}^{LL'}$, let us find the Green's function G in the coordinate representation. It is more convenient to express $G(\mathbf{r},\mathbf{r}';E)$ in terms of regular and irregular solutions $Z_L^i(\mathbf{r},E)$ and $J_L^i(\mathbf{r},E)$, of the Schroedinger equation for an isolated potential $v_i(r)$. $Z_L^i(\mathbf{r},E)$ and $J_L^i(\mathbf{r},E)$ are normalized such that fot $r \geq r_m$

$$Z_L^i(\mathbf{r},E) = \kappa Y_L(\mathbf{r}_i)[n_l(\kappa r_i) - \cot\delta_l^i j_l(\kappa r_i)] \tag{23}$$

and

$$J_L^i(\mathbf{r},E) = Y_L(\mathbf{r}_i)j_l(\kappa r_i) \tag{24}$$

where j_l and n_l are spherical Bessel and Neumann functions respectively. In terms of these functions $G(\mathbf{r},\mathbf{r}';E)$ can be expressed as[14]

$$G(\mathbf{r}_i,\mathbf{r}_j';E) = \sum_{LL'} Z_L^i(\mathbf{r}_i,E)T_{ij}^{LL'}(\kappa)Z_L^i(\mathbf{r}_i,E) - \sum_L Z_L^i(\mathbf{r}_i,E)J_L^i(\mathbf{r}_i;E) \tag{25}$$

Eq.(25) is exact within the muffin-tin approximation and can be used to evaluate the Green's function for a cluster of atoms. However, for an infinite medium it requires the inversion of an infinite matrix in Eq. (22). Therefore, to proceed further we require some approximation. In the KKR-CPA, the disordered alloy A_xB_{1-x} is replaced by an ordered effective medium which is determined by the requirement that if a particular site is replaced by an A or B atom, the average scattering from this site with respect to the effective medium is zero i.e.

$$xT_{00}^A + (1-x)T_{00}^B = T_{00}^C \tag{26}$$

where T_{00}^A and T_{00}^B are the restricted site averages of T_{00} when 0th site is occupied by an A or B atom. T_{00}^C is the path operator for the KKR-CPA medium. Eq. (26) can further be simplified to more familiar and computationally simpler form. These are given by

$$T_{00}^C = \frac{1}{N}\sum_{\mathbf{k}}[t_C^{-1} - B(\mathbf{k},E)]^{-1} \tag{27}$$

$$T_{00}^{A(B)} = [1 - T_{00}^C(t_C^{-1} - t_{A(B)}^{-1})]^{-1}T_{00}^C \tag{28}$$

and

$$t_C^{-1} = xt_A^{-1} + (1-x)t_B^{-1} + (t_A^{-1} - t_C^{-1})T_{00}^C(t_B^{-1} - t_C^{-1}). \tag{29}$$

Here t_C is the energy shell matrix element associated with the effective scatterer and is obtained by solving Eq. (29) self-consistently. Note that this is a nontrivial task as it involves a Brillouin zone integration at each step. This was a big hurdle at the early stages of the implementation of the theory which was overcome by using a simple method for integration called special direction method[15]. Although this method is computatinally very efficient, it is not as accurate as the tetrahedron method which has been recently generalized for the random alloys[16].

For full charge self-consistent calculation, one needs charge densities in A and B cells, which are obtained from the restricted site averages of the Green's function (25) denoted by $\langle G(\mathbf{r},\mathbf{r};E)\rangle_{A(B)}$

$$\rho_{A(B)}(\mathbf{r}) = -\frac{1}{\pi}\int_{-\infty}^{E_F} Im\langle G(\mathbf{r},\mathbf{r};E)\rangle_{A(B)}dE \tag{30}$$

where $\rho_{A(B)}(\mathbf{r})$ is the charge density associated with an A(B) cell.The Fermi energy is generally calculated from the Lloyd's formula for the integrated density of states given by[1]

$$N(E) = N_0(E) + \frac{2}{\pi}Im[x\log\|t_C^{-1} - t_B^{-1}\| + y\log\|t_C^{-1} - t_A^{-1}\|] - \frac{2}{\pi N}\sum_{\mathbf{k}} Im\log\|t_C^{-1} - B(\mathbf{k},E)\| \tag{31}$$

where $N(E)$ is the integrated density of states for the alloy and $N_0(E)$ is for the free electrons. Although this formula is very convenient to use, it sometimes gives unphysical jumps. Kaprzyk and Bansil[17] have recently derived a generalized formula which does not give unphysical jumps.

The component density of states can be calculated from

$$\rho_{A(B)}(E) = -\frac{1}{\pi}\int Im\langle G(\mathbf{r},\mathbf{r};E)\rangle_{A(B)}d^3r$$

and the average density of states can be calculated from Eq. (5). The calculation of charge densities from Eq. (30) using integration along real energy is slow and time consuming. To speed up the calculations, the complex energy method is now being used[18]. The **k**-space integration too becomes much faster for the complex energy as the integrand becomes smoother.

For the charge self-consistent KKR-CPA, the CPA equation (29) is first solved for some initial potentials v_A and v_B. The new charge densities are then computed using Eq. (30). From these charge densities one can determine new potentials $v_{A(B)}(\mathbf{r})$ using the local density functional theory and iterate the whole procedure till the self-consistency is achieved. At this point the final results do not depend on the initial choice of the potentials in A or B cells. At this level this theory is a first-principles parameter-free theory of the random alloys. In $x \to 0$ limit (perfect limit) it reduces to the standard KKR formulation of the band theory and describes the impurity limit correctly. Thus, it is now possible to treat pure metals, single impurity and concentrated alloys at the same footing. In the following, I shall discuss some applications of the KKR-CPA.

2.2 Density of States

To highlight the effect of charge self-consistency, Fig. 1 shows a comparison of charge self-consistent density of states for $Nb_{0.5}Mo_{0.5}$ by Rajput *et. al*[19] with non-charge self-consistent calculations of Donato *et al*[20] and Giuliano *et al*[21]. It is seen that while the results of Donato *et. al* agree reasonably well with the charge self-consistent calculation, the calculation of Giuliano *et al* shows large differences particularly above E_F. This could be attributed to their different choice of potentials. Fig. 2 shows a comparison of charge self-consistent component density of states for Nb and Mo in $Nb_{0.5}Mo_{0.5}$ with soft x-ray experimental results[20]. There seems to be a good agreement between the theory and the experiment, considering that matrix elements etc. have been neglected in the theory.

The KKR-CPA has been applied to several disordered alloys such as Cu-Ge (Ref. 22), Cu-Ni (Ref. 23), Cu-Zn (Ref. 24), Cu-Al (Ref. 25), Cu-Pd (Ref. 26), Cu-Au (Ref. 27), Ag-Pd (Ref. 28) etc. Recently it has been applied to very complex systems such as high T_C superconductors[29].

2.3 Complex Energy Bands

The concept of energy bands can easily be generalized to random alloys and is very useful in interpreting various experimental results[3,23]. For the ordered A material the energy bands are given by the KKR secular equation

$$||t_A^{-1} - B(\mathbf{k}, E)|| = 0 \tag{32}$$

For the alloy, the energy bands are given by the secular equation

$$||t_C^{-1} - B(\mathbf{k}, E)|| = 0 \tag{33}$$

For pure A(B) material, $t_C = t_{A(B)}$ and Eq. (33) reduces to the KKR equation (32). One generally solves these equations for $E(\mathbf{k})$ by fixing $\mathbf{k}$. $E(\mathbf{k})$ are real for perfect A or B material but turn out to be complex in alloys. This is an important alloying effect and is due to the disorder scattering.

For a long time it was thought that the energy bands and energy gaps were the consequences of the long range order in solids. However, the random alloys too can have well defined bands, provided the disorder smearing $E(\mathbf{k})$ is much smaller compared to their separation. This can be seen in Fig. 3, which shows complex energy bands for $Cu_{0.9}Ge_{0.1}$ alloy along ΓX direction[22]. The most striking feature of this figure is the appearance of a new band around -0.2 Ry due to Ge impurities. Note that the disorder smearings for different bands differ considerably and in general, are different for different $\mathbf{k}$. The existence of energy bands and band gaps has been confirmed by a variety of experimental techniques. Using the angle-resolved

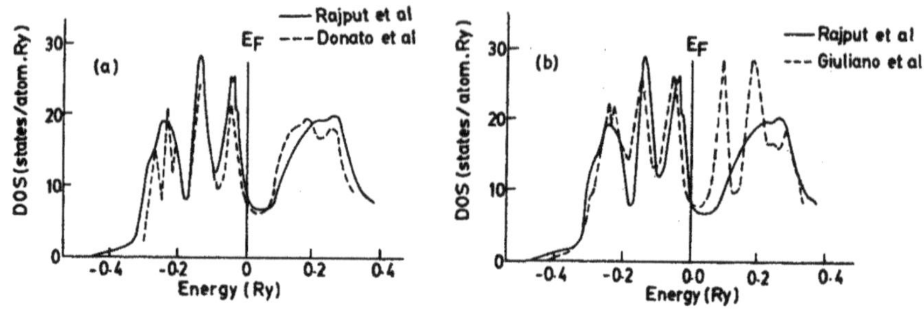

Fig. 1. Density of states for $Nb_{0.5}Mo_{0.5}$ alloy using the charge self-consistent KKR-CPA (full curve) and non-charge self-consistent calculations (dashed curves). After Rajput *et al* [Ref. 19].

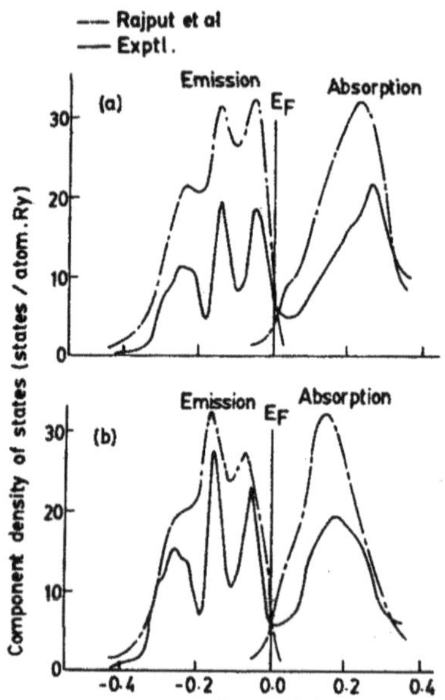

Fig. 2. Component density of states for $Nb_{0.5}Mo_{0.5}$ alloy (a) Nb component, (b) Mo component. The dashed curve shows the experimental results of Ref. 20. After Rajput *et al* [Ref. 19].

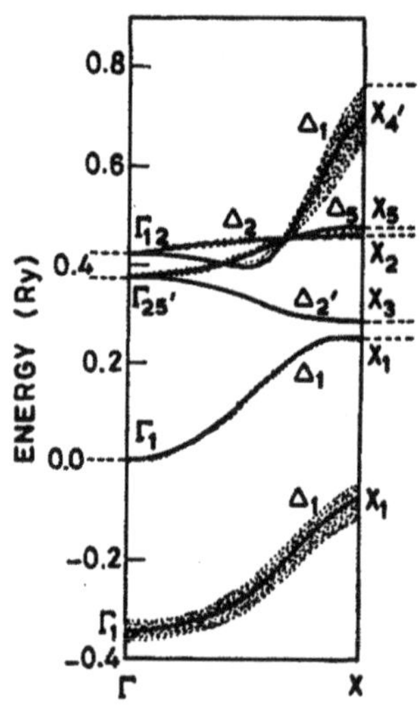

Fig. 3. Complex energy bands along ΓX direction for $Cu_{0.9}Ge_{0.1}$. The vertical length of shading around the levels equals $2ImE(\mathbf{k})$ and dashed lines on the symmetry points mark the energy levels in pure Cu. After Prasad and Bansil [Ref. 22].

photoemission technique, the alloy energy bands have been measured for a number of Cu-based alloys and were found in good agreement with the predictions of the KKR-CPA (Refs. 25 and 30). The energy band gaps for various alloys have been measured by optical absorption, differential reflectivity and piezo-reflectance techniques. The composition dependence of various gaps deduced from these experiments was found to be different for different Cu-based alloy systems, in agreement with the KKR-CPA results[31]. Thus the concept of complex energy band is very useful in interpreting various experiments and to see how the bands for the alloy evolve as a function of alloying.

2.4 Fermi Surface

If the band intersecting the Fermi energy is well defined for an alloy, it should have a well defined Fermi surface. In the early seventies there were doubts if a Fermi surface could be defined for a random alloy. Now it has been established experimentally that ,in general, concentrated alloys do have well defined Fermi surfaces. This can be seen from Fig. 4, which shows few typical cross-sections of the Fermi surface of $Cu_{0.7}Zn_{0.3}$ alloy and Cu (Ref. 24). The theoretical results are seen to be in excellent agreement with the results obtained from 2D-ACAR experiment[32]. Note that the alloy Fermi surface has a half-width because of the disorder and is not sharp as for pure metals. Using the 2D-ACAR technique, Fermi surface dimensions have been measured for several disordered alloys such as Cu-Zn (Ref. 32), Cu-Ge (Ref. 33), Cu-Pd (Ref. 34), Nb-Mo (Ref. 35) and Li-Mg (Ref. 36). The KKR-CPA does extremely well in predicting the Fermi surface of several alloys for which such calculations have been done.

2.5 Surface States

Although there exists a lot of work on the bulk electronic structure of disordered alloys, not much work has been done on their surface electronic structure. The presence of a surface can give rise to surface states which are localised in the surface region and can be probed by angle-resolved photoemission technique. Several such states have been observed in several alloys. Recently Prasad *et al*[37] have developed a technique based on Green's functiom matching method of Inglesfield[38] to investigate Shockley-type surface states (SSS) associated with the semi-infinite surface of a random alloy. In this method the average Green's function of the semi-infinite alloy is constructed by matching the vacuum and bulk Green's functions on the bounding surface. The bulk Green's function is obtained by using the KKR-CPA and the surface is modelled as an abrupt potential step. Although this is a rather idealized model of a surface, such an approach has given reasonable results for surface electronic structure of various pure metals and disordered alloys. Among the alloys which have been studied using this technique are Cu-Al, Cu-Zn, Cu-Ge, Cu-Ni, Cu-Pd, and Cu-Au (Refs. 37 and 39). Fig. 5 shows results for the $\bar{Y}$-centered Shockley-type surface state on (110) surface of $Cu_{0.9}Al_{0.1}$ alloy and Cu using this

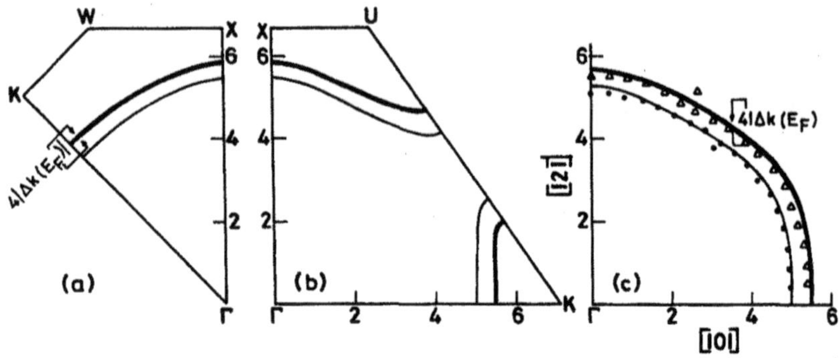

Fig. 4. Cross-sections of the Fermi surface of Cu (light solid) and $Cu_{0.7}Zn0.3$ (thick solid) in three different planes in the Brillouin zone. The thickness of the alloy curves shows the disorder smearing of the alloy Fermi surface. Experimental data for Cu (filled circles) and $Cu_{0.7}Zn_{0.3}$ (triangles) are from Ref. 32. After Prasad *et al* [Ref. 24].

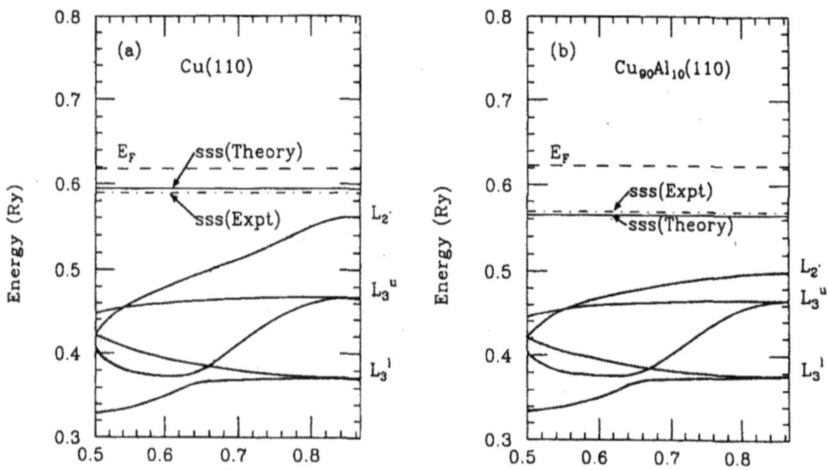

Fig. 5. Bulk energy bands along the straight line joining $\mathbf{k} = (0, 0, 0.5)$ and the symmetry point L in the Brillouin zone in (a) Cu and (b) $Cu_{0.9}Al_{0.1}$ relevant to the $\bar{Y}$-centered Shockley surface states (SSS) on the (110) surface. The bands in the alloy are shown without disorder induced widths for simplicity. The theoretical and experimental values of binding energies of the SSS are shown by horizontal lines. After Prasad *et al* [Ref. 37].

technique. Also shown in the figure are corresponding results obtained from the angle-resolved photoemission experiment. The agreement between the theory and the experiment is good.

3. The KKR-CCPA and its Applications

I shall first briefly discuss the main ideas of the augmented space formalism (ASF) which will be needed in developing the KKR-CCPA.

3.1 The Augmented Space Formalism

The ASF was developed by Mookerjee[40,41] as a tool for configurational averaging. Suppose we want the configurational average of a quantity $A(\{n_i\})$ which is a function of random variables $\{n_i\}$ with given probability distribution $p(\{n_i\})$. The configurational average of $A(\{n_i\})$ can be written as

$$\langle A\rangle = \int A(\{n_i\})p(\{n_i\})\prod_i dn_i. \tag{34}$$

For a random substitutional alloy $p(\{n_i\})$ can be expressed as a product of the individual single-site probability distributions $p_i(n_i)$ i.e.

$$p(\{n_i\}) = \prod_i p_i(n_i). \tag{35}$$

If $p_i(n_i)$ has finite moments to all orders, it can be written as the matrix element

$$p_i(n_i) = -\frac{1}{\pi} Im\langle f_0^i|(n_i I - M^i)^{-1}|f_0^i\rangle, \tag{36}$$

where M^i is an operator in the configuration space ϕ_i of rank m, spanned by $\{|f_j^i\rangle\}, j = 1, 2, \ldots, m-1$, with m being the number of components of the alloy. The configuration space ϕ_i contains all the possible configurations of the site i and $|f_0^i\rangle$ is referred to as the ground state in ϕ_i. The basis $\{|f_j^i\rangle\}$ can be defined if $p_i(n_i)$ is known explicitly. For a random binary alloy $A_x B_y, p_i(n_i)$ can be written as

$$p_i(n_i) = x\delta(n_i - 1) + y\delta(n_i), \tag{37}$$

where

$$n_i = \begin{cases} 1 & \text{for } i = A \\ 0 & \text{for } i = B \end{cases}.$$

For this $p_i(n_i), M^i$ is a tridiagonal matrix in the space ϕ_i of rank 2 sapnned by $|f_0^i\rangle$ and $|f_1^i\rangle$ with a representation

$$M^i = \begin{pmatrix} x & \sqrt{xy} \\ \sqrt{xy} & y \end{pmatrix}. \tag{38}$$

It is easily seen that

$$M^i|f_0^i\rangle = x|f_0^i\rangle + \sqrt{xy}|f_1^i\rangle, \tag{39}$$

and

$$M^i|f_1^i\rangle = y|f_1^i\rangle + \sqrt{xy}|f_0^i\rangle. \tag{40}$$

The bases $|f_0^i\rangle$ and $|f_1^i\rangle$ which will be called as the ground state and the excited state in the configuration space respectively are given by

$$|f_0^i\rangle = \sqrt{x}|A\rangle_i + \sqrt{y}|B\rangle_i, \tag{41}$$

and

$$|f_1^i\rangle = \sqrt{y}|A\rangle_i - \sqrt{x}|B\rangle_i. \tag{42}$$

The bases $|A\rangle_i$ and $|B\rangle_i$ indicate the occupancy of the ith site. By the augmented-space theorem[41] we get

$$\langle A\rangle = \langle F|\tilde{A}(\{M^i\})|F\rangle, \tag{43}$$

where $|F\rangle = \prod_i |f_0^i\rangle$ and represents a state in the configurartion space $\Phi = \prod_i \phi_i$. The remaining bases in the configuration space are those for which there are excited states $|f_1^i\rangle$ on one or more sites. These are conventionally written as $|F_s\rangle$, where s is a composite index of one or more sites on which we have $|f_1^i\rangle$ while on the rest of the sites we have $|f_0^i\rangle$. For example, a state $|F_{jk..l}\rangle$ is given by

$$|F_{jk..l}\rangle = \prod_i |f_n^i\rangle \; with\; n = \begin{cases} 0 & \text{for } i \neq j, k, \ldots, l \\ 1 & \text{for } i = j, k, \ldots, l \end{cases}. \tag{44}$$

$\tilde{A}(\{M^i\})$ is an operator in the augmented space $\Psi = H \otimes \Phi$, where H is the Hilbert space spanned by $\{|i\rangle\}$. The augmented space is spanned by $\{|i, F_s\rangle\}$ and its rank is $N \times 2^N$, where N is the total number of atoms in the solid. The operator function $\tilde{A}(\{M^i\})$ is the same function of M^i as $A(\{n_i\})$ is of $\{n_i\}$. Note that Eq. (43) is exact but cannot be used for computational purpose because the rank of the augmented space is infinite for a solid. Therefore, an approximation such as cluster CPA is used to reduce the rank of the augmented space and will be discussed in the next section.

3.2 The KKR-CCPA

In this section I shall discuss the application of ASF to derive KKR-CCPA (Refs. 8 and 10). In this approximation a small cluster C is chosen out of $\{|i\rangle\}$ and the complement of the cluster is chosen as an effective medium. The randomness is present only within C and, therefore, configurational averaging is done only over all the configurations of the cluster. The self-consistent KKR-CCPA equations are obtained from the condition that the average Green's function obtained from this technique is equal to the Green's function of the effective medium. Depending on the size of the cluster, one can generate various kinds of effective mediums. Thus, this formulation offers a convenient and systematic way of going beyond the KKR-CPA.

Our starting equation is Eq. (22) for the on-the energy shell matrix element for the path operator which can be rewritten as

$$T_{ij}^{LL'} = (A^{-1})_{ij}^{LL'}, \tag{45}$$

where

$$A_{ij}^{LL'} = C_i^{LL'}\delta_{ij} - B_{ij}^{LL'}, \tag{46}$$

and $C_i^{LL'}$ is the inverse of the on-the-shell single scatterer t-matrix. Supressing the angular momentum indices we can write

$$T_{ij} = [(C_i I - B)^{-1}]_{ij}. \tag{47}$$

Note that only C_i carries information about randomness. The non-randomness of the off-diagonal terms B_{ij} in the KKR framework, in contrast to the tight-binding framework, where randomness appears in both diagonal as well as in off-diagonal terms, makes the application of the ASF much simpler. The first step towards implementing ASF to the problem is to write C_i in terms of the random variable n_i,

$$C_i = C_B + \delta C n_i, \tag{48}$$

where

$$\delta C = C_A - C_B. \tag{49}$$

From Eq. (46) , the matrix A can be written as

$$\begin{aligned} A &= \sum_i C_i |i\rangle\langle i| - \sum_{i,j(i\neq j)} B_{ij}|i\rangle\langle j| \\ &= C_B \sum_i |i\rangle\langle i| + \delta C \sum_i |i\rangle\langle i| n_i - \sum_{i,j(i\neq j)} B_{ij}|i\rangle\langle j|. \end{aligned} \tag{50}$$

Our aim is to find configuration-averaged T_{ij}, which by the augmented space theorem[41], is given by

$$\langle T_{ij}\rangle = \langle i, F|\tilde{A}^{-1}|j, F\rangle, \tag{51}$$

where

$$\tilde{A} = C_B \sum_i |i\rangle\langle i| \otimes I + \delta C \sum_i |i\rangle\langle i| \otimes M^i - \sum_{i,j(i\neq j)} B_{ij}|i\rangle\langle j| \otimes I. \tag{52}$$

Here I and M^i are the operators in the configuration space Φ. Now the augmented space Ψ is partitioned into subspaces labelled I and II; the subspace I is spanned by

$\{|i, F_s\rangle\}, i \in C$, where C is the chosen cluster and $\{|F_s\rangle\}$ span the configuration space of the cluster. For a cluster of size m, the subspace I is of rank $m \times 2^m$. Now the mean-field approximation is used and the subspace II is replaced by an effective medium. Thus the subspace II has only one configuration, namely, the ground state $|F\rangle$, and hence, is spanned by $\{|j, F\rangle\}, j \in C'$, where C' is the complement of the cluster C and is of rank N-m.

Since we need the $\langle i, F| \ldots |j, F\rangle$ element of $\tilde{A}^{-1}$, we partition $\tilde{A}$ as follows:

$$\tilde{A} = \begin{pmatrix} A_I & A' \\ A'^T & A_{II} \end{pmatrix}, \tag{53}$$

where A_I is in subspace I and A_{II} is in subspace II. By partition theorem we get the inverse of $\tilde{A}$ in subspace I as

$$[\tilde{A}^{-1}]_I = (A_I - A' A_{II}^{-1} A'^T)^{-1} = \hat{A}^{-1}. \tag{54}$$

The four constituent matrices of A are given by

$$A_I = C_B \sum_{i \in C} |i\rangle\langle i| \otimes \sum_{F_s \in \Phi} |F_s\rangle\langle F_s| + \delta C \sum_{i \in C} |i\rangle\langle i| \otimes M^i - \sum_{i,j \in C} B_{ij} |i\rangle\langle j| \otimes \sum_{F_s \in \Phi} |F_s\rangle\langle F_s|, \tag{55}$$

$$A_{II} = \left(C_{eff} \sum_{k \in C'} |k\rangle\langle k| - \sum_{k,l \in C'} b_{kl} |k\rangle\langle l| - \sum_{k,l \in C'} B_{kl} |k\rangle\langle l| \right) \otimes \sum_{F_s \in \Phi} |F_s\rangle\langle F_s|, \tag{56}$$

$$A' = -\left(\sum_{i \in C} \sum_{k \in C'} b_{ik} |i\rangle\langle k| + \sum_{i \in C} \sum_{k \in C'} B_{ik} |i\rangle\langle k| \right) \otimes \sum_{F_s \in \Phi} |F_s\rangle\langle F_s|. \tag{57}$$

and A'^T is transpose of A'. In Eqs. (56) and (57) b_{kl} are the off-diagonal corrections in C, the diagonal corrections already contained in C_{eff}. Note that A_I does not contain b_{ij} , because subspace I contains the real atoms and their configurations, while A_{II} has these off-diagonal corrections because subspace II contains only effective atoms. For convenience, we add b_{kl} to B_{kl}, which are also off-diagonal in site indices and define B^{eff} as

$$B_{kl}^{eff} = B_{kl} + b_{kl}. \tag{58}$$

Thus Eqs. (56) and (57) can now be written as

$$A_{II} = \left(C_{eff} \sum_{k \in C'} |k\rangle\langle k| - \sum_{k,l \in C'} B_{kl}^{eff} |k\rangle\langle l| \right) \otimes \sum_{F_s \in \Phi} |F_s\rangle\langle F_s|, \tag{59}$$

and

$$A' = -\left(\sum_{i \in C} \sum_{k \in C'} B_{ik}^{eff} |i\rangle\langle k| \right) \otimes \sum_{F_s \in \Phi} |F_s\rangle\langle F_s|. \tag{60}$$

The elements of $\hat{A}$ in Eq. (54) can be found if we can evaluate the tripple product $(A' A_{II}^{-1} A'^T)$. Since A_{II} is the matrix in the effective medium with the cluster C removed, we have

$$A_{II}^{-1} = T^{eff(C)}. \tag{61}$$

The superscript C indicates that $T^{eff(C)}$ is the path operator matrix of the effective medium with the cluster removed from the medium. Though it is not possible to calculate this quantity, it can be completely eliminated from the calculations. From Eqs. (59) and (60) we get

$$\begin{aligned} A' A_{II}^{-1} A'^T &= \sum_{i,j \in C} \sum_{k,l \in C'} B_{ik}^{eff} T_{kl}^{eff(C)} B_{lj}^{eff} |i\rangle\langle j| \otimes \sum_{F_s \in \Phi} |F_s\rangle\langle F_s| \\ &= \sum_{i,j \in C} \xi_{ij}^C |i\rangle\langle j| \otimes \sum_{F_s \in \Phi} |F_s\rangle\langle F_s|, \end{aligned} \tag{62}$$

where

$$\xi_{ij}^C = \sum_{k,l \in C'} B_{ik}^{eff} T_{kl}^{eff(C)} B_{lj}^{eff} \; for \; i, j \in C. \tag{63}$$

It is clear from Eq. (62) that the matrix $A' A_{II}^{-1} A'^T$ is diagonal in the configuration space Φ. The off-diagonal terms in $\hat{A}$ thus come from A_I only. From Eqs. (54), (55), and (62) we get

$$\hat{A} = \Big(\sum_{i \in C} (C_B - \xi_{ii}^C |i\rangle\langle i| - \sum_{i,j \in C} (B_{ij} + \xi_{ij}^C)|i\rangle\langle j| \Big) \otimes \sum_{F_s \in \Phi} |F_s\rangle\langle F_s| + \delta C \sum_{i \in C} |i\rangle\langle i| \otimes M^i. \tag{64}$$

Next important step is to partition $\hat{A}$ as follows:

$$\hat{A} = \begin{pmatrix} A_1 & A_{12} \\ A_{21} & A_2 \end{pmatrix}, \tag{65}$$

where A_1 is in the subspace 1 spanned by $|i, F\rangle, i \in C$ and is of rank m. The subspace 2 is the complement of subspace 1 and has rank $m \times (2^m - 1)$. The inverse of $\hat{A}$ in subspace 1 is then given by

$$[\hat{A}^{-1}]_1 = (A_1 - A_{12} A_2^{-1} A_{21})^{-1}. \tag{66}$$

The matrices A_1, A_{12}, A_{21}, and A_2 have been evaluated in Refs. (8) and (10). For example A_1 is given by

$$A_1 = \sum_{i \in C} (\bar{C} - \xi_{ii}^C)|i, F\rangle\langle i, F| - \sum_{i,j \in C} (B_{ij} + \xi_{ij}^C)|i, F\rangle\langle j, F|, \tag{67}$$

where

$$\bar{C} = x C_A + y C_B. \tag{68}$$

From Eq. (51) we get

$$\langle T_{ij}\rangle = \langle i, F|(A_1 - \omega\Gamma\omega)^{-1}|j, F\rangle, \tag{69}$$

where

$$\omega = (xy)^{1/2}\delta C, \tag{70}$$

and

$$\Gamma = \sum_{i,j\in C} |i, F\rangle\langle i, F_i|A_2^{-1}|j, F_j\rangle\langle j, F|. \tag{71}$$

For the translationally invariant effective medium, Eq. (50) can be written as

$$A_{eff} = C_{eff}\sum_i |i\rangle\langle i| - \sum_{i,j} B_{ij}^{eff}|i\rangle\langle j|. \tag{72}$$

We get the effective path operators T_{ij}^{eff}, for the sites inside the cluster, from Eq. (72) by partitioning technique as

$$T_{ij}^{eff} = \langle i|\left(\sum_{k\in C}(C_{eff} - \xi_{kk}^C)|k\rangle\langle k| - \sum_{k,l\in C}(B_{kl}^{eff} + \xi_{kl}^C)|k\rangle\langle l|\right)^{-1}|j\rangle. \tag{73}$$

The selfconsistency condition $\langle T_{ij}\rangle = T_{ij}^{eff}$ implies that

$$\langle F|(A_1 - \omega\Gamma\omega)^{-1}|F\rangle = \sum_{i\in C}(C_{eff} - \xi_{ii}^C)|i\rangle\langle i| - \sum_{i,j\in C}(B_{ij}^{eff} + \xi_{ij}^C)|i\rangle\langle j|. \tag{74}$$

The KKR-CCPA equations follow directly from Eq. (74) as

$$C_{eff} = \bar{C} - \omega\langle i, F_i|A_2^{-1}|i, F_i\rangle\omega, \tag{75}$$

$$b_{ij} = \omega\langle i, F_i|A_2^{-1}|j, F_j\rangle\omega, \; for\; i, j \in C. \tag{76}$$

Note that these equations are self-consitent equations, which involve C_{eff}, and b_{ij} on the right hand side as well and can be solved iteratively. For one atom cluster the KKR-CCPA reduces to the KKR-CPA giving the correct limit. Also, like KKR-CPA it is exact in the limit when the concentration of the either constituent vanishes, and also in the limit when the difference between the scattering strengths of the two constituents is small.

The site-diagonal Green's function for a particular configuration of the alloy can be written from Eq. (25) as

$$G_{SD}(\mathbf{r}, \mathbf{r}') = Tr(T_{ii}F^i) - \sum_L Z_L^i(\mathbf{r}_i)J_L^i(\mathbf{r}_i'), \tag{77}$$

where Tr stands for trace over L and

$$F^i_{LL'} = Z^i_L(\mathbf{r}_i)Z^i_L(\mathbf{r}'_i). \tag{78}$$

Configurational average of this Green's function has been calculated by Razee and Prasad[10] within the KKR-CCPA and is given by

$$\langle G_{SD}(\mathbf{r},\mathbf{r}')\rangle = Tr(T^{eff}_{00}[\bar{F} + (C_{eff} - \bar{C})(\delta C)^{-1}\delta F] - \sum_{j\in C} T^{eff}_{ij} b_{ji}(\delta C)^{-1}\delta F) -$$

$$\sum_L [xZ^A_L(\mathbf{r})J^A_L(\mathbf{r}') + yZ^B_L(\mathbf{r})J^A_L(\mathbf{r}')], \tag{79}$$

where

$$\bar{F} = xF^A + yF^B, \tag{80}$$

and

$$\delta F = F^A - F^B. \tag{81}$$

The average density of states within the KKR-CCPA can be calculated from this Green's function using Eq. (5). The charge density in A or B cell, $\rho_{A(B)}(\mathbf{r})$ is given by

$$\rho_{A(B)}(\mathbf{r}) = -\frac{1}{\pi} Tr\ Im \int_{-\infty}^{E_F} F^{A(B)} D_{A(B)} T^{eff}_{00}\, dE, \tag{82}$$

where

$$D_{A(B)} = [I + (C_{A(B)} - C_{eff})T^{eff}_{00}]^{-1}. \tag{83}$$

One can calculate $v_{A(B)}(\mathbf{r})$ from $\rho_{A(B)}(\mathbf{r})$ using the local density approximation and can achieve charge self-consistency.

3.3 Application to One-dimensional Muffin-tin Model

The KKR-CCPA has been applied to a one-dimensional muffin-tin model[42], which retains many features of the three dimensional model but reduces computational effort considerably. Various expressions regarding the wavefunctions, scattering matrices and the method for calculating path operators for this model have been discussed in Ref. (8). The KKR-CCPA has been implementd for a two-atom cluster on this model. Fig. 6(a) shows a comparison of the KKR-CCPA and KKR-CPA densities of states for x= 0.1. It can be seen that in the first majority band, there is no apparent difference between the two results. However, in the impurity band there are two peaks in the KKR-CCPA result, in contrast to a smooth KKR-CPA result. The extra structure in the KKR-CCPA density of states arises due to the correlated scattering from two-atom clusters. This can be seen from Fig. 6(b) which shows the local density of states on site A when single A and two A impurities are embedded in the pure B medium. A comparison of the two figures shows that the structure in the impurity band arises approximately around the impurity levels due to A and AA impurities. In Ref. 8 it is shown that for x = 0.5 also the KKR-CCPA gives few extra structures in each band which are due to various clusters embedded in the KKR-CPA medium.

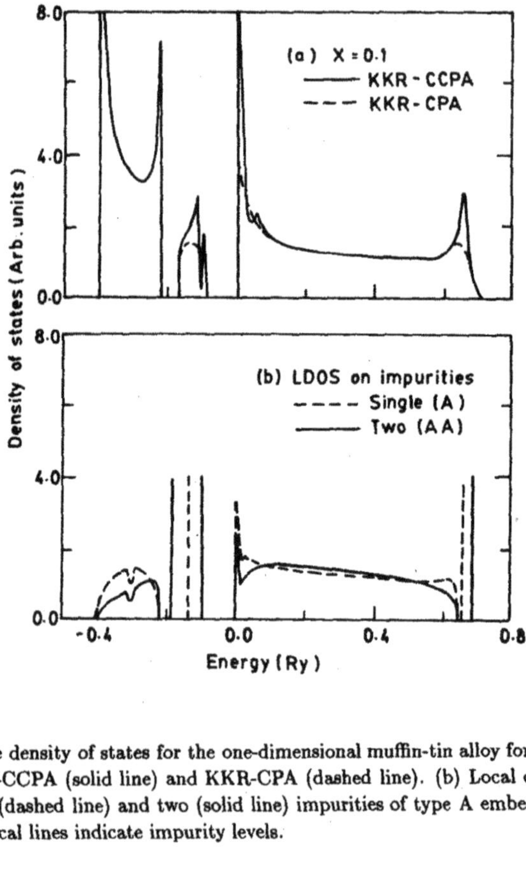

Fig. 6. (a) Average density of states for the one-dimensional muffin-tin alloy for $x = 0.1$ calculated by using the KKR-CCPA (solid line) and KKR-CPA (dashed line). (b) Local density of states on A site for a single (dashed line) and two (solid line) impurities of type A embedded in the pure B medium. The vertical lines indicate impurity levels.

Recently, the above formulation has been generalized to disorderd alloys with short-range order[11]. It has been found that the introduction of the short-range order can induce large changes in the density of states.

4. Acknowledgements

The author would like to thank Drs. A. Mookerjee, V. Kumar and S. S. A. Razee for helpful discussions. This work was supported by the Department of Science and Technology, New Delhi, India, through Grant No.SP/S2/M-39/87.

5. References

1. H. Ehrenreich and L. Schwartz, in *Solid State Physics*, ed. F. Seitz and D. Turnbull (Academic, New York, 1976), vol. 31, p. 149.
2. J. S. Faulkner, in *Progress in Materials Science*, **27**, 1 (1982).
3. A. Bansil, in *Electronic Band Structure and its Applications*, Lecture Note Series, Vol. 283 (Springer-Verlag, Heidelberg, 1987).
4. R. Prasad, *Indian J. Pure and Appl. Phys.* **29**, 255 (1992).
5. P. Soven, *Phys. Rev.* **B 2**, 4715 (1970).
6. For the early history see, for example, Ref. 1.
7. H. Wright *et al, Phys. Rev.* **B 35**, 519 (1987); D. van der Marel *et al, Phys. Rev.* **B 32**, 6331 (1985); J. Banhart *et al, Phys. Rev.* **B 37**, 6027 (1988).
8. S. S. A. Razee, S. S. Rajput,R. Prasad, and A. Mookerjee, *Phys. Rev.* **B 42**, 9391 (1990).
9. S. S. A. Razee, A. Mookerjee, and R. Prasad, *J. Phys: Condensed Matter* **3**, 3301 (1991).
10. S. S. A. Razee and R. Prasad, *Phys. Rev.* **B 45**, 3265 (1992).
11. S. S. A. Razee and R. Prasad, *Phys. Rev.* **B 48**, 1349 (1993).
12. S. S. A. Razee and R. Prasad, *Phys. Rev.* **B 48**, 1361(1993).
13. B. L. Gyorffy, *Phys. Rev.* **B 5**, 2382 (1972).
14. J. S. Faulkner and G. M. Stocks, *Phys. Rev.* **B 21**, 3222 (1980).
15. R. Prasad and A. Bansil, *Phys. Rev.* **B 21**, 496 (1980).
16. S. Kaprzyk and P. E. Mijnarends, *J. Phys.* **C 19**, 1283 (1986); S. Kaprzyk and A. Bansil (unpublished).
17. S. Kaprzyk and A. Bansil, *Phys. Rev.* **B 42**, 7358 (1990).
18. R. Zeller, J. Dentz, and P. H. Dederichs, *Solid State Commun.* **44**, 993 (1982); D. D. Johnson, F. J. Pinski, and G. M. Stocks, *Phys. Rev.* **B 30**, 5508 (1984).
19. S. S. Rajput, R. Prasad, S. Kaprzyk, and A. Bansil, to be published.
20. E. Donato *et al, Phys. Stat. Sol. (b)* **95, K** 37 (1979).
21. E. S. Giuliano *et al*, in *Transition metals*, ed. M. J. G. Lee *et al* (Institute of Physics, Bristol, 1977), p. 40.
22. R. Prasad and A. Bansil, *Phys. Rev. Lett.* **48**, 113 (1982).
23. A. Bansil, *Phys. Rev.* **B 20**, 4035 (1979).

24. R. Prasad, S. C. Papadopoulos, and A. Bansil, *Phys. Rev.* **B 23**, 2607 (1981).
25. H. Asonen *et al*, *Phys. Rev.* **B 25**, 7075 (1982).
26. R. S. Rao *et al*, *Phys. Rev.* **B 29**, 1713 (1984); H. Winter *et al*, *Phys. Rev.* **B 33**, 2370 (1986).
27. R. S. Rao *et al*, *Phys. Rev.* **B 31**, 3245 (1985).
28. H. Winter and G. M. Stocks, *Phys. Rev.* **B 27**, 882 (1983).
29. S. Kaprzyk and A. Bansil, *Phys. Rev.* **B 43**, 10335 (1991).
30. R. S. Rao and M. Pessa, in Ref. 3.
31. R. S. Rao, R. Prasad, and A. Bansil, *Phys. Rev.* **B 28**, 5762 (1983).
32. M. Haghooie, S. Berko, and U. Mizutani, in *Positron Annihilation*, eds. R. Hasiguti and K. Fujiwara (Japan Institute of Metals, Sendai, 1979).
33. P. E. Mijnarends *et al*, *Phys. Rev. Lett.* **59**, 720 (1987).
34. L. C. Smedskjaer *et al*, *Phys. Rev. Lett.* **59**, 2479 (1987).
35. C. R. Bull *et al*, *Phys. Rev.* **B 29**, 6378 (1984).
36. W. Triftshauser *et al*, *Materials Science Forum*, **105-110**, 501 (1992); S. S. Rajput *et al*, *J. Phys: Condensed Matter*, **5**, 6419 (1993).
37. R. Prasad, A. Y. Serageldin, and A. Bansil, *J. Phys: Condensed Matter* **3**, 801 (1991).
38. J. E. Inglesfield, *Surf. Sci.* **76**, 355 (1978); *Rep. Prog. Phys.* **45**, 223 (1982).
39. A. Y. Serageldin, R. Prasad, and A. Bansil, *Modern Phys. Lett.* **B 7**, 25 (1993).
40. A. Mookerjee, *J. Phys.* **C 6**, L205 (1973); **C 6**, 1340 (1973).
41. A. Mookerjee, in this volume and in Ref. 3.
42. W. H. Butler, *Phys. Rev.* **B 14**, 468 (1976).

SELF-CONSISTENT GREEN'S FUNCTION METHOD FOR RANDOM ALLOYS AND THEIR SURFACES

J. KUDRNOVSKÝ
Institute of Physics, Academy of Sciences of Czech Republic,
CZ-180 40 Praha 8, Czech Republic
and
Institute of Technical Electrochemistry,
Technical University, A-1060 Wien, Austria

I. TUREK
Institute of Physical Metallurgy, Academy of Sciences of Czech Republic,
CZ-616 62 Brno, Czech Republic

and

V. DRCHAL
Institute of Physics, Academy of Sciences of Czech Republic,
CZ-180 40 Praha 8, Czech Republic

ABSTRACT

An efficient self-consistent Green's function method developed for calculation of the electronic properties of disordered alloys and their surfaces within the local density approximation of the density functional theory is reviewed. The electronic states are described by the first-principles tight-binding linear muffin-tin orbital method. The semi-infinite nature of the system is incorporated within the surface Green's function approach. A generalization of the coherent potential approximation method is used to treat the effect of disorder in alloys with complex lattices as well as in alloys with varying concentration profiles at the surface. The potentials are treated within the atomic sphere approximation, but for their construction both the monopole and dipole components of the charge density are included in the case of surfaces. We discuss as well some computational details concerning the numerical implementation of the method.

1. Introduction

The knowledge of the electronic properties of a system is of vital importance for understanding of the studied physical processes on the microscopic ground. In the past this led to efficient schemes to determine the electronic structure of crystals and ordered alloys, which are based on the density-functional formalism (see for example Refs. 1 and 2). The last decade is characterized by a fast development of the material science with emphasis on the materials engineering and the preparation of new materials, often with unique properties. We just mention the semiconductor and metallic superlattices, magnetic overlayers on both magnetic and non-magnetic substrates, or the high-temperature superconductors. Surfaces of crystals and alloys are as well of great interest from both theoretical and technological points of view, as many of material properties depend critically on the properties of the surface region. The catalysis, chemisorption, crystal growth or the segregation of atoms to the surface are just a few such examples. From the theoretical point of view, a deep insight into the physical origins of various processes in solids can be obtained by studying the physical properties as functions of alloy composition or coverage of a clean substrate by foreign atoms.

The presence of the structural and compositional inhomogeneities requires the use of techniques different from those employed in the conventional band structure calculations. The Green's functions correspond to quantum-mechanical probability, while the wave functions correspond to its amplitude. The observables depend linearly on Green's functions, but are bilinear forms in wave functions. Consequently, the Green's functions are the relevant physical quantities to be evaluated for materials without the translational symmetry like random alloys and their surfaces. Once the Green's function of the studied system is found, any physical quantity of interest can be determined. It should be noted, that the conventional band structure methods can still be used to study some specific problems for surfaces or random alloys by modelling them via suitably chosen slabs or supercells. By construction, the Green's function depends rather on the charge density than directly on the wave function, and it is thus much less sensitive to the imposed boundary condition. In addition, the Green's function language allows to understand the nature of the physical problems and approximations used in a systematic way.

In this paper we give a brief account of an efficient self-consistent Green's function method for the calculation of electronic properties of disordered alloys and their surfaces. The main features of the method can be summarized as follows: (i) the application of the all-electron tight-binding linear muffin-tin orbital (TB-LMTO) method[1,2,3] within the local density approximation (LDA) to describe the electronic structure from the first-principles; (ii) the description of the semi-infinite geometry of the system using the surface Green's function (SGF) formalism;[4,5] (iii) the use of the coherent potential approximation (CPA) approach[6,7,8] extended to inhomogeneous systems like surfaces[9,10] or complex lattices;[11,12,13] (iv) the characterization of the vacuum region by empty spheres which represent the continuation of the semi-infinite solid lattice to infinity,[14] (v) the description of the atomic potentials in the atomic sphere approximation (ASA), and (vi) the inclusion of both monopole and dipole terms of the charge density in the calculation of the Madelung potential[15] at the solid surface.

The advantage of the present approach is the unified treatment of both bulk alloys and their surfaces. This is achieved by using the same starting Hamiltonian in both cases, which in the orthogonal LMTO representation γ has the following form[1,2]

$$H^{\gamma}_{RL,R'L'} = C_{RL}\delta_{RR'}\,\delta_{LL'} + \Delta^{1/2}_{RL}\,S^{\gamma}_{RL,R'L'}\,\Delta^{1/2}_{R'L'}, \quad O^{\gamma}_{RL,R'L'} = \delta_{RR'}\,\delta_{LL'}. \tag{1}$$

In Eq. (1), H^{γ} and O^{γ} are, respectively, the Hamiltonian and overlap matrices. Further, R and R' are the site indices while L and L' refer to the orbital indices. The geometry of the problem enters the Hamiltonian via the structure constant matrix S^γ expressed in terms of the canonical structure constant matrix S^0, which is known analytically[10]

$$S^\gamma_{RL,R'L'} = [\, S^0\,(1-\gamma\, S^0)^{-1}\,]_{RL,R'L'}\,. \tag{2}$$

The elements $S^0_{RL,R'L'}$ depend only on $(R-R')/w$, where R and R' are the atomic positions, and w is the average Wigner-Seitz radius of a solid. Due to the semi-infinite nature of the surface problem, all layers parallel to the surface could have different local physical properties. In order to overcome this difficulty, we assume that from a certain layer on, the electronic properties of all subsequent layers are identical to those of the corresponding infinite systems, namely either to a homogeneous alloy or to the vacuum. The solid is thus considered to be divided into three regions: (i) a homogeneous semi-infinite bulk alloy, (ii) a homogeneous vacuum region represented by empty spheres and characterized by flat potentials, and (iii) an intermediate region consisting of several layers where all inhomogeneities (chemical or electronic) are concentrated, and which contains also a few layers of empty spheres of the vacuum-solid interface. Here we shall consider the simplest case, namely an ideal periodic lattice randomly occupied by atoms A and B. We thus neglect the lattice relaxations in bulk which are connected with different sizes of alloy constituents (see Sec. 6.5), as well as possible inward or outward shifts of surface layers. We can then use the ideal bulk structure constant S^0 for both the bulk and surface problems.

The properties of individual atoms occupying the ideal lattice sites are characterized by the potential parameters X_{RL} ($X = C, \Delta$, and γ), which are randomly X^A_{RL} and X^B_{RL} in a solid, and X^v_{RL} for empty spheres in the vacuum region. Note that the potential parameters, in general, depend also on the site index R due to the charge self-consistency. They are the same for each group $\{R_p\}$ of geometrically and electronically equivalent sites. For example, $\{R_p\}$ can be sites of one of inequivalent sublattices of a complex lattice, or a layer of atoms close to the surface with properties different from those in the bulk. On a simple lattice, e.g. fcc lattice, there is only one set of potential parameters. The potential parameters have a simple physical interpretation:[1,16] they describe the positions C_{RL}, the widths Δ_{RL}, and the distortions γ_{RL} of the 'pure' RL-bands. The concentrations c^Q of alloy constituents $Q = A, B$ could as well depend on the site index $R \in \{R_p\}$, $c^Q_R \equiv c^Q_p$. The set of numbers c^Q_R thus defines the concentration profile in the solid. In a homogeneous bulk alloy all c^Q_R are independent of the site index R. In random alloys with a complex lattice structure, like quasibinary semiconductor alloys with the sphalerite, fluorite, or rocksalt structures,[11] in the transition metal alloys with partial order,[12] or in the magnetic alloys with the DO_3 structure,[13] the local concentrations c^Q_R can have two or more different values on a lattice. The most complex is the case of the segregation of one of the alloy components at the solid surface, where the concentrations c^Q_R may differ from the bulk values in several surface layers.

The potential parameters are obtained from the solutions of the radial scalar-relativistic Schrődinger equation for the LDA potential in a corresponding atomic sphere. The potential parameters can be combined into the potential function matrix

$$P^0_{RL}(z) = \frac{z - C_{RL}}{\Delta_{RL} + \gamma_{RL}(z - C_{RL})}, \tag{3}$$

which plays a central role in the TB-LMTO formalism.[1,2] In Eqs. (1-3), $L = (lm)$ is the orbital index with the angular momentum $l \leq 2$, which allows to describe all transition metals and most of interesting semiconductors (the basis set consists of s-, p-, and d-orbitals on each site). The relativistic effects (like the spin-orbit coupling) and the higher angular momenta ($l > 2$) can be included, but are omitted in this paper for simplicity.

In conclusion of this Section, we mention one important feature of the LMTO basis set, namely its transformation property. The starting Hamiltonian was defined in the so-called orthogonal LMTO representation γ whose name is self-explanatory: the basis set orbitals are mutually orthogonal (see Eq. (1)). This representation is convenient to formulate the problem and to evaluate the physical quantities of interest here, as the overlap matrix O^γ is the unit matrix. The orthogonal LMTO basis set can be transformed exactly into many other basis sets whose properties can be tailored to a specific problem. For example, it is convenient to perform the configurational averaging[7,8] or to introduce the surface[9,10,14,15] in the so-called tight-binding LMTO representation, or to transform to a specific LMTO representation in which the downfolding to a minimal basis set[11] can be performed. The idea of the LMTO representations, introduced by Andersen,[3] is a powerful tool for the applications of the LMTO formalism to random alloys and their surfaces. In the next Section we shall give a brief account of this idea in the language of the Green's functions suitable for further applications.

2. Green's Functions in the LMTO Theory

The Green's function matrix corresponding to the Hamiltonian (1) is defined as

$$G_{RL,R'L'}(z) = [(z - H^\gamma)^{-1}]_{RL,R'L'}. \tag{4}$$

For systems with broken translational symmetry it is, however, advantageous to reformulate the theory by using different LMTO representations. A general LMTO representation α is specified by a set α_{Rl} of the so-called screening constants,[3] which define new structure constant matrices $S^\alpha_{RL,R'L'}$ and corresponding potential functions $P^\alpha_{RL}(z)$ by means of the following transformations

$$S^\alpha_{RL,R'L'} = [S^0(1 - \alpha S^0)^{-1}]_{RL,R'L'}, \tag{5}$$

and

$$P^\alpha_{RL}(z) = \frac{P^0_{RL}(z)}{1 - \alpha_{RL}P^0_{RL}(z)} = \frac{z - C_{RL}}{\Delta_{RL} + (\gamma_{RL} - \alpha_{RL})(z - C_{RL})}. \tag{6}$$

We note that for spherically symmetric potentials, the potential parameters C_{RL}, Δ_{RL}, γ_{RL} and the potential functions $P^\alpha_{RL}(z)$ can be represented by the site- and orbital-diagonal matrices C, Δ, γ and $P^\alpha(z)$. In fact, they depend on the orbital quantum number l rather then on the orbital index L itself. The off-diagonal matrices S^α are defined in terms of the original, canonical structure constant matrix S^0.

The Green's function matrix $G(z)$ corresponding to the Hamiltonian (1) can be expressed using the LMTO representation α as

$$G_{RL,R'L'}(z) = \lambda^\alpha_{RL}(z) + \mu^\alpha_{RL}(z)\, g^\alpha_{RL,R'L'}(z)\, \mu^\alpha_{R'L'}(z), \tag{7}$$

where the auxiliary Green's function $g^\alpha(z)$ is defined by

$$g^\alpha_{RL,R'L'}(z) = [(P^\alpha(z) - S^\alpha)^{-1}]_{RL,R'L'}. \tag{8}$$

The site-diagonal matrices $\lambda^\alpha(z)$ and $\mu^\alpha(z)$ are given by

$$\begin{aligned} \lambda^\alpha_{RL}(z) &= -\frac{1}{2}\frac{\ddot{P}^\alpha_{RL}(z)}{\dot{P}^\alpha_{RL}(z)} = \frac{\gamma_{RL} - \alpha_{RL}}{\Delta_{RL} + (\gamma_{RL} - \alpha_{RL})(z - C_{RL})}, \\ \mu^\alpha_{RL}(z) &= [\dot{P}^\alpha_{RL}(z)]^{1/2} = \frac{\Delta^{1/2}_{RL}}{\Delta_{RL} + (\gamma_{RL} - \alpha_{RL})(z - C_{RL})}. \end{aligned} \tag{9}$$

In Eqs. (9), the symbols $\dot{P}^\alpha(z)$ and $\ddot{P}^\alpha(z)$ denote, respectively, the first and second derivative with respect to the variable z. We remark that the expressions containing the potential function $P^\alpha(z)$, Eqs. (6) and (9), are valid within the ASA,[17] while the parametrized forms using potential parameters are consistent with the Hamiltonian H^γ, Eq. (1).

The advantage of the above Green's function formulation lies in the possibility to choose the LMTO representation α[18] which is the most suitable for a particular physical problem. The transformation between two representations involves a re-scaling of the auxiliary Green's function $g^\alpha(z)$ and a trivial change of the potential functions. The only non-trivial change concerns the transformation of the structure constants. This, however, can be done once for a given problem. Clearly, the full Green's function $G(z)$ is independent of the particular choice of the LMTO representation α.

The most important LMTO representations are: (i) the canonical representation with $\alpha \equiv 0$, (ii) the orthogonal representation with $\alpha \equiv \gamma$, and (iii) the tight-binding representation with $\alpha_{Rl} \equiv \beta_l$ given by the site independent but l-dependent particular values,[1,2,3] which lead to the shortest spatial extent of the structure constants S^β. In the orthogonal representation (α=γ) the scaling factors $\lambda^\gamma(z)$ and $\mu^\gamma(z)$ become trivial and the true resolvent $G(z)$ differs from the auxiliary one $g^\gamma(z)$ only by an energy independent re-scaling. The structure constant S^γ depends on the actual values of the potential parameters γ_{Rl}. This leads, in random alloys, to the off-diagonal randomness in the alloy Hamiltonian (1) which cannot be treated within the CPA unless additional approximations are adopted. This undesirable feature can be avoided using either the canonical or tight-binding representations ($\alpha = 0$ or β). In this way a complete separation of the structural and atom-dependent parts can be achieved. The problem of the chemical disorder can be then reduced to the configurational averaging of the auxiliary resolvent $g^\alpha(z)$ by means of the CPA since the off-diagonal structure matrix S^{α} is non-random and the random quantities $P^{\alpha}(z)$ are site-diagonal.

The detailed discussion of the CPA approach will be given in Sec. 3, but the convenience of the Green's function approach can be illustrated in the case of an isolated substitutional impurity.[17] We neglect for simplicity the lattice relaxations and the change of potential parameters in the host near the impurity. The auxiliary Green's function $g^{\alpha}(z)$ for this problem satisfies the Dyson equation with the perturbation confined to the impurity site, which can be solved easily by inverting the local Green's function matrix. On the other hand, the perturbation in the Hamiltonian (1) extends, in principle, over the whole crystal due to the extended nature of the structure constants S^{γ}.

The main advantage of the tight-binding representation over the canonical one for the case of random substitutional alloys is in the numerical feasibility of the evaluation of the Fourier transform of the structure constants. The elements $S^{0}_{RL,R'L'}$ of the canonical structure constant behave like $d^{-(l+l'+1)}$ (where $d = |R-R'|$) as contrasted with the exponential-like decay of the tight-binding structure constants in the real space.[18] Consequently, the lattice sums must be done using the Ewald technique in the canonical representation. In the tight-binding representation, the lattice sums extend essentially up to the second nearest-neighbor shell for close-packed lattices and are easy to perform.

A profound difference between the canonical and tight-binding representations occurs in the surface-related problems.[5,9,10,14,15] The short-range character of the tight-binding structure constants allows to view the semi-infinite solid as a stack of the so-called principal layers with only nearest-neighbor interactions between them (see Sec. 4). As the principal layer contains only a few atomic layers, and as the electronic structure changes are usually confined to a few top surface layers, the problem can be described by the finite matrices. On the contrary, the large spatial extent of the canonical structure constants requires the use of specific, numerically highly involved, techniques.[19]

In the LMTO method, it is customary to use s-, p-, and d-orbitals at each lattice site. The use of all orbitals is sometimes undesirable from the interpretational and computational points of view. A minimal basis set of the physically relevant orbitals suffices for an accurate description provided that remaining states are not neglected but included approximately by means of the Lőwdin downfolding technique.[1,11] We now briefly describe this approach in the language of the Green's functions and LMTO representations. We divide the set of RL-orbitals into two subsets, the lower ($\mathcal{L}$) and the higher ($\mathcal{H}$), according to the values of potential parameters. This division depends on the type of atoms, the lattice in question, and on the energy range of interest. In the energy region of interest the potential functions of $\mathcal{H}$-orbitals should acquire large magnitudes as the corresponding $\mathcal{H}$-resonances are located far away. This forms a basis for a simplified linearization scheme which allows to determine the electronic structure of a solid from the $\mathcal{LL}$ block of the auxiliary resolvent $g^{\alpha}(z)$. Here and below we use the block-matrix notation: $\mathcal{L}$ and $\mathcal{H}$ denotes the group of orbitals with $(R, L) \in \mathcal{L}, \mathcal{H}$, respectively. We express $G(z)$ and

$g^\alpha(z)$ in the following, still exact form, using Löwdin downfolding

$$\begin{aligned}
G_{SS'}(z) &= \lambda_S^\alpha(z)\,\delta_{SS'} + \mu_S^\alpha(z)\,g_{SS'}^\alpha(z)\,\mu_{S'}^\alpha(z), \quad (S,S'=\mathcal{L} \text{ or } \mathcal{H}),\\
g_{\mathcal{LL}}^\alpha(z) &= [P_\mathcal{L}^\alpha(z) - S_{\mathcal{LL}}^\alpha - S_{\mathcal{LH}}^\alpha F_{\mathcal{HH}}^\alpha(z) S_{\mathcal{HL}}^\alpha]^{-1},\\
g_{\mathcal{LH}}^\alpha(z) &= g_{\mathcal{LL}}^\alpha(z) S_{\mathcal{LH}}^\alpha F_{\mathcal{HH}}^\alpha(z), \quad g_{\mathcal{HL}}^\alpha(z) = F_{\mathcal{HH}}^\alpha(z) S_{\mathcal{HL}}^\alpha g_{\mathcal{LL}}^\alpha(z),\\
g_{\mathcal{HH}}^\alpha(z) &= F_{\mathcal{HH}}^\alpha(z) + F_{\mathcal{HH}}^\alpha(z) S_{\mathcal{HL}}^\alpha g_{\mathcal{LL}}^\alpha(z) S_{\mathcal{LH}}^\alpha F_{\mathcal{HH}}^\alpha(z),
\end{aligned} \tag{10}$$

where

$$F_{\mathcal{HH}}^\alpha(z) = [P_\mathcal{H}^\alpha(z) - S_{\mathcal{HH}}^\alpha]^{-1}. \tag{11}$$

The physically unrelevant orbitals now enter via the quantity $F_{\mathcal{HH}}^\alpha(z)$, which acts only in the $\mathcal{H}$ subspace. Note that the norm $\|F_{\mathcal{HH}}^\alpha\|$ depends on the LMTO representation $\alpha_\mathcal{H}$ of $\mathcal{H}$-orbitals. The idea is to choose such a representation $\alpha = (\alpha_\mathcal{L}, \alpha_\mathcal{H})$ that $\|P_\mathcal{H}^\alpha(z)\| \gg \|S_{\mathcal{HH}}^\alpha\|$ holds in the energy range of interest centered around an energy ε_0. We thus require $(P_\mathcal{H}^\alpha(\varepsilon_0))^{-1} = 0$, which is the equation determining the unknown $\alpha_\mathcal{H}$ representation. Then it is possible to make an approximation $F_{\mathcal{HH}}^\alpha(z) \approx (P_\mathcal{H}^\alpha(z))^{-1}$ in the vicinity of the energy ε_0. This approximate value of the quantity $F_{\mathcal{HH}}^\alpha(z)$ can be then used in the block-partitioned equations (10). In this way a consistent approximation for each block as well as for the full resolvent $G(z)$ can be obtained. The numerical advantage comes from the fact that instead of inversion in the full $\mathcal{L}+\mathcal{H}$ space, only the inversion in the subspace $\mathcal{L}$ must be performed, while the approximate inversion in the $\mathcal{HH}$ block, Eq. (11), is now a simple diagonal operation. As concerns the LMTO representation for a minimal basis set in the $\mathcal{L}$ subspace, it can be chosen arbitrarily. For example, if the surfaces or random alloys are concerned, the tight-binding LMTO representation $\beta_\mathcal{L}$ is a suitable choice. We shall discuss some specific features connected with the application to random alloys in the next Section.

Here we mention examples of the downfolding technique. In open structures like semiconductors it is common to introduce empty spheres in the interstitial sites with their full sp^3d^5 basis although only the s-orbitals play a significant role for these sites. Consequently, p- and d-orbitals of empty spheres can be downfolded which significantly reduces the dimension of the minimal basis set subspace $\mathcal{L}$.[11] Another example is represented by a substitutional sp-impurity or vacancy in a transition metal host, where the impurity d-orbitals or vacancy p- and d-orbitals can be downfolded. Similarly, the f-orbitals in transition metals can be treated in this way as well as the d-orbitals of elements like Zn, Cd, or Hg near the Fermi level in simple metals or semiconductor alloys. We remark that the downfolding technique is especially efficient for solids with complex lattices, where the dimension of the subspace $\mathcal{L}$ can be much smaller than that of the full space $\mathcal{L}+\mathcal{H}$.

Summarizing, the notion of the physical Green's function $G(z)$ and the auxiliary Green's function $g^\alpha(z)$ expressed in different LMTO representations α leads to a unified framework for the treatment of isolated impurities, concentrated disordered alloys as well as their surfaces. In all these cases, the auxiliary resolvent $g^\alpha(z)$ is the quantity that can be easily evaluated and configurationally averaged or treated via the downfolding. The scaling character of the connection between the auxiliary

Green's function $g^\alpha(z)$ and the physical Green's function $G(z)$ is preserved also after the configurational averaging for the case of random alloys[7,8] and their surfaces.[9] This allows to calculate all physically relevant quantities like the local densities of states or the Bloch spectral densities for bulk alloys (Sec. 3) or surfaces (Sec. 4). The local atom- and layer-dependent Green's functions $G_{RL,RL'}(z)$ are also necessary for electronic charge and spin densities to achieve the complete LDA self-consistency of the theory (Sec. 5).

3. Configurational Averaging

In this Section we shall describe the configurational averaging of the Green's function $G(z)$, Eq. (7), within the CPA and will discuss its basic properties. The CPA is the effective medium theory and its rigorous derivation is based on the multiple-scattering formalism.[6] We present here its simple, intuitive derivation. After the configurational averaging the translational symmetry of a system is recovered. The scattering properties of atoms can be now characterized by the effective coherent potential function $\mathcal{P}^\beta_R(z)$ in a direct analogy with the potential functions $P^\beta_R(z)$ characterizing the scattering properties of the constituent atoms $Q = A$ and B. All calculations will be done in the tight-binding LMTO representation, hence the superscript β. The coherent potential functions are the same in each group $\{R_p\}$ of geometrically and electronically equivalent sites, $\mathcal{P}^\beta_R(z) \equiv \mathcal{P}^\beta_p(z)$, $R \in \{R_p\}$, $p = 1, 2, .., M$. These different groups $\{R_p\}$ refer either to different sublattices of a complex bulk solid, or to different layers in the intermediate region in surface problems. The corresponding configurationally averaged Green's function, or equivalently, the Green's function of effective medium, is

$$\bar{g}^\beta_{RL,R'L'}(z) \equiv \langle g^\beta(z) \rangle_{RL,R'L'} = [(\mathcal{P}^\beta(z) - S^\beta)^{-1}]_{RL,R'L'}. \tag{12}$$

The barred quantity and the symbol $\langle \ldots \rangle$ denote the configurational averaging. The unknown $\mathcal{P}^\beta_p(z)$ can be determined within the single-site approximation in the following manner. We embed an atom of the type Q ($Q = A, B$) on a given site $R \in \{R_p\}$ into the effective medium $\mathcal{P}^\beta$. The scattering from such a single impurity object is described by a single-site T-matrix $t^{\beta,Q}_p(z)$ defined as

$$\begin{aligned} t^{\beta,Q}_{p,LL'}(z) &= \{(P^{\beta,Q}_p(z) - \mathcal{P}^\beta_p(z))[1 + \Phi^\beta_p(z)(P^{\beta,Q}_p(z) - \mathcal{P}^\beta_p(z))]^{-1}\}_{LL'} \\ \Phi^\beta_{p,LL'}(z) &\equiv \bar{g}^\beta_{RR}(z) = \frac{1}{N_p} \sum_{k_p} \{[(\mathcal{P}^\beta(z) - S^\beta(k_p))^{-1}]_{pp}\}_{LL'}. \end{aligned} \tag{13}$$

Here, $P^{\beta,Q}_p(z)$ denotes the potential function of a given atom, Eq. (6), which is generally different on inequivalent sites as a result of the charge self-consistency. It is important to note that a simple form of the scattering matrix, Eq. (13), is due to the single-site character of alloy scattering in the tight-binding LMTO representation, as discussed in Sec. 2. The quantity $\Phi^\beta_p(z)$ is the on-site element of the configurationally averaged auxiliary Green's function $\bar{g}^\beta(z)$, Eq. (12), corresponding to a given subset $\{R_p\}$ of lattice sites. In Eq. (13), N_p is the number of the lattice sites in the subset

$\{R_p\}$, and the sum runs over the Brillouin zone of k-vectors k_p corresponding to this subset. We note that $\Phi_p^\beta(z)$ depends on all inequivalent $\mathcal{P}_q^\beta(z)$, $(q = 1, 2, ..., M)$. The determination of $\Phi_p^\beta(z)$ is of central importance for the CPA. Its evaluation is rather straightforward for the bulk case, but nontrivial for surfaces. This will be discussed in the next Section.

The configurational averaging within the single-site approximation is equivalent to the requirement of vanishing of the resulting scattering. This requirement leads to the following equation for each inequivalent subset $\{R_p\}$

$$\sum_{Q=A,B} c_p^Q \tau_{p,LL'}^{\beta,Q}(z) = 0, \quad (p = 1, 2, ..., M). \tag{14}$$

In Eq. (14), c_p^Q denote the concentrations of atoms Q on a subset of sites $\{R_p\}$ ($c_p^A + c_p^B = 1$ for each p). The conditions (14) are the CPA equations for generally inhomogeneous medium, and its numerical solution will be discussed in Sec. 6.1. As $\Phi_p^\beta(z)$ depends on all inequivalent $\mathcal{P}_q^\beta(z)$, Eqs. (14) represent the set of M coupled CPA equations for unknown $\mathcal{P}_p^\beta(z)$. Once $\mathcal{P}^\beta(z)$ and $\bar{g}^\beta(z)$ are determined, we go over to the physical, orthogonal LMTO representation γ, to evaluate the quantities of interest. To this end, the configurationally averaged $\bar{G}(z)$ must be determined. We introduce the occupation index η_R^Q: $\eta_R^Q = 1$ if an atom of the type Q is at the site R, and $\eta_R^Q = 0$ otherwise. The configurationally averaged physical resolvent $\langle G(z)\rangle = \bar{G}(z)$ can be expressed as[8]

$$\begin{aligned} \bar{G}_{RL,R'L'}(z) &= \sum_Q \lambda_{RL}^{\beta,Q}(z)\langle \eta_R^Q \rangle \delta_{RR'}\,\delta_{LL'} + \sum_Q \sum_{Q'} \mu_{RL}^{\beta,Q}(z)\,\bar{g}_{RL,R'L'}^{\beta,QQ'}(z)\,\mu_{R'L'}^{\beta,Q'}(z), \\ \bar{g}_{RL,R'L'}^{\beta,QQ'}(z) &= \langle \eta_R^Q\, g_{RL,R'L'}^\beta(z)\, \eta_{R'}^{Q'} \rangle. \end{aligned} \tag{15}$$

The site-diagonal term, proportional to η_R^Q, averages trivially

$$\sum_Q \lambda_{RL}^{\beta,Q}(z)\langle \eta_R^Q \rangle = \sum_Q c_p^Q\, \lambda_{pL}^{\beta,Q}(z) \equiv \langle \lambda_{pL}^\beta(z) \rangle. \tag{16}$$

In Eq. (16), we assumed that $R \in \{R_p\}$. The evaluation of $\bar{g}^{\beta,QQ'}(z)$ is non-trivial, but it is facilitated by the multiplicative character of quantities η_R^Q in this product. The derivation can be performed using the formalism developed in Appendix of Ref. 8. Assuming that $R \in \{R_p\}$ and $R' \in \{R_q\}$, the final result with an explicit separation of site diagonal and site off-diagonal terms can be summarized as [1]

$$\begin{aligned} \bar{g}_{RL,R'L'}^{\beta,QQ'}(z) &= c_p^Q\,[\bar{g}_{RR}^\beta(z)\, f_p^{\beta,Q}(z)]_{LL'}\,\delta_{RR'}\,\delta_{QQ'} \\ &\quad + c_p^Q c_q^{Q'}\,[\tilde{f}_p^{\beta,Q}(z)\,\bar{g}_{RR'}^\beta(z)\, f_q^{\beta,Q'}(z)]_{LL'}\,(1-\delta_{RR'}), \end{aligned}$$

$$\begin{aligned} f_{q,LL'}^{\beta,Q}(z) &= [\{1 + (P_q^{\beta,Q}(z) - \mathcal{P}_q^\beta(z))\,\Phi_q^\beta(z)\}^{-1}]_{LL'}, \\ \tilde{f}_{p,LL'}^{\beta,Q}(z) &= [\{1 + \Phi_p^\beta(z)\,(P_p^{\beta,Q}(z) - \mathcal{P}_p^\beta(z))\}^{-1}]_{LL'}. \end{aligned} \tag{17}$$

[1]The scaling factors $f^A(z), f^B(z)$ can be expressed differently, in a direct relation to notation used in Ref. 8 as: $c^A f^A(z) = -[P^B(z) - \mathcal{P}(z)]\,(\Delta P(z))^{-1}$, and $c^B f^B(z) = [P^A(z) - \mathcal{P}(z)]\,(\Delta P(z))^{-1}$. Here, $\Delta P(z) = P^A(z) - P^B(z)$. Similar relations hold for transposed $\tilde{f}^Q(z)$. The subscripts and superscripts were omitted.

In Eq. (17) and below, the symbol $\tilde{X}$ denotes the transposed matrix to the matrix X. The site-diagonal scaling factors $f_p^{\beta,Q}(z)$ also enter the expressions for the single-site T-matrix, Eq. (13), and the local conditionally averaged resolvent (see Eq. (22) below). We note that quantities $\bar{g}^{\beta,QQ'}(z)$ are all expressed solely via $\bar{g}^{\beta}(z)$, Eq. (12). Combining Eqs. (15 - 17), and assuming again that $R \in \{R_p\}$ and $R' \in \{R_q\}$ we can derive for the physical configurationally averaged resolvent $\bar{G}(z)$ the following expression

$$
\begin{aligned}
\bar{G}_{RL,R'L'}(z) &= \{\Lambda_p^{\beta}(z)\,\delta_{RR'} + \tilde{M}_p^{\beta}(z)\,\bar{g}_{RR'}^{\beta}(z)\,M_q^{\beta}(z)\}_{LL'}, \\
\Lambda_{p,LL'}^{\beta}(z) &= \langle \lambda_{pL}^{\beta}(z)\rangle\,\delta_{LL'} \\
&\quad + \sum_{Q\neq Q'} c_p^Q c_p^{Q'} \left[\mu_p^{\beta,Q}(z)\,\tilde{f}_p^{\beta,Q}(z)\,\Phi_p^{\beta}(z)\,f_p^{\beta,Q'}(z)\,(\,\mu_p^{\beta,Q}(z) - \mu_p^{\beta,Q'}(z))\right]_{LL'}, \\
M_{q,LL'}^{\beta}(z) &= \sum_{Q} c_q^Q\, f_{q,LL'}^{\beta,Q}(z)\,\mu_{qL'}^{\beta,Q}(z), \\
\tilde{M}_{p,LL'}^{\beta}(z) &= \sum_{Q} c_p^Q\, \mu_{pL}^{\beta,Q}(z)\,\tilde{f}_{p,LL'}^{\beta,Q}(z).
\end{aligned}
\qquad (18)
$$

Eqs. (15 - 18) are the main result of this Section. They give the expression for the full configurationally averaged Green's function of disordered bulk alloy and its surface, from which the physical quantities of interest can be obtained. This result is as well attractive from the formal point of view, since the configurationally averaged Green's function $\langle G(z)\rangle$, Eq. (18), has the same scaling structure as the original Green's function $G(z)$, Eq. (7). This formal analogy goes even further, as it is possible to put the expression for $\langle G(z)\rangle$ into another, physically transparent form[8]

$$
\begin{aligned}
\langle G(z)\rangle_{RL,R'L'} &= [(z - H^{eff}(z))^{-1}]_{RL,R'L'}, \\
H_{RL,R'L'}^{eff}(z) &= [\tilde{C}_p(z)\,\delta_{RR'} + \tilde{\Delta}_p^{1/2}(z)\,S_{RR'}^{\gamma}(z)\,\tilde{\Delta}_q^{1/2}(z)]_{LL'}, \\
S_{RL,R'L'}^{\gamma}(z) &= [S^0\,(1 - \tilde{\gamma}(z)\,S^0)^{-1}]_{RL,R'L'}.
\end{aligned}
\qquad (19)
$$

We assumed again that $R \in \{R_p\}$ and $R' \in \{R_q\}$. The site-diagonal quantities $\tilde{C}_p(z)$, $\tilde{\Delta}_p$, and $\tilde{\gamma}_p(z)$ are functions of $\mathcal{P}_p^{\beta}(z)$, $\Lambda_p^{\beta}(z)$, and $M_p^{\beta}(z)$ (see Ref. 8 for details). The energy-dependent translationally invariant effective Hamiltonian $H_{RL,R'L'}^{eff}(z)$ represents just the intuitively expected result of the configurational averaging applied to the Hamiltonian (1). The random potential parameters C_{RL}, Δ_{RL}, and γ_{RL} are substituted on a given subset $\{R_p\}$ by their coherent-potential counterparts: $\tilde{C}_{p,LL'}(z)$, the effective atomic levels; $\tilde{\Delta}_{p,LL'}(z)$, the effective bandwidths; and $\tilde{\gamma}_{p,LL'}(z)$, the effective band distortions. The common tight-binding CPA theories[6] are able to treat correctly only the diagonal disorder (different atomic levels). The off-diagonal disorder (different bandwidths and bandshapes) can be treated within these theories only if one adopts further approximations, which may introduce uncontrollable errors.

A detailed information on behavior of electrons in random solids can be obtained from their self-energy $\Sigma(z)$. Its definition is rather straightforward for alloys with the diagonal disorder.[6] We generalize here this notion to the case of Hamiltonian (1) in the following manner:

$$\langle G(z) \rangle = [z - H^{ref} - \Sigma(z)]^{-1}, \tag{20}$$

where H^{ref} is a reference Hamiltonian with respect to which $\Sigma(z)$ is defined, e.g. the concentration-weighted average of the Hamiltonian (1). The self-energy $\Sigma(z)$ is then obtained from Eqs. (19,20) as

$$\Sigma_{RL,R'L'}(z) = H^{eff}_{RL,R'L'}(z) - H^{ref}_{RL,R'L'}. \tag{21}$$

Contrary to simple empirical tight-binding CPA theories,[6] the self-energy $\Sigma(z)$ is, generally, off-diagonal with respect to the site indices as the Hamiltonian (1) exhibits the off-diagonal randomness. Alternatively, its Bloch transform $\Sigma(\mathbf{k}, z)$ is a $\mathbf{k}$-dependent quantity.

To get charge densities, the local quantities related to the conditionally averaged physical Green's function are necessary. For this reason we give below the expression for the atom- and orbital-resolved density of states matrix $D^Q_{p,LL'}(E)$ for sites $R \in \{R_p\}$

$$\begin{aligned}
D^Q_{p,LL'}(E) &= -\frac{1}{\pi}\,\mathrm{Im}\,\mathcal{D}^Q_{p,LL'}(E + i0), \\
\mathcal{D}^Q_{p,LL'}(z) &= \lambda^{\beta,Q}_{pL}(z)\,\delta_{LL'} + [\dot{P}^{\beta,Q}_{pL}(z)]^{1/2}\,\Phi^{\beta,Q}_{p,LL'}(z)\,[\dot{P}^{\beta,Q}_{p'L}(z)]^{1/2}, \\
\Phi^{\beta,Q}_{p,LL'}(z) &= \left\{\Phi^{\beta}_{p}(z)\,\left[1 + (P^{\beta,Q}_{p}(z) - \mathcal{P}^{\beta}_{p}(z))\,\Phi^{\beta}_{p}(z)\right]^{-1}\right\}_{LL'}.
\end{aligned} \tag{22}$$

The first term in the expression for $\mathcal{D}^Q_p(z)$ cancels the unphysical poles of the potential function $P^{\beta,Q}_p(z)$. The quantities $\mathcal{D}^Q_p(z)$ and $\Phi^{\beta,Q}_p(z)$ are, respectively, the local conditionally averaged physical and auxiliary Green's functions.[6,8,9] The L-diagonal components of $D^Q_p(E)$ enter the expression for the spheridized charge density while the parts off-diagonal in L are needed for the non-spheridized charge density (see Sec. 5).

Similarly, the spectral density $A_p(\mathbf{k}, E)$, or the $\mathbf{k}$-resolved density of states, corresponding to the subset $\{R_p\}$ of lattice points, is related to both site diagonal and site off-diagonal elements of the configurationally averaged physical and auxiliary Green's functions, Eq. (18) and (12), respectively. The result is given by

$$\begin{aligned}
A_p(\mathbf{k}, E) &= -\frac{1}{\pi}\,\mathrm{Im}\sum_L \mathcal{A}_{p,L}(\mathbf{k}, E + i0), \\
\mathcal{A}_{p,L}(\mathbf{k}, z) &= \Lambda^{\beta}_{p,LL}(z) + [\tilde{M}^{\beta}_p(z)\,\bar{g}^{\beta}_{pp}(\mathbf{k}, z)\,M^{\beta}_p(z)]_{LL},
\end{aligned} \tag{23}$$

where $\bar{g}^{\beta}_{pp}(\mathbf{k}, z)$ is the Fourier transform of $\bar{g}^{\beta}_{RR'}(z)$, over the Brillouin zone corresponding to a subset $\{R_p\}$ of lattice points R, R'. The notion of the spectral density substitutes in random alloys and the surface problems the notion of energy bands in the conventional band theory.

Some remarks concerning the downfolding technique and the CPA are now in order. First, the set of downfolded orbitals must be the same for all atoms occupying randomly a given subset $\{R_p\}$ of lattice sites, e.g. it must be the same for Pb and Cd atoms on the cation sublattice in the (Pb,Cd)F_2 semiconductor alloy. This is obvious from the division of the full space into $\mathcal{L}$ and $\mathcal{H}$ blocks. Second, the condition for determination of the LMTO representation of the $\mathcal{H}$-orbitals is generalized to $[\mathcal{P}^{\alpha}_{\mathcal{H}}(\varepsilon_0)]^{-1}=0$ (compare Sec. 2). The coherent potential function $\mathcal{P}^{\alpha}_{\mathcal{H}}(z)$, contrary to the potential function $P^{\alpha}_{\mathcal{H}}(z)$, is known only numerically, after the CPA equations are solved. Fortunately, the $\mathcal{H}$-resonances lie, by definition, well outside the energy region of interest. The inverse of the coherent potential function $\mathcal{P}^{\alpha}_{\mathcal{H}}(z)$ has near ε_0 only a small imaginary part induced by a weak hybridization with the $\mathcal{L}$-orbitals, while its real part approaches the virtual crystal-like limit $\langle[P^{\alpha}_{\mathcal{H}}(z)]^{-1}\rangle = c_A[P^{\alpha,A}_{\mathcal{H}}(z)]^{-1} + c_B[P^{\alpha,B}_{\mathcal{H}}(z)]^{-1}$. The solution of the equation $\langle[P^{\alpha}_{\mathcal{H}}(\varepsilon_0)]^{-1}\rangle = 0$ is $\alpha_{\mathcal{H}} = \langle\gamma_{\mathcal{H}} + \Delta_{\mathcal{H}}/(\varepsilon_0 - C_{\mathcal{H}})\rangle$, expressed only via the potential parameters of the alloy constituents on a subset $\{R_p\} \in \mathcal{H}$. We can write then $F^{\alpha}_{\mathcal{H}\mathcal{H}}(z) \approx (V^{\alpha}_{\mathcal{H}} - z)/\Gamma^{\alpha}_{\mathcal{H}}$, where $V^{\alpha}_{\mathcal{H}}$ and $\Gamma^{\alpha}_{\mathcal{H}}$ are the site- and orbital-diagonal matrices expressed via the potential parameters. The final result for the auxiliary resolvent of the minimal basis set orbitals $\langle g^{\alpha}(z)\rangle_{\mathcal{L}\mathcal{L}}$, Eq. (10), is

$$\begin{aligned} \langle g^{\alpha}(z)\rangle_{\mathcal{L}\mathcal{L}} &= [\mathcal{P}^{\alpha}_{\mathcal{L}}(z) - S^{\alpha,eff}_{\mathcal{L}\mathcal{L}}(z)]^{-1}, \\ S^{\alpha,eff}_{\mathcal{L}\mathcal{L}}(z) &= S^{\alpha}_{\mathcal{L}\mathcal{L}} + S^{\alpha}_{\mathcal{L}\mathcal{H}}\,[(V^{\alpha}_{\mathcal{H}} - z)/\Gamma^{\alpha}_{\mathcal{H}})]\,S^{\alpha}_{\mathcal{H}\mathcal{L}}. \end{aligned} \tag{24}$$

This result has a simple physical interpretation for the case of ordered solids: in the original LMTO representation γ, the terms $S^{\alpha}_{\mathcal{L}\mathcal{H}}\,(V^{\alpha}_{\mathcal{H}}/\Gamma^{\alpha}_{\mathcal{H}})\,S^{\alpha}_{\mathcal{H}\mathcal{L}}$ and $S^{\alpha}_{\mathcal{L}\mathcal{H}}\,(1/\Gamma^{\alpha}_{\mathcal{H}})\,S^{\alpha}_{\mathcal{H}\mathcal{L}}$ give, respectively, corrections to the Hamiltonian and overlap matrices of the truncated minimal basis space $\mathcal{L}$. The corrections to the overlap matrix are decisive for accurate band structures obtained from the minimal basis set orbitals. We just note that the bands can be obtained from the poles of the related auxiliary Green's function $\langle g^{\alpha}(z)\rangle_{\mathcal{L}\mathcal{L}}$, Eq. (24). In conclusion of this paragraph we give one specific example of the downfolding.[11] The conduction and valence states around the common gap in random $Cd_{1-x}Hg_xTe$ zincblende semiconductor alloy are well described by $\mathcal{L} =$ M(sp^3)Te(p^3)E_1(s)E_2(s), where M=Cd or Hg, and E_1(s), E_2(s) are, respectively, the s-orbitals at the empty tetrahedral sites. Other examples and the numerical studies can be found in Refs. 20, 21, and 22.

We conclude this Section by some remarks concerning the symmetry of the CPA-related quantities with respect to the orbital index L. We repeat that for spherical potentials, the potential parameters and related quantities λ^{β}, μ^{β}, and potential functions $P^{\beta}(z)$ are diagonal with respect to the orbital momentum index L. On the other hand, all the quantities $\Lambda^{\beta}(z)$, $M^{\beta}(z)$, the coherent potential function $\mathcal{P}^{\beta}(z)$, and the Green's function $\Phi^{\beta}(z)$ have the symmetry of the lattice. They are, therefore, generally non-diagonal matrices with respect to indices L, L'. Specifically, for cubic lattices and $l \le 2$, they are still diagonal with respect to L. For cubic lattices with $l > 2$, as well as for surfaces where the symmetry of the lattice is lowered, they are already non-diagonal matrices with respect to L, L'. Consequently, the group analysis must be done for each specific case. As concerns the self-energy $\Sigma(z)$, Eq. (21), it is generally non-diagonal with respect to L, L' even for cubic lattices and

$l \leq 2$, where $\Phi^{\beta}(z)$ is diagonal.

4. Surface Green's Functions

In this Section we will discuss the evaluation of $\bar{g}^{\beta}_{pp}(z)$, Eq. (12), for the case of surfaces of random alloys[9] or random overlayers on non-random substrates.[23,10] For the surface-related problems, the subset $\{R_p\}$ of structurally and electronically equivalent sites is simply a layer of atoms parallel to the solid surface. We shall start with the definition ot the principal layer[4] (PL) mentioned in Sec. 1. The semi-infinite solid can be partitioned into PLs such that only nearest-neighbor PLs are coupled by the structure constants. A principal layer can include one or more atomic layers depending on the solid face, the lattice type, and the spatial extent of the structure constants S^{α}. It defines the characteristic dimension D of the problem, $D=n(l_{max}+1)^2$, where n is the number of atoms in the two-dimensional primitive cell of the PL and l_{max} is the maximal angular momentum. The advantage of the tight-binding representation, with the structure constants S^{β} having the fastest decay with respect to interatomic distances, is thus obvious. We shall limit ourselves to the simplest case, when the PL consists of one atomic layer with one atom per primitive cell. This includes a number of important low-index surfaces, e.g. the fcc(001) and fcc(111) faces within the first nearest-neighbor terms in S^{β}, or bcc(110) face within the first and second nearest-neighbor terms in S^{β}. The generalization to more complex PLs can be also made.[24] We can further simplify the problem by neglecting possible outward or inward relaxations of surface layers, as well as lattice relaxations due to different sizes of alloy constituents. Accordingly, $S^{\beta}_{RL,R'L'}$ has the block tri-diagonal form with respect to layer indices p and q, $R \in \{R_p\}, R' \in \{R_q\}$. Employing further the translational symmetry parallel to the surface, one gets

$$S^{\beta}_{pp}(\mathbf{k}_{\parallel}) = S^{\beta}_{00}(\mathbf{k}_{\parallel}), \quad S^{\beta}_{pq}(\mathbf{k}_{\parallel}) = S^{\beta}_{01}(\mathbf{k}_{\parallel})\,\delta_{p+1,q} + S^{\beta}_{10}(\mathbf{k}_{\parallel})\,\delta_{p-1,q}, \tag{25}$$

where

$$S^{\beta}_{pq}(\mathbf{k}_{\parallel}) = \sum_{R\in\{R_{pq}\}} \exp(i\mathbf{k}_{\parallel}\cdot R)\, S^{\beta}(R). \tag{26}$$

Here, $\mathbf{k}_{\parallel}$ is a vector from the surface Brillouin zone (SBZ), and the symbol R_{pq} denotes the set of vectors that connect one lattice site in the p-th layer with all lattice sites in the q-th layer. In Eqs. (25, 26) we made use of the fact that the bulk structure constants S^{β} depend only on the difference vector $R = R_p - R_q$.

The coherent potential function $\mathcal{P}^{\beta}(z)$ is the site-diagonal matrix, which is layer-dependent due to the charge self-consistency and possible concentration inhomogeneities

$$\mathcal{P}^{\beta}(z) = \begin{cases} \mathcal{P}^{\beta,v}(z) & \text{for layers in the vacuum region,} \\ \mathcal{P}^{\beta}_{p}(z) & \text{for layers in the intermediate region } (p = 1, 2, ..., M), \\ \mathcal{P}^{\beta,a}(z) & \text{for layers in the bulk alloy region.} \end{cases} \tag{27}$$

The subscripts v, and a refer to the vacuum and bulk alloy regions, respectively, as discussed in Sec. 1. The values $\mathcal{P}^{\beta}_{p}(z)$ are obtained from the solution of the set of coupled CPA equations (13,14) for M layers in the intermediate region. On the other hand, the quantities $\mathcal{P}^{\beta,v}(z)$ and $\mathcal{P}^{\beta,a}(z)$ are found from the separate calculations for corresponding infinite systems, namely for the homogeneous vacuum represented by empty spheres, or the homogeneous bulk alloy. In the latter case, the value $\mathcal{P}^{\beta,a}(z)$ is obtained from the charge self-consistent bulk LMTO-CPA theory (see Sec. 5).

The configurationally averaged Green's function matrix $\bar{g}^{\beta}_{pq}(k_{\parallel},z)$, which is the Fourier transform of $\bar{g}^{\beta}_{RR'}(z)$, Eq. (12), $R \in \{R_p\}$, $R' \in \{R_q\}$, is represented by an inverted infinite tridiagonal matrix (similarly as the structure constant $S^{\beta}_{pq}(k_{\parallel})$) with respect to the PL indices p,q. The effect of the homogeneous semi-infinite systems (bulk alloy or vacuum) with the physical properties identical for each layer can be characterized by a single quantity, the surface Green's function (SGF).[4,5] By definition, the SGF is the top PL projection of the Green's function of the homogeneous semi-infinite bulk alloy or vacuum, and its knowledge is of central importance for the present formalism. The SGF can be determined directly in a real space by the technique developed in Refs. 26 and 5, which avoids the use of the bulk Green's function and related integration over the surface-normal component $k_{\perp}$ of the k-vector in the Brillouin zone, common to other approaches.[14,15,27,28] The basic idea is nearly self-evident: by adding (removing) a PL of atoms to (from) the surface we recover the same semi-infinite solid. The above condition is easily expressed mathematically, and one obtains for the SGFs of the bulk alloy $\mathcal{G}^{\beta,a}(k_{\parallel},z)$ and of the vacuum $\mathcal{G}^{\beta,v}(k_{\parallel},z)$ the following equations

$$\begin{aligned} \mathcal{G}^{\beta,a}(k_{\parallel},z) &= (\mathcal{P}^{\beta,a}(z) - S^{\beta}_{00}(k_{\parallel}) - S^{\beta}_{01}(k_{\parallel})\,\mathcal{G}^{\beta,a}(k_{\parallel},z)\,S^{\beta}_{10}(k_{\parallel}))^{-1}, \\ \mathcal{G}^{\beta,v}(k_{\parallel},z) &= (\mathcal{P}^{\beta,v}(z) - S^{\beta}_{00}(k_{\parallel}) - S^{\beta}_{10}(k_{\parallel})\,\mathcal{G}^{\beta,v}(k_{\parallel},z)\,S^{\beta}_{01}(k_{\parallel}))^{-1}, \end{aligned} \tag{28}$$

which have to be solved self-consistently for each $k_{\parallel}$ and energy z. These SGFs provide the necessary coupling of the intermediate region to the semi-infinite bulk alloy and vacuum. The equations for the SGFs (28) have a simple physical interpretation: the first two terms in the brackets of Eqs. (28) describe the inverse Green's function of the isolated layer of atoms, $[\mathcal{G}^{\beta,l}(k_{\parallel},z)]^{-1} = \mathcal{P}^{\beta}(z) - S^{\beta}_{00}(k_{\parallel})$, which is coupled to the semi-infinite solids characterized by corresponding SGFs, $\mathcal{G}^{\beta,a}(z)$ and $\mathcal{G}^{\beta,v}(z)$. This coupling is mediated by the interlayer structure constants S^{β}_{01} and S^{β}_{10}. The resulting semi-infinite solid is the same because it consists of an infinite number of PLs. The numerical solution of Eqs. (28) will be discussed in Sec. 6.3.

For the (p,q) block $(1 \le p,q \le M)$ of the inverse configurationally averaged Green's function $\bar{g}^{\beta}(k_{\parallel},z)$ one thus gets

$$\{(\bar{g}^{\beta}(k_{\parallel},z))^{-1}\}_{pq} = \{\mathcal{P}^{\beta}_{p}(z) - S^{\beta}_{00}(k_{\parallel}) - \Gamma^{\beta}_{p}(k_{\parallel},z)\}\,\delta_{pq} - S^{\beta}_{01}(k_{\parallel})\,\delta_{p+1,q} - S^{\beta}_{10}(k_{\parallel})\,\delta_{p-1,q}, \tag{29}$$

where

$$\begin{aligned} \Gamma^{\beta}_{1}(k_{\parallel},z) &= S^{\beta}_{10}(k_{\parallel})\,\mathcal{G}^{\beta,v}(k_{\parallel},z)\,S^{\beta}_{01}(k_{\parallel}), \\ \Gamma^{\beta}_{p}(k_{\parallel},z) &= 0 \qquad \text{for } p = 2,3,\ldots,M-1, \\ \Gamma^{\beta}_{M}(k_{\parallel},z) &= S^{\beta}_{01}(k_{\parallel})\,\mathcal{G}^{\beta,a}(k_{\parallel},z)\,S^{\beta}_{10}(k_{\parallel}). \end{aligned} \tag{30}$$

The quantities Γ^{β}_{1} and Γ^{β}_{M} have thus the meaning of embedding potentials that couple the intermediate region to the vacuum region and to the bulk alloy, respectively. In other words, the concept of the SGF allows one to reduce the original problem of the infinite order in the PL indices to an effective problem of the finite order M in the PL indices.

The desired quantity $\Phi_p^\beta(z)$, Eq. (13), which enters both the CPA equations (14) and the expression for the density of states matrix, Eq. (22), needed for the charge self-consistency within the LDA, is obtained by integrating over the SBZ of the (p,p) block of $\bar{g}^\beta(\boldsymbol{k}_\parallel, z)$, namely

$$\Phi_p^\beta(z) = \frac{1}{N_\parallel} \sum_{\boldsymbol{k}_\parallel} \bar{g}_{pp}^\beta(\boldsymbol{k}_\parallel, z). \tag{31}$$

In this equation, $N_\parallel$ is the number of atoms in a given layer. Due to the block tri-diagonal structure of $\{\bar{g}^\beta(\boldsymbol{k}_\parallel, z)\}^{-1}$, Eq. (29), the explicit evaluation of the block diagonal parts of $\bar{g}^\beta(\boldsymbol{k}_\parallel, z)$ is possible using the partitioning technique.[25] The result is[9,23]

$$\bar{g}_{pp}^\beta(\boldsymbol{k}_\parallel, z) = [\mathcal{P}_p^\beta(z) - S_{00}^\beta(\boldsymbol{k}_\parallel) - \Pi_p^\beta(\boldsymbol{k}_\parallel, z) - \tilde{\Pi}_p^\beta(\boldsymbol{k}_\parallel, z)]^{-1}, \tag{32}$$

where

$$\begin{aligned} \Pi_p^\beta(\boldsymbol{k}_\parallel, z) &= S_{01}^\beta(\boldsymbol{k}_\parallel)\,[\mathcal{P}_{p+1}^\beta(z) - S_{00}^\beta(\boldsymbol{k}_\parallel) - \Pi_{p+1}^\beta(\boldsymbol{k}_\parallel, z)]^{-1}\, S_{10}^\beta(\boldsymbol{k}_\parallel), \\ \tilde{\Pi}_p^\beta(\boldsymbol{k}_\parallel, z) &= S_{10}^\beta(\boldsymbol{k}_\parallel)\,[\mathcal{P}_{p-1}^\beta(z) - S_{00}^\beta(\boldsymbol{k}_\parallel) - \tilde{\Pi}_{p-1}^\beta(\boldsymbol{k}_\parallel, z)]^{-1}\, S_{01}^\beta(\boldsymbol{k}_\parallel). \end{aligned} \tag{33}$$

The set of above recursive equations for unknowns Π_p^β and $\tilde{\Pi}_p^\beta$ is terminated by the following conditions at the boundaries of the intermediate region

$$\begin{aligned} \Pi_M^\beta(\boldsymbol{k}_\parallel, z) &= S_{01}^\beta(\boldsymbol{k}_\parallel)\, \mathcal{G}^{\beta,a}(\boldsymbol{k}_\parallel, z)\, S_{10}^\beta(\boldsymbol{k}_\parallel), \\ \tilde{\Pi}_1^\beta(\boldsymbol{k}_\parallel, z) &= S_{10}^\beta(\boldsymbol{k}_\parallel)\, \mathcal{G}^{\beta,v}(\boldsymbol{k}_\parallel, z)\, S_{10}^\beta(\boldsymbol{k}_\parallel), \end{aligned} \tag{34}$$

which are nothing else than the embedding potentials Γ_M^β and Γ_1^β of the intermediate region, Eqs. (30).

In conclusion of this Section, we give another example illustrating a deep connection between the Green's functions of bulk and its surface. Namely, we determine the Green's function of the bulk layer $\mathcal{G}^{\beta,b}(z)$, i.e. the layer situated infinitely deep below the solid surface, from the SGF. We couple via S_{01}^β and S_{10}^β the isolated layer of bulk alloy atoms to the ideal 'left' and 'right' semi-infinite alloy subspaces characterized by the corresponding SGFs. The result is, of course, the infinite bulk alloy. Formally, it holds

$$\mathcal{G}^{\beta,b}(\boldsymbol{k}_\parallel, z) = [\mathcal{P}^{\beta,a}(z) - S_{00}^\beta(\boldsymbol{k}_\parallel) - S_{01}^\beta(\boldsymbol{k}_\parallel)\mathcal{G}^{\beta,a}(\boldsymbol{k}_\parallel, z)\, S_{10}^\beta(\boldsymbol{k}_\parallel) - S_{10}^\beta(\boldsymbol{k}_\parallel)\tilde{\mathcal{G}}^{\beta,a}(\boldsymbol{k}_\parallel, z)\, S_{01}^\beta(\boldsymbol{k}_\parallel)]^{-1}. \tag{35}$$

Here, $\tilde{\mathcal{G}}^{\beta,a}(\boldsymbol{k}_\parallel, z)$ is the SGF of the conjugated semi-infinite subspace of the random alloy (compare Eq. (28)). In this way, the maximum internal consistency for the bulk and surface calculations can be obtained, which is of importance when charge self-consistent calculations are performed.

5. Charge Self-Consistency

The all-electron charge density within the individual atomic sphere centered at the site $R \in \{R_p\}$ and occupied by an atom of the type Q can be evaluated according to

$$\varrho_p^Q(r) = \sum_{LL'} \int_{-\infty}^{E_F} \psi_{pL}^Q(r,E)\, D_{p,LL'}^Q(E)\, \psi_{pL'}^Q(r,E)\, dE \;+\; \varrho_p^{Q,core}(r). \tag{36}$$

The density of states matrix $D_{p,LL'}^Q(E)$ is given by Eq. (22), while the wave functions $\psi_{pL}^Q(r,E) = R_{pl}^Q(|r|,E)Y_L(\hat{r})$ are found from the radial solutions of the Schrődinger equation $R_{pl}^Q(|r|,E)$ (normalized to unity within the corresponding atomic sphere) and from the real spherical harmonics $Y_L(\hat{r})$. The second term in Eq. (36) is the spherically symmetric density due to the core electrons. The spherically averaged charge density can be obtained by a simpler formula

$$\varrho_p^Q(|r|) = \frac{1}{4\pi} \sum_{L} \int_{-\infty}^{E_F} D_{p,LL}^Q(E)\, [R_{pl}^Q(|r|,E)]^2\, dE \;+\; \varrho_p^{Q,core}(|r|), \tag{37}$$

containing the L-diagonal elements $D_{p,LL}^Q(E)$ of the local density of states matrix, Eq. (22).

The wave functions in Eqs. (36, 37) obey the Schrődinger equation with the effective one-electron potential given by

$$V_p^Q(|r|) = -\frac{2Z^Q}{|r|} + V_p^{Q,H}(\varrho_p^Q(|r|)) + V_p^{Q,xc}(\varrho_p^Q(|r|)) + V_p^{Mad}, \tag{38}$$

which depends both on the site $R \in \{R_p\}$ and on the type of atom Q. The first three terms in Eq. (38) have the standard form and meaning common to all LDA calculations using spherical charge density (see e.g. Refs. 1, 2, 16, and 18). The first term describes the Coulomb attraction to the nucleus with the charge Z^Q, the second term is the Hartree potential due to the spheridized electronic charge density $\varrho_p^Q(|r|)$, and the third term is the exchange-correlation contribution. The last term in Eq. (38), the Madelung term, describes the spherically averaged parts of the Hartree potential coming from the full, non-spheridized charge density $\varrho_p^Q(r)$ in other atomic spheres. Quite generally, this term can be expressed as

$$V_p^{Mad} = \sum_{L} \sum_{q} M_{pq}^{sL}\, \bar{Q}_q^L, \tag{39}$$

where

$$\bar{Q}_q^L = \sum_{Q} c_q^Q \left\{ \frac{\sqrt{4\pi}}{2l+1} \int_0^{s^Q} |r|^l\, Y_L(\hat{r})\, \varrho_q^Q(r)\, dr - Z^Q \delta_{l,0} \right\}. \tag{40}$$

We note that the Madelung term at a site $R \in \{R_p\}$ is a linear combination of the configurationally averaged multipole moments $\bar{Q}_q^L$ at all nonequivalent sites $R' \in \{R_q\}$. The quantities M_{pq}^{sL} are called Madelung constants. In the following we discuss the meaning of the Madelung term in different cases.

In a bulk system consisting of several sublattices with one or more sublattices occupied randomly (e.g. in $Cd_{1-x}Hg_xTe$ quasibinary alloys), the averaged net charges $\bar{Q}_p^{l=0}$ will be nonzero thus describing a possible charge transfer among different sublattices with different electronic, magnetic, and chemical nature. Since the configurationally averaged environment of each site usually has a certain degree of the symmetry (e.g. tetrahedral), it can be expected that the higher multipole moments can be neglected and that the monopole-monopole interaction (term $L \equiv s$ $(l = m = 0)$ in Eq. (40)) will play a decisive role.

In bulk systems with all sites equivalent (e.g. in substitutionally disordered fcc or bcc alloys), the above arguments together with the condition of overall charge neutrality lead to the vanishing Madelung term, since the averaged charge transfer at each site is equal to zero. However, one can go a step further and include the effect of the local fluctuations of the electronic charge. The first possibility is to change the atomic sphere radii s^Q so that each atomic sphere is charge neutral while preserving the total volume of the random alloy[12,18] (see also Sec. 6.5). In this way not only the average value of the Madelung term, Eq. (40), is zero but also the local fluctuations around this value are minimized. The absence of the Madelung contribution to the one-electron potentials, Eq. (39), is compensated by the different values s^Q up to which the radial Schrődinger equation is integrated thus leading to modified potential parameters. The second possibility is based on an assumption that in a densely-packed metallic alloy the local charge fluctuations are effectively screened on distances greater than the nearest-neighbor distance.[29] In other words, it is supposed that each site together with its nearest-neighbor shell of atoms form a charge neutral cluster. This leads to the Madelung term of the form

$$V^{Q,Mad} = -\frac{2}{d_1}(N^Q - Z^Q), \tag{41}$$

where N^Q is the the average total number of electrons at the atom of the type Q and d_1 denotes the average nearest-neighbor distance in the random alloy. Note that this Madelung contribution, which accounts for possible local charge fluctuations, is atom-dependent. This should be contrasted with the general form of Eqs. (39, 40) which corresponds to averaged electrostatic interaction among different sublattices (or layers), and is therefore p-dependent. Both approaches can be applied to the case of complex metallic lattices provided that s^Q and $V^{Q,Mad}$, Eq. (41), will depend on the sublattice index p.

In the surface-related problems, the Madelung term must describe not only the charge redistribution among different layers (due to the presence of the surface and possible concentration inhomogeneities), but also the strong deviations from the spherical symmetry of the charge density at the surface. It has been shown[15,10] that the inclusion of the term $L \equiv p_z$ $(l = 1, m = 0)$ (the dipole moment and the quantization axis perpendicular to the surface) is necessary to get results comparable to the full-potential slab calculations. In addition to the Madelung term in Eq. (40), the nonzero net charges and dipole moments in the surface layers lead to the electrostatic potential barrier across the solid-vacuum interface. This quantity (and the closely related work function) is also very sensitive to the proper treatment of the charge nonsphericity.[15,10]

At this stage it is worthwhile mentioning some differences between bulk and surface alloys. The most notable difference is the pinning of the Fermi level at the surface by the underlying bulk. Consequently, the overall charge neutrality cannot be achieved by shifting the Fermi level to accommodate valence electrons as in the bulk case but via a reconstruction of the electronic structure in the spatial region of the solid-vacuum interface (the intermediate region). As a result, as already mentioned, the properties of a top few layers differ from each other as well as from those deeper in the solid (the infinite medium).

6. Computational Implementation

In this Section, we present some details concerning the numerical implementation of the formalism developed in Secs. 3 to 5. We omit completely the discussion of results of numerical calculations, but for the reader's convenience we give below reference to calculations done for particular systems. The bulk properties of transition metal alloys with the fcc lattice structure were studied in Refs. 7, 8, 30, and 31, with the bcc lattice structure in Refs. 8, 30, and 32, and with the hcp lattice structure in Ref. 33. The magnetic properties of transition metal alloys were calculated in Refs. 13, 34, and 35. The properties of random alloys with complex lattice structures (DO_3-based alloys, Heusler-like alloys, and alloys with Cu_3Au structure exhibiting ordering tendencies) were the object of study in Refs. 12, 13, and 36. The semiconductor alloys with the sphalerite,[11] fluorite,[20,21] and rocksalt[22] lattice structures were already mentioned in Sec. 3. Random overlayers of transition metals on the non-random transition metal substrate were studied in Refs. 10 and 23 including the study of surface ordering,[37] while the surfaces of random transition metal alloys were the subject of study in Refs. 9 and 38. The formalism was applied also to the semiconductor surfaces.[24] We refer the reader interested in a specific application to the above papers.

6.1. Solution of the CPA Equations

The CPA equations (13, 14) are a set of coupled nonlinear equations for the coherent potential functions $\mathcal{P}_p^\beta(z)$. The physical solution must preserve the Herglotz property of the Green's function $\Phi^\beta(z)$. The form of CPA equations as expressed by (14) is not suitable for numerical implementation as it does not converge in the strong scattering case. Therefore, we rewrite the system of CPA equations to the numerically more stable form by introducing the notion of the coherent interactor $\Omega_p^\beta(z)$.[8] It is defined with help of the configurationally averaged Green's function $\Phi_p^\beta(z)$, Eq. (13), as follows

$$\begin{aligned}\Phi^\beta_{p,LL'}(z) &= \{[\mathcal{P}_p^\beta(z) - \Omega_p^\beta(z)]^{-1}\}_{LL'} \\ &= \frac{1}{N_p}\sum_{k_p}\{[(\mathcal{P}^\beta(z) - S^\beta(k_p))^{-1}]_{pp}\}_{LL'}.\end{aligned} \tag{42}$$

The physical meaning of $\Omega_p^\beta(z)$ is simple: it describes an energy-dependent coupling of an effective atom in a random alloy, as characterized by the coherent potential function $\mathcal{P}^\beta(z)$, to all other sites. Using the definition (42), the expression for the local conditionally averaged Green's function $\Phi_p^{\beta,Q}(z)$, Eq. (22), becomes surprisingly simple and physically transparent

$$\Phi_{p,LL'}^{\beta,Q}(z) = \{[P_p^{\beta,Q}(z) - \Omega_p^{\beta}(z)]^{-1}\}_{LL'}. \tag{43}$$

It has formally the same structure as the averaged Green's function $\Phi_p^{\beta}(z)$, but the coherent potential function $\mathcal{P}^{\beta}(z)$ is substituted by the potential function $P_p^{\beta,Q}(z)$ of a given atom $Q = A, B$. This result is just in the spirit of the single-site approximation. The CPA equations (14) can be then put into the following form

$$\Phi_{p,LL'}^{\beta}(z) = \sum_Q c_p^Q \, \Phi_{p,LL'}^{\beta,Q}(z). \tag{44}$$

The set of equations (42-44) is suitable for numerical implementation. It is solved by simple iterations starting with the virtual-crystal value of $P_p^{\beta}(z)$ (Eq. (6) with concentration weighted values of species potential parameters) for the coherent potential function $\mathcal{P}^{\beta}(z)$. The above system of CPA equations converges reliably for both bulk and surface cases, and independently for each energy. This 'parallel' feature can be used in future, together with an obvious parallelism inherent to the k-space Brillouin zone (BZ) integration, to significantly speed up calculations. The k-space integration in Eq. (42) takes advantage of energy variables with finite imaginary parts, thereby smoothing out the integrand and speeding up the iteration process. The desired quantities are found by the numerical analytical continuation back to the real axis[39] in the last step of the numerical calculations.

6.2. Fast LMTO-CPA Method for Bulk Alloys

The bottleneck of the charge self-consistent calculations for random alloys is repeated evaluation of $\Phi_p^{\beta}(z)$, Eq. (42), which requires k-integrations over the irreducible part of the bulk BZ. The basic idea how to overcome this difficulty was proposed recently by Akai[40], and employed later on by other authors[41,29] as well by us when writing computer codes. In a given LDA-CPA iteration $(n+1)$, we use the following approximate form of $\Phi_p^{\beta\,(n+1)}(z)$ (we suppress for simplicity of writing the subset index p):

$$\begin{aligned}
\Phi_{LL'}^{\beta\,(n+1)}(z) &= (1 - r_{n+1})\,\Phi_{LL'}^{\beta\,(n)}(z) + r_{n+1}\,\{[\mathcal{P}^{\beta\,(n)}(z) - S^{\beta}(k_{\parallel}^{(n)})]^{-1}\}_{LL'}, \\
\Phi_{LL'}^{\beta\,(0)}(z) &= \frac{1}{N_0}\sum_{i=1}^{N_0}\{[\mathcal{P}^{\beta\,(0)}(z) - S^{\beta}(k^{(i)})]^{-1}\}_{LL'}.
\end{aligned} \tag{45}$$

Here, $\Phi^{\beta\,(n)}(z)$ is the value of $\Phi^{\beta}(z)$ from the previous iteration, and $k^{(n)}$ is a k-vector from the irreducible BZ chosen at random. The value $\mathcal{P}^{\beta\,(n)}(z)$ is updated in each iteration from corresponding CPA equations. The starting value of $\Phi_{LL'}^{\beta}(z)$, the quantity $\Phi_{LL'}^{\beta\,(0)}(z)$, is obtained by summing up a certain number N_0 of k-points in the BZ with the virtual-crystal value $\mathcal{P}^{\beta\,(0)}(z)$ for the coherent potential function (typically $N_0 = 50$ - 60 k-points is used for fcc or bcc lattices). It is obvious from Eq. (45), that the first k-points with unconverged values of the coherent potential function are contained in the BZ sum for $\Phi^{\beta}(z)$ for any other successive iteration, although their influence can be relaxed to some extent by a suitable choice of the mixing factor r_n (the simplest choice $r_n = 1/n$ corresponds to equal weights for each k-point in the final sum[40]). However, for a sufficiently large number of k-points used during the iteration process, Eq. (45), the influence of the 'bad memory' of these first points can be significantly reduced (typically, $N_{BZ} = 500$ - 600 of k-points suffices for an accurate self-consistent solution for fcc- or bcc-lattices). Another important requirement for the success of this method is the choice of the $\{k^{(n)}\}$ mesh. It is obvious from the description of this method that this mesh must be a random set, not a regular mesh as in the conventional BZ integration. Akai[40] proposes to use the so-called Weyl distribution instead of the Monte Carlo choice, because the BZ integral converges to its limiting value as N_{BZ}^{-1} rather than $N_{BZ}^{-1/2}$ as in the Monte Carlo method. All our calculations are done in the complex energy plane and deconvoluted at the end to real axis as mentioned in the previous subsection. The method proves to be very efficient and our experience is that the charge self-consistent electronic structure calculations for random alloys with fcc- or bcc-lattice structures can be performed even on advanced personal computers.

6.3. Solution of the Equation for SGF

The natural way of solving the nonlinear equation (28) for the SGF seems to be a simple iteration of Eq. (28), using the layer Green's function $\mathcal{G}^{\beta\,(0)}(k_\parallel, z) = [(\mathcal{P}^\beta(z) - S^\beta_{00}(k_\parallel)]^{-1}$, as a starting value. The next iteration adds to an initial layer another one, so that after n iterations we have a slab consisting of $(n+1)$ layers. At some $(k_\parallel, z)$ points hundreds of such iterations are needed, particularly near singularities, to obtain the converged value. This problem may be overcome by the renormalization-decimation technique proposed in Ref. 42. The basic idea of this method, which we use to solve Eqs. (28), is the following. We replace the original set of PLs by an effective one with doubled lattice constant, where a given PL and its two nearest neighbor PLs are replaced by an effective layer in a new set. These new layers are coupled to each other by effective energy-dependent interlayer structure constants with weaker interactions than those in the original set. This replacement can be repeated iteratively until the effective interlayer interactions vanish. After n iterations one thus represents a semi-infinite solid by a stack of $(2^n + 1)$ layers. Clearly, this method represents a significant improvement over the simple iteration scheme described at the beginning of this subsection.

It exists, however, a simpler and even more effective solution of Eqs. (28). One can still use a simple iteration procedure, but the starting value $\mathcal{G}^{\beta\,(0)}(k_\parallel, z)$ should be chosen as close to the exact solution as possible. We can use the known solution in the vicinity of the energy point z, namely $\mathcal{G}^{\beta\,(0)}(k_\parallel, z) = \mathcal{G}^\beta(k_\parallel, z - \delta z)$. For the initial energy z_0 outside the band spectrum, where the Green's function is a smooth function without singularities, it is sufficient to use a simple iteration scheme as described at the beginning of this subsection. The calculations are performed in the complex plane and deconvoluted to real axis at the end of calculations. This method in its efficiency supersedes the renormalization-decimation technique, and in addition it is much simpler to program. It converges reliably provided the energy points z and $z-\delta z$ are sufficiently close. In other words, we pay for the simplicity of this method by loosing the 'parallel' property with respect to the energy (not to the $k_\parallel$-vector) which, on the other hand, is preserved for the renormalization-decimation technique.

6.4. Solution of the LDA Equations

The problem of the charge self-consistency is essentially the same for both bulk disordered alloys and their surfaces. The basic quantity for the charge density calculations is the local density of states matrix $D^Q_{p,LL'}(E)$, Eq. (22), which must be calculated using the methods described in Secs. 2 to 4. It is this quantity which contains the information on the mutual positions of atoms in a solid and on their occupations by component atoms Q. On the other hand, the radial wave functions $R^Q_{pl}(|\mathbf{r}|, E)$ for both the valence and core orbitals can be found for each atomic sphere separately. The evaluation of the spheridized electron charge densities $\varrho^Q_p(|\mathbf{r}|)$ according to Eq. (37) therefore represents only a negligible numerical effort compared to the determination of the density of states matrix $D^Q_{p,LL'}(E)$. The charge densities are expressed in terms of the energy-moments $m^{Q,k}_{pL}$ ($k = 0, 1, 2$) of the local atom- and momentum- resolved density of states[1,2,16]

$$m^{Q,k}_{p,L} = \int_{-\infty}^{E_F} (E - E^Q_{\nu pl})^k \, D^Q_{p,LL}(E) \, dE. \tag{46}$$

These moments appear as a result of approximating the energy dependence of the radial wave function $R^Q_{pl}(|\mathbf{r}|, E)$ by three terms of its Taylor expansion around a chosen value $E^Q_{\nu pl}$. The non-spheridized charge density, Eq. (36), need not be evaluated throughout the whole atomic sphere, since the nonspherical parts of the density enter the one-electron potentials (38) only via the multipole moments, Eq. (40), which are integral characteristics. These integrals are expressed using the moments $m^{Q,kk'}_{p,LL'}$ ($k, k' = 0, 1, 2$) of the local density of states matrix, Eq. (22), as

$$m^{Q,kk'}_{p,LL'} = \int_{-\infty}^{E_F} (E - E^Q_{\nu pl})^k \, D^Q_{p,LL'}(E) \, (E - E^Q_{\nu pl'})^{k'} \, dE. \tag{47}$$

The resulting multipole moments are thus reduced to radial integrals up to the Wigner-Seitz radius s^Q and angular integrals over the unit sphere.[15]

The charge density of core electrons $\varrho^{Q,core}_p(|\mathbf{r}|)$ is recalculated at each LDA iteration instead of the usual frozen-core approximation.[15,16] This has only negligible effect on the resulting valence charge densities, but the corresponding changes in the core orbitals and their eigenvalues are of importance for the core-level shifts[10] or hyperfine magnetic fields.

To achieve the self-consistency within the ASA, the effective one-electron potentials (38) are needed. The Madelung constants, due to the long range nature of the electrostatic interactions, must be evaluated by means of the Ewald technique. For complex lattices consisting of several sublattices this is a standard problem.[16] For surfaces, the two-dimensional Ewald techniques must be used.[15,19] Moreover, the problem of the charge neutrality of the finite number of self-consistently treated layers must be overcome. One possibility is to introduce additional energy shifts to potentials in the intermediate region which restores the perfect charge neutrality.[15] Another possibility is to redefine reference points for electrostatic potentials:[10] instead of reference points at infinite distances from the surface, one can choose reference points in the first bulk and vacuum layers which are not treated self-consistently.

The calculation of the potential parameters X_{pL} $(X = C, \Delta, \gamma)$ from the wave functions and their radial derivatives at $|r| = s^Q$ is again a procedure well documented in the literature[16] and is omitted here. The input of these potential parameters into the Green's function problem (Secs. 2 to 4) closes up the loop and makes the theory fully self-consistent in the CPA and LDA sense.

As mentioned in Sec. 5, the Fermi level is a bulk property even for surface problems. Its evaluation requires the knowledge of the total bulk density of states, which is found by integrating over the subset of k-points corresponding to an infinite solid (the bulk Brillouin zone). Similarly, the layer-resolved densities of states needed for the charge self-consistent solution within the LDA are obtained by integrating over the subset of k-points corresponding to a layer (the surface Brillouin zone). The internal consistency between both calculations can be achieved by evaluating the bulk Green's function (and the Fermi level) by integrating the Green's function of a bulk layer, Eq. (35), over the surface Brillouin zone. The Green's function determined in this way then enters the self-consistent TB-LMTO-CPA codes.

6.5. The Effect of Lattice Relaxations

Alloying of atoms leads, in general, to the off-diagonal randomness in the Hamiltonian (1) for random alloys. The off-diagonal randomness has essentially two reasons: (i) the different positions of constituent atoms in the Periodic Table, and (ii) the different sizes of atoms. These effects, in fact, cannot be separated from each other but they should be considered together. From the point of view of the electronic structure, the effect of the off-diagonal randomness manifests itself in different widths and shapes of species bands, and enters the theory via the different values of the potential parameters Δ_{RL} and γ_{RL}. We have seen that such randomness can be included in the CPA without additional approximations using the present formalism. The case of ordered alloys with different sizes of atoms can be treated without problems,[16] as the structure constants $S^{\beta}_{RR'}$ are still related to an ideal lattice. The LMTO method is rather insensitive to the choice of sphere radii s^Q used in the solution of the Schrődinger equation provided that the spheres fill completely the space, do not overlap very much (the criterion can be found in Refs. 1, 2, and 18), and the electrostatic contributions from possible charge transfers are included properly via the Madelung terms in the LDA potentials. This observation forms a basis for an approximate treatment of the charge self-consistency for both ordered[18] and disordered[8,12] alloys. It is based on the choice of the sphere radii s^Q in such a manner that atoms are approximately neutral. These conclusions were verified by comparison with corresponding charge self-consistent calculations for ordered[24,12] and disordered[43] alloys. The flexibility in the choice of atomic radii can be employed to develop a simple, approximate theory describing the influence of lattice relaxations on the electronic structure of random alloys with different sizes of constituent atoms.[8,21,22,44]

The different sizes of atoms cause some structural deformations in random alloys, so that the structure constant $S^{Q}_{RL,R'L'}$ is, in general, random and dependent on the occupation of sites R and R' by atoms Q and Q', respectively. Strictly speaking, the above statement already contains an approximation, since $S^{Q}_{RR'}$ actually depends on the configuration $\{R\}$ of all atoms in a random alloy. But as a first step, this is an acceptable assumption. To account for such distortions approximately, we set

$$S^{0}_{RL,R'L'}(\{R\}) \to S^{0,QQ'}_{RL,R'L'} \approx \lambda^{Q}_{L}\, S^{0}_{RL,R'L'}\, \lambda^{Q'}_{L'}, \tag{48}$$

where $S^{0}_{RL,R'L'}$ is the nonrandom structure constant of a perfect, unrelaxed lattice in question. In deriving Eq. (48), we have neglected possible angle distortions and assumed that they are locally isotropic. The multiplicative form of the random factors λ^{Q}_{L} in Eq. (48) is justified for small distortions, when it holds $(s^{Q}+s^{Q'})/(2w^{alloy}) \approx (s^{Q}s^{Q'})^{1/2}/w^{alloy}$ (w^{alloy} is the Wigner-Seitz radius of the random alloy). Because $S^{0}_{RL,R'L'}$ depends on the distance $d=|R-R'|$ as $(w^{alloy}/d)^{l+l'+1}$, one can choose $\lambda^{Q}_{L} \approx (w^{alloy}/s^{Q})^{l+1/2}$. This choice assumes additional approximation concerning hybridization terms ($l \neq l'$). In fact, we need an approximate form for $S^{\beta}_{RR'}$ rather then for $S^{0}_{RR'}$. But for the d-states, for which the size effect is the most relevant, it holds $S^{\beta}_{RR'} \approx S^{0}_{RR'}$, and above approximations are plausible. By combining the Eqs. (48), (1) and (2), one can find that the description of lattice relaxations can be reformulated as follows: the structure constants remain unchanged, while the potential parameters Δ^{Q}_{L} and γ^{Q}_{L} are rescaled according to $\Delta^{Q}_{L} \to \lambda^{Q}_{L}\Delta^{Q}_{L}\lambda^{Q}_{L}$ and $\gamma^{Q}_{L} \to \lambda^{Q}_{L}\gamma^{Q}_{L}\lambda^{Q}_{L}$. This gives rise to additional off-diagonal disorder, which can be treated by the present formalism as demonstrated before. This result is in the spirit of the tight-binding language: bond length variations are modelled by variations of hopping matrix elements. The above approximate theory has thus very clear and simple physical interpretation. The implementation of this simple approach to random transition metal alloys with different sizes of atoms, like CuPd[45] or CuAu,[12] as well as to semiconductor alloys like $(Cd,Pb)F_2$[21] improved the agreement of theoretical calculations with experiment compared to the theories completely neglecting the effect of lattice relaxations. The results for CuPd and CuAu alloys are in agreement with calculations performed by the modified supercell method[46] (see Ref. 47). It is clear, however, that a more sophisticated theory of lattice relaxations in random alloys should be developed in future.

6.6. Comparison with Other Approaches

In this subsection we shall compare the present method for the evaluation of the electronic structure of disordered alloys and their surfaces with other approaches available in the literature.

We first mention the special quasirandom structures method.[47] It is essentially a supercell method with such an arrangement of atoms that it mimics the first few, physically most relevant radial correlation functions of a random alloy. A reasonably small size of supercells is obtained for compositions like $A_{75}B_{25}$, $A_{25}B_{75}$, and especially for $A_{50}B_{50}$. The method describes reasonably well only local properties of random alloy. On the other hand, it is able to include to some extent the local environment effects, and any standard band structure technique, including the full-potential ones, can be used. Its applicability to complex lattices and random surfaces remains open.

Closely related to our approach for the case of bulk alloys is the Korringa-Kohn-Rostoker (KKR) method combined with the CPA approach[48] (KKR CPA), and its ASA version, formulated recently in the tight-binding representation.[49] A deep internal relation between the KKR and LMTO methods is discussed in Ref. 2. Roughly speaking, the LMTO is a linearized version of the KKR method, which shares advantages of both the fixed basis set methods (the variational principle, the straightforward way how to include various perturbations or how to evaluate the physical quantities of interest, etc.) and of the partial waves methods (the small basis set, the possibility to treat each element of the Periodic Table on the same footing, etc.).

As already mentioned, the applications of the supercell or slab methods to random surfaces is of a limited use. From methods which treat the true semi-infinite solid similarly as done in the present paper, we mention the embedded Green's function method[50], and the self-consistent Green's function methods for ordered semiconductor surfaces[27] and for localized defects on surfaces.[51] The generalization of above methods to random surfaces is not obvious, but they are able to treat atomic potentials with no shape approximations. Closely related to the present approach are the LMTO-SGF method for clean surfaces,[15] and the layer KKR method for ordered[19] and disordered[52] interfaces. Only monopole terms in the Madelung potential were included in Refs. 19 and 52, which is justified for the case of interfaces with a smooth change of atomic potentials across the interface. The straightforward way of determining the SGF in our method as compared with those in Refs. 15, 19, and 52 is a great advantage of the present approach, both numerically and conceptually.

7. Summary and Conclusions

We have given a brief review of an efficient self-consistent Green's function method to calculate properties of disordered alloys and their surfaces. The approach is based on the local density approximation within the all-electron first-principles tight-binding linear muffin-tin orbital method. In this approach both the intralayer and interlayer interactions are short ranged, which greatly simplifies the evaluation of the surface Green's function needed for a proper description of the semi-infinite systems, like surfaces and overlayers. A simple technique was described which allows to evaluate the surface Green's function directly in a real space without the knowledge of the bulk resolvent. The potentials of constituent atoms are treated as spherical ones within the atomic sphere approximation. In the case of surface-related problems, we include both the monopole and dipole terms in the multipole expansion of the charge density, which accounts for its nonspherical shape at the vacuum-solid or solid-solid interface and which yields the correct value of the electrostatic surface barrier, the work function, and of the surface energy. The method makes possible to study simple metals, transition metals, or semiconductors and insulators on the same footing. In the latter case, the concept of empty interstitial spheres to describe electron states of open-structure solids is used together with the downfolding technique to take full advantage of the minimal basis set. The effect of disorder in the bulk or at the surface is included via the coherent potential approximation capable of describing various situations occurring in the experiment: ideal crystal and its surface, completely disordered alloy as well as the case of the partial ordering on one sublattice of a complex lattice, various stages of adsorption ranging from a low coverage limit to the case of adsorbed multilayers, as well as such complex systems as the surface of disordered alloy with a possible segregation of one of alloy component at the surface. The theory can be improved in some aspects, of which we mention the development of the fully-relativistic versions for bulk and surface cases, the inclusion of inward or outward layer shifts at the surface, the extension of the formalism to the spin-polarized case, the effect of electron correlations not accounted properly within the local density approximation, the chemisorption of simple gases on the transition metal alloy surfaces, or the study of more complex structures like interfaces, sandwiches, and superlattices. Further, we want to mention that in conjunction with the generalized perturbation method[53] or with the concentration fluctuation method,[54] the present approach can supply the effective cluster interactions[55,38] needed for the construction of the bulk and surface phase diagrams and the study of the surface segregation from first principles.[55,56] Finally, the present formalism is compatible with a recent treatment of charge and spin fluctuations about the local spin density approximation.[57,58] Specifically, the disordered local moment state above the Curie temperature can be studied systematically not only for the bulk systems but also for their surfaces.

Acknowledgements

The financial support for this work was provided by the Academy of Sciences of Czech Republic (Project No. 11015) and the Austrian Science Foundation (P8918). Two of the authors (J.K. and V.D.) thank Prof. P. Weinberger for creating a good working atmosphere at the Institute of Technical Electrochemistry, Technical University of Vienna, where this review was completed.

References

1. O. K. Andersen, O. Jepsen, and D. Glötzel, in *Highlights of Condensed-Matter Theory*, eds. F. Bassani, F. Fumi, and M. P. Tosi (North-Holland, New York, 1985), p. 59.

2. O. K. Andersen, these *Proceedings.*

3. O. K. Andersen and O. Jepsen, *Phys. Rev. Lett.* **53** (1984) 2571.

4. F. Garcia-Moliner and V. R. Velasco, *Prog. Surf. Sci.* **21** (1986) 93.

5. B. Wenzien, J. Kudrnovský, V. Drchal, and M. Šob, *J. Phys. Condens. Matter* **1** (1989) 9893.

6. B. Velický, S. Kirkpatrick, and H. Ehrenreich, *Phys. Rev.* **175** (1968) 747.

7. J. Kudrnovský, V. Drchal, and J. Mašek, *Phys. Rev.* **B35** (1987) 2487.

8. J. Kudrnovský and V. Drchal, *Phys. Rev.* **B41** (1990) 7515.

9. J. Kudrnovský, P. Weinberger, and V. Drchal, *Phys. Rev.* **B44** (1991) 6410.

10. J. Kudrnovský, I. Turek, V. Drchal, P. Weinberger, N. E. Christensen, and S. K. Bose, *Phys. Rev.* **B46** (1992) 4222.

11. J. Kudrnovský, V. Drchal, M. Šob, N. E. Christensen, and O. K. Andersen, *Phys. Rev.* **B40** (1989) 10029.

12. J. Kudrnovský, S. K. Bose, and O. K. Andersen, *Phys. Rev.* **B43** (1991) 4613.

13. J. Kudrnovský, N. E. Christensen, and O. K. Andersen, *Phys. Rev.* **B43** (1991) 5924.

14. W. Lambrecht and O. K. Andersen, *Surf. Sci.* **178** (1986) 256.

15. H. L. Skriver and N. M. Rosengaard, *Phys. Rev.* **B43** (1991) 9538.

16. H. L. Skriver, *The LMTO Method* (Springer, Berlin, 1984).

17. O. Gunnarsson, O. Jepsen, and O. K. Andersen, *Phys. Rev.* **B27** (1983) 7144.

18. O. K. Andersen, O. Jepsen, and M. Šob, in *Electronic Band Structure and its Applications*, ed. M. Yussouff (Springer, Berlin, 1987), p. 1.

19. J. M. Mac Laren, S. Crampin, D. D. Vvedensky, and J. B. Pendry, *Phys. Rev.* **B40** (1989) 12164.

20. J. Kudrnovský and N. E. Christensen, *Solid State Commun.* **78** (1991) 153.

21. J. Kudrnovský, N. E. Christensen, and J. Mašek, *Phys. Rev.* **B43** (1991) 12597.

22. J. Kudrnovský and N. E. Christensen, *Phys. Rev.* **B43** (1991) 9758.

23. J. Kudrnovský, B. Wenzien, V. Drchal, and P. Weinberger, *Phys. Rev.* **B44** (1991) 4068.

24. M. Šob and J. Kudrnovský, in *MRS Symposia Proceedings* **209**: *Defects in Materials*, eds. P. D. Bristowe, J. E. Epperson, J. E. Griffith, and Z. Liliental-Weber (Materials Research Society, Pittsburgh, 1991), p. 177.

25. P. O. Lőwdin, *J. Chem. Phys.* **19** (1951) 1396.

26. J. Kudrnovský and V. Drchal, in *Studies in Surface Science and Catalysis* **36**, ed. J. Koukal (Elsevier, Amsterdam, 1988), p. 74.

27. J. Pollmann and S. T. Pantelides, *Phys. Rev.* **B18** (1978) 5524.

28. V. M. Tapilin, *Surf. Sci.* **206** (1988) 405.

29. I. A. Abrikosov, Yu. H. Vekilov, P. A. Korzhavyi, A. V. Ruban, and L. E. Shilkrot, *Solid State Commun.* **83** (1992) 867.

30. J. Kudrnovský, V. Drchal, and I. Turek, *J. Phys.* **F17** (1987) L283.

31. J. Kudrnovský and V. Drchal, *phys. stat. sol. (b)* **148** (1988) K23.

32. J. Kudrnovský and V. Drchal, *Z. Phys.* **B73** (1989) 489.

33. J. Kudrnovský, V. Drchal, M. Šob, and O. Jepsen, *Phys. Rev.* **B41** (1990) 10459; and *Phys. Rev.* **B43** (1991) 4622.

34. S. K. Bose, J. Kudrnovský, M. van Schilfgaarde, P. Blőchl, O. Jepsen, M. Methfessel, and T. Paxton, *J. Magn. Magn. Mater.* **87** (1990) 97.

35. J. Kudrnovský, S. K. Bose, and O. K. Andersen, *J. Phys. Soc. Japan* **59** (1990) 4511.

36. J. Kudrnovský, S. K. Bose, and O. Jepsen, *Phys. Rev.* **B43** (1991) 14409.

37. J. Kudrnovský, S. K. Bose, and V. Drchal, *Phys. Rev. Lett.* **69** (1992) 308.

38. V. Drchal, J. Kudrnovský, L. Udvardi, P. Weinberger, and A. Pasturel, *Phys. Rev.* **B45** (1992) 14328.

39. K. C. Hass, B. Velický, and H. Ehrenreich, *Phys. Rev.* **B29** (1984) 3697.

40. H. Akai, *J. Phys. Condens. Matter* **1** (1989) 8045.

41. I. A. Abrikosov, Yu. H. Vekilov, and A. V. Ruban, *Phys. Lett.* **A154** (1991) 407.

42. M. P. López Sancho, J. M. López Sancho, and J. Rubio, *J. Phys.* **F15** (1985) 851.

43. J. Kudrnovský, I. Turek, and V. Drchal, (unpublished).

44. J. Mašek and J. Kudrnovský, *Solid State Commun.* **58** (1986) 67.

45. J. Kudrnovský and V. Drchal, *Solid State Commun.* **70** (1989) 577.

46. Z. W. Lu, S.-H. Wei, and A. Zunger, *Phys. Rev.* **B44** (1991) 3387; and *Phys. Rev.* **B45** (1992) 10314.

47. A. Zunger, S.-H. Wei, L. G. Ferreira, and J. E. Bernard, *Phys. Rev. Lett.* **65** (1990) 353.

48. H. Winter and G. M. Stocks, *Phys. Rev.* **B27** (1983) 882.

49. P. P. Singh, D. de Fontaine, and A. Gonis, *Phys. Rev.* **B44** (1991) 8578.

50. J. E. Inglesfield and G. A. Benesh, *Phys. Rev.* **B37** (1988) 6682.

51. M. Scheffler, C. Droste, A. Fleszar, F. Máca, G. Wachutka, and G. Barzel, *Physica* **B172** (1991) 143.

52. S. Crampin, R. Monnier, T. Schulthess, G. H. Schadler, and D. D. Vvedensky, *Phys. Rev.* **B45** (1992) 464.

53. F. Ducastelle and F. Gautier, *J. Phys.* **F6** (1976) 2039; G. Trèglia, B. Le-

grand, and F. Ducastelle, *Europhys. Lett.* **7** (1988) 575.

54. B. L. Gyorffy and G. M. Stocks, *Phys. Rev. Lett.* **50** (1983) 374.
55. F. Ducastelle, *Order and Phase Stability*, (North Holland, Amsterdam, 1991).
56. A. Pasturel, V. Drchal, J. Kudrnovský, and P. Weinberger, *Phys. Rev.* **B** (submitted).
57. B. L. Gyorffy, A. J. Pindor, J. Staunton, G. M. Stocks, and H. Winter, *J. Phys.* **F15** (1985) 1337.
58. B. L. Győrffy, A. Barbieri, J. B. Staunton, W. A. Shelton, and G. M. Stocks, *Physica* **B172** (1991) 35.

MAGNETISM AND COMPOSITIONAL ORDER IN TRANSITION METAL ALLOYS

J.B.STAUNTON
Department of Physics, University of Warwick,
Coventry CV4 7AL, U.K.
B.L. GYORFFY
H.H.Wills' Physics Laboratory, University of Bristol,
Tyndall Avenue, Bristol BS8 1TL, U.K.
D.D. JOHNSON
Sandia National Laboratories,Livermore,CA,U.S.A.
F.J. PINSKI
Department of Physics, University of Cincinnati
Cincinnati OH 45221, U.S.A.
G.M. STOCKS
Metals and Ceramics Division, Oak Ridge National Laboratory,
Oak Ridge TN 37830, U.S.A.

Abstract

We describe our work on the theory of magnetism and compositional order in transition metal alloys. The self-consistent field, Korringa Kohn Rostoker - Coherent Potential Approximation (SCF-KKR-CPA) formal background to the work is reviewed and then the theory for magnetic and compositional correlations in alloys at high temperatures is outlined. The theory is illustrated in terms of several case studies. Recent developments in the theory of itinerant magnetism at finite temperatures are also reported.

1 Introduction

It is well known that the most important source of magnetic interactions comes from electrostatic electron-electron exchange interactions. In magnetic insulators it is possible to associate electrons with particular atomic sites, define site 'spin' operators $\hat{s}_i$ and the Heisenberg Hamiltonian,for example, can be used to describe the magnetic behaviour of such systems. In metallic systems, however, the situation is very different. It is not possible to allocate their itinerant electrons in this way and such pairwise interactions are inapplicable. In metals, magnetism is a complicated many electron phenomenon.

For such systems magnetism is subtly connected with other properties. For example, some materials show a small thermal expansion coefficient below the Curie temperature, T_c, a large forced increase in volume when an external magnetic field is applied, a sharp decrease of spontaneous magnetism and of T_c when pressure is applied and large changes in the elastic constants on going through T_c. These are the famous invar effects, so called because they occur in the f.c.c. invar alloys Fe-Ni

(65% Fe), Fe-Pd and Fe-Pt. Another example, of particular relevance to this article, is the connection between compositional ordering in an alloy and its magnetic state. When the states of magnetic and compositional order are strongly coupled, physically interesting and technologically important new phenomena may arise. Some alloys, such as Fe-Ni, develop directional chemical order when annealed in a magnetic field [1]. The magnetic properties of many alloys are sensitive to the local environment. For example, ordered Ni-Pt (50%) is an anti-ferromagnetic alloy [2] whereas its disordered counterpart is ferromagnetic.

For several years it has generally been agreed that spin polarised band theory, established within the Spin Density Functional Formulism [3], is able to provide a reliable description of magnetic properties of transition metal systems at absolute zero temperature. This is essentially the modern version of the Stoner-Wohlfarth theory [4,5], in which the magnetic moments are assumed to originate from itinerant d-electrons whose spins are aligned by exchange interactions. It provides a mechanism for non-integer moments together with a plausible account of the many body nature of magnetic moment formation at $T = 0K$ [6]. This theory maps the many electron problem onto one of the independent electrons moving in the effective potentials and magnetic fields which themselves depend on all the other electrons. This is the self-consistent spin polarised band theory picture.

Whilst the modern version of the Stoner -Wohlfarth theory adequately describes the magnetic ground state of transition metals, its straightforward generalisation to finite temperatures fails miserably. The predicted Curie temperature, T_c, is too high typically by a factor of five and there are no moments and no Curie Weiss law above T_c. The theory seems to miss the dominant thermal fluctuation of the magnetisation. Its structure allows the thermally averaged magnetisation, $\bar{M}$, to vanish only as the exchange splitting between the spin polarised bands collapses together with the local magnetisation in each unit cell. Clearly an improved theory must allow thermal fluctuations of the orientations of these 'local moments', attributes set up by the collective behaviour of all the electrons. A discussion of the implementation of this sort of theory is included later on in this article. For the time being we consider metallic systems in their magnetic states at temperatures well below their Curie temperatures.

2 Spin polarised electronic structure of disordered alloys.

We first discuss the parameter free, modern way of formulating the Stoner theory and start with a brief description of Spin Density Functional theory (for detailed reviews [3,7]. Consider a many electron system in an external potential, $V^{ext.}$ and external magnetic field $\mathbf{B}^{ext.}$

$$
\begin{aligned}
V^{ext.} &= \sum_i v_i^{ext.}(\mathbf{r}_i) \\
\mathbf{B}^{ext.} &= \sum_i \mathbf{b}_i^{ext.}(\mathbf{r}_i) \\
\mathbf{r}_i &= \mathbf{r} - \mathbf{R}_i
\end{aligned}
$$

where $\mathbf{R}_i$ denotes the position of a lattice site. The many body Hamiltonian is

$$
\begin{aligned}
\hat{H} &= -(\hbar^2/2m)Tr.\int d\mathbf{r}\nabla\psi^{\dagger}(\mathbf{r},t)\cdot\nabla\psi(\mathbf{r},t) \\
&+Tr.\int d\mathbf{r}\psi^{\dagger}(\mathbf{r},t)(V^{ext.}(\mathbf{r}) + \mathbf{B}^{ext.}(\mathbf{r})\cdot\tilde{\boldsymbol{\sigma}}\psi(\mathbf{r},t) \\
&+(e^2/2)Tr.\int\int d\mathbf{r}\, d\mathbf{r}'\psi^{\dagger}(\mathbf{r},t)\psi^{\dagger}(\mathbf{r}',t)\psi(\mathbf{r}',t)\psi(\mathbf{r},t)/|\mathbf{r}-\mathbf{r}'| \qquad (1)
\end{aligned}
$$

in terms of creation and annihilation operators $\psi^{\dagger}$, ψ ($Tr.$ denotes trace and $\tilde{}$ a 2×2 matrix). It is possible to prove that in the grand canonical ensemble at a given temperature, T, and chemical potential ν, the equilibrium charge $n(\mathbf{r})$ and magnetisation density $\mathbf{m}(\mathbf{r})$ are determined by the external potential and magnetic field. The correct equilibrium charge and magnetisation densities minimise the Gibbs grand potential Ω

$$
\begin{aligned}
\Omega[n,\mathbf{m}] &= \int d\mathbf{r}V^{ext.}(\mathbf{r})n(\mathbf{r}) + \int d\mathbf{r}\mathbf{B}^{ext.}(\mathbf{r})\cdot\mathbf{m}(\mathbf{r}) \\
&+(e^2/2)\int\int d\mathbf{r}\, d\mathbf{r}'n(\mathbf{r})n(\mathbf{r}')/|\mathbf{r}-\mathbf{r}'| \\
&-\nu\int d\mathbf{r}n(\mathbf{r}) + G[n,\mathbf{m}] \qquad (2)
\end{aligned}
$$

where G is a unique functional of charge and magnetisation densities at a given T, ν. The variational principle now states that Ω is a minimum for the equilibrium n, $\mathbf{m}$ and G can be written

$$
G[n,\mathbf{m}] = T_s[n,\mathbf{m}] - TS_s[n,\mathbf{m}] + \Omega_{xc}[n,\mathbf{m}] \qquad (3)
$$

with T_s, S_s being respectively the kinetic energy and entropy of a system of non-interacting electrons with densities n, $\mathbf{m}$ at a temperature T. Ω_{xc} is the exchange and correlation contribution to the Gibbs Free Energy. The minimum principle can be shown to be identical to the corresponding equation for a system of non-interacting electrons moving in an effective potential $\tilde{W}^{eff.}$

$$
\begin{aligned}
\tilde{W}^{eff.}[n,\mathbf{m}] &= V^{ext.}\tilde{1} + \mathbf{B}^{ext.}\cdot\tilde{\boldsymbol{\sigma}} + e^2\tilde{1}\int d\mathbf{r}'n(\mathbf{r}')/|\mathbf{r}-\mathbf{r}'| \\
&+\tilde{1}(\delta\Omega_{xc}/\delta n(\mathbf{r})) + (\delta\Omega_{xc}/\delta\mathbf{m}(\mathbf{r}))\cdot\tilde{\boldsymbol{\sigma}} \qquad (4)
\end{aligned}
$$

which satisfy the following set of equations

$$\{(\epsilon + (\hbar^2/2m)\nabla^2)\tilde{1} - \tilde{W}^{eff.}\}\tilde{G}(\mathbf{r},\mathbf{r}';\epsilon) = \tilde{1}\delta(\mathbf{r}-\mathbf{r}') \tag{5}$$

$$n(\mathbf{r}) = -Im(1/\pi)\int d\epsilon f(\epsilon-\nu)Tr.\tilde{G}(\mathbf{r},\mathbf{r};\epsilon) \tag{6}$$

$$\mathbf{m}(\mathbf{r}) = -Im(1/\pi)\int d\epsilon f(\epsilon-\nu)Tr.\tilde{\sigma}\tilde{G}(\mathbf{r},\mathbf{r};\epsilon) \tag{7}$$

$\tilde{G}$ represents an effective single electron Green function. It is found that

$$\begin{aligned}\Omega &= \int d\epsilon f(\epsilon-\nu)\epsilon n(\epsilon) - (e^2/2)\int\int d\mathbf{r}\,d\mathbf{r}' n(\mathbf{r})n(\mathbf{r}')/|\mathbf{r}-\mathbf{r}'| \\ &\quad +\Omega_{xc} - \int d\mathbf{r}\{(\delta\Omega_{xc}/\delta n(\mathbf{r}))n(\mathbf{r}) + (\delta\Omega_{xc}/\delta\mathbf{m}(\mathbf{r}))\cdot\mathbf{m}(\mathbf{r})\}\end{aligned} \tag{8}$$

where the single particle density of states is

$$n(\epsilon) = -Im(1/\pi)\int d\mathbf{r}Tr.\tilde{G}(\mathbf{r},\mathbf{r};\epsilon) \tag{9}$$

These equations demonstrate the spin polarised, self-consistent band structure basis to the many electron phenomenon. The scheme is exact but in practice approximations are made for the form of Ω_{xc} , the most common being

$$\Omega_{xc} \simeq \int d\mathbf{r}\Omega^0_{xc}(n,m)n(\mathbf{r}) \tag{10}$$

where Ω^0_{xc} is the exchange-correlation contribution to the Free Energy of a homogeneous electron gas. this is the 'local density' approximation and justification for its use are given in detail by Gunnarsson and Lundqvist (8) and others. This theory accurately describes a range of equilibrium properties of transition metals including whether a material is a magnet, the size of its magnetic moment per site, its cohesive energy etc. The underlying spin polarised bands are found, for example, in the book by Moruzzi, Janak and Williams [9].

Stocks and Winter [10] and Johnson et al.[11] have discussed in detail how this formulism may be applied to randomly disordered paramagnetic and ferromagnetic alloys A_cB_{1-c}. But,before summarising the key aspects, we need to consider how to treat accurately the problem of an electron moving through a lattice whose sites are randomly occupied by one atomic species or another. A reliable way of going about this is to use the Coherent Potential Approximation (C.P.A.)[12] and multiple scattering theory (KKR-CPA),[10,11].

The effective potentials, of the form described above appropriate to A and B atoms, are spherically symmetric. In multiple scattering language, the single particle Green function $G(\mathbf{r},\mathbf{r}';\epsilon)$ is written [14]

$$\tilde{G}(\mathbf{r}_i,\mathbf{r}'_j;\epsilon) = \sum_{L,L'}\{Z_L(\mathbf{r}_i)\tau^{ij}_{L,L'}Z^{\dagger}_{L'}(\mathbf{r}'_j) - Z_L(\mathbf{r}_i)J^{\dagger}_L(\mathbf{r}'_i)\delta_{ij}\delta_{LL'}\} \tag{11}$$

in which $Z_L(\mathbf{r}_i)$ is the regular solution of the single site Schrödinger equation appropriate to the ith lattice site, J_L is the irregular solution. $\tau^{\tilde{ij}}$ represents the scattering path operator[15] and when it operates on the wave incident at the site at $\mathbf{R}_j$ it gives the scattered wave coming from the site at $\mathbf{R}_i$, including all the effects of scattering in between, i.e.

$$\tau^{\tilde{ij}}_{L,L'} = t^{\tilde{i}}_L \delta_{ij}\delta_{L,L'} + \sum_{k\neq i}\sum_{L''} t^{\tilde{i}}_L G^0_{L,L''}(\mathbf{R}_i - \mathbf{R}_k;\epsilon)\tilde{1}\tau^{\tilde{kj}}_{L'',L'} \tag{12}$$

where $t^{\tilde{i}}_L$ describes the Lth (l,m) channel single site scattering from site i and $G^0_{L,L'}$ the structure constants appropriate to the crystal lattice.

The idea behind the C.P.A. is to retrieve the Bloch theorem and construct a lattice of effective complex C.P.A. potentials such that the motion of an electron through this lattice approximates the motion of an electron, *on the average*, through the randomly disordered lattice. A C.P.A. potential described by a scattering t-matrix $t_{\tilde{c},L}$ is placed on every site and the multiple scattering in this ordered lattice specified by $\tau^{\tilde{c},ij}_{L,L'}$. An A atom impurity in this C.P.A. host can be described by an operator $\tau^{\tilde{A},ij}_{L,L'}$, similarly a B atom impurity by $\tau^{\tilde{B},ij}_{L,L'}$. The C.P.A. condition is then

$$c\tau^{\tilde{A},ij}_{L,L'} + (1-c)\tau^{\tilde{B},ij}_{L,L'} = \tau^{\tilde{c},ij}_{L,L'} \tag{13}$$

where c is the probability of any site being occupied by an A atom. This means that the optimum coherent potential is chosen by insisting that the substitution of a single site of the C.P.A. lattice by either an A or B site potential produces no further scattering, *on the average*. In practice $t_{\tilde{c},L}$ can be found by solving the site diagonal version of the above equation only and for a ferro- or paramagnetic alloy in which the axis of spin polarisation is the same on every site this is written

$$c\tau^{A,00}_{L\sigma,L'\sigma} + (1-c)\tau^{B,00}_{L\sigma,L'\sigma} = \tau^{c,00}_{L\sigma,L'\sigma} \tag{14}$$

$$\tau^{c,00}_{L\sigma,L'\sigma} = (1/V_{BZ})\int d\mathbf{k}(t^{-1}_{c,\sigma} - G^0(\mathbf{k}))^{-1}_{LL'} \tag{15}$$

$$\tau^{A(B),00}_{L\sigma,L'\sigma} = \{[1 + (t^{-1}_{A(B),\sigma} - t^{-1}_{c,\sigma})\tau^{c,00}_{\sigma\sigma}]^{-1}\tau^{c,00}_{\sigma\sigma}\}_{LL'} = \sum_{L''} D^{A(B)}_{L\sigma,L''\sigma}\tau^{c,00}_{L''\sigma,L'\sigma} \tag{16}$$

(a quantity without L,L' labels denotes a matrix whose elements are labelled by angular momentum quantum numbers $L = l,m$). Observables are calculated by taking configurational averages estimated by considering fluctuations about this C.P.A. medium[16]. For example, partially averaged charge and magnetisation densities $\langle n\rangle_{i,A(B)}$, $\langle\mu\rangle_{i,A(B)}$ on a site i in an A_cB_{1-c} binary alloy are given by

$$\langle n(\mathbf{r}_i)\rangle_{i,A(B)} = -(1/\pi)Im\int d\epsilon f(\epsilon-\nu)[\langle G_{++}(\mathbf{r}_i,\mathbf{r}_i;\epsilon)\rangle_{i,A(B)} + \langle G_{--}(\mathbf{r}_i,\mathbf{r}_i;\epsilon)\rangle_{i,A(B)}] \tag{17}$$

$$\langle\mu(\mathbf{r}_i)\rangle_{i,A(B)} = -(1/\pi)Im\int d\epsilon f(\epsilon-\nu)[\langle G_{++}(\mathbf{r}_i,\mathbf{r}_i;\epsilon)\rangle_{i,A(B)} - \langle G_{--}(\mathbf{r}_i,\mathbf{r}_i;\epsilon)\rangle_{i,A(B)}] \tag{18}$$

where

$$\langle G_{\sigma\sigma}(\mathbf{r}_i,\mathbf{r}_i;\epsilon)\rangle \simeq \sum_L \{Z^{A(B)}_{L\sigma}(\mathbf{r}_i)\tau^{A(B),00}_{L\sigma,L\sigma}Z^{A(B)}_{L\sigma}(\mathbf{r}_i) - Z^{A(B)}_{L\sigma}(\mathbf{r}_i)J^{A(B)}_{L\sigma}(\mathbf{r}_i\} \tag{19}$$

and $\langle\cdots\rangle_{i\alpha,j\beta,\ldots}$ represents the configurational average with the restriction that the ith site is occupied by an α atom, the jth site by a β atom etc. The Bloch spectral function [13,14], $A^B_\sigma(\mathbf{k},\epsilon)$ is related to restricted averages of the non-site diagonal Green function which can be similarly expressed by considering fluctuations about the C.P.A. medium in terms of $Z^{A(B)}$, t_c and τ^c. It is a quantity through which it is possible to investigate the averaged single particle (quasiparticle) eigenvalue spectrum in both wave-vector and energy dependent detail.

A formal version of Spin Density Functional theory applied to an alloy of concentration c would require consideration of each possible configuration, the minimisation of the Gibbs Free Energy of each, finding the equilibrium charge and magnetisation densities, followed by the appropriate averaging. These steps are of course intractable. It is necessary to reverse the minimisation involved in the SDF theory with the configurational averaging in an approximate way which is consistent with the mean field spirit of the C.P.A. This is achieved by minimising the configurationally averaged Free Energy with respect to the partially averaged charge and magnetisation densities $\langle n\rangle_{i,A(B)} = \bar{n}_{iA(B)}$, $\langle\mu\rangle_{i,A(B)} = \bar{\mu}_{iA(B)}$ [11]. The average Free Energy is written

$$\langle\Omega\rangle = c\langle\Omega\rangle_{i,A} + (1-c)\langle\Omega\rangle_{i,B} \tag{20}$$

$$\begin{aligned}
\langle\Omega\rangle_{i\alpha} = {} & Q_\alpha\nu - \int d\epsilon f(\epsilon-\nu)\sum_\sigma\langle N(\epsilon)\rangle_{i\alpha} \\
& -e^2/2\{\int\int d\mathbf{r}_i d\mathbf{r}'_i \bar{n}_{i\alpha}(\mathbf{r}_i)\bar{n}_{i\alpha}(\mathbf{r}'_i)/|\mathbf{r}_i-\mathbf{r}'_i| \\
& +\sum_{j\neq i}\int\int d\mathbf{r}_i d\mathbf{r}'_j \bar{n}_{i\alpha}(\mathbf{r}_i)\bar{n}_j(\mathbf{r}'_j)/|\mathbf{r}_i-\mathbf{r}'_j| \\
& +\sum_{j,k\neq i}\int\int d\mathbf{r}_j d\mathbf{r}'_k \bar{n}_j(\mathbf{r}_j)\bar{n}_k(\mathbf{r}'_k)/|\mathbf{r}_j-\mathbf{r}'_k|\} \\
& -\int d\mathbf{r}_i\{(\delta\Omega_{xc}/\delta\bar{n}_{i\alpha}(\mathbf{r}_i) - \Omega^0_{xc}(\bar{n}_{i\alpha},\bar{\mu}_{i\alpha}))\bar{n}_{i\alpha}(\mathbf{r}_i) + (\delta\Omega_{xc}/\delta\bar{\mu}_{i\alpha}(\mathbf{r}_i))\bar{\mu}_{i\alpha}(\mathbf{r}_i)\} \\
& -\sum_{j\neq i}\sum_{\beta=A,B} c_\beta\int d\mathbf{r}_j\{(\delta\Omega_{xc}/\delta\bar{n}_{j\beta}(\mathbf{r}_j) - \Omega^0_{xc}(\bar{n}_{j\beta},\bar{\mu}_{j\beta}))\bar{n}_{j\beta}(\mathbf{r}_j) \\
& +(\delta\Omega_{xc}/\delta\bar{\mu}_{j\beta}(\mathbf{r}_j))\bar{\mu}_{j\beta}(\mathbf{r}_j)\}
\end{aligned} \tag{21}$$

in which $N_\sigma(\epsilon)$ is the integrated single particle density of states and may be written using the Lloyd determinant [13].

$$N_\sigma(\epsilon) = N_0(\epsilon) - Im(1/\pi)\ln\|t_\sigma^{-1} - G^0(\epsilon)\| \tag{22}$$

and Q_α is the charge appropriate to the $A(B)$ site. The minimisation

$$\delta\langle\Omega\rangle_{i\alpha}/\delta\bar{n}_{i\alpha} = \delta\langle\Omega\rangle_{i\alpha}/\delta\bar{\mu}_{i\alpha} = 0 \tag{23}$$

causes the resulting $\bar{n}_{i\alpha}$ and $\bar{\mu}_{i\alpha}$ to set up consistent effective potentials associated with an A ar B site which are written

$$\begin{aligned}\tilde{W}_{i\alpha}[\bar{n}_{i\alpha},\bar{\mu}_{i\alpha};\mathbf{r}_i] \;=\; & \{-(Q^\alpha e^2/r_i) \\ & +e^2\int d\mathbf{r}'_i\bar{n}_{i\alpha}(\mathbf{r}'_i)/|\mathbf{r}_i-\mathbf{r}'_i| - \sum_{j\neq i}\bar{Q}e^2/|\mathbf{r}_i-\mathbf{R}_j+\mathbf{R}_i| \\ & +\sum_{j\neq i}e^2\int d\mathbf{r}'_j\bar{n}_j(\mathbf{r}'_j)/|\mathbf{r}_i-\mathbf{r}'_j| \\ & +\delta\Omega_{xc}/\delta\bar{n}_{i\alpha}(\mathbf{r}_i)\}\tilde{1}+\delta\Omega_{xc}/\delta\bar{\mu}_{i\alpha}(\mathbf{r}_i)\tilde{\sigma}_z\end{aligned} \tag{24}$$

where Q^α is the nuclear charge on an α site,$\bar{Q}$ the average.

To summarise, for an alloy of composition A_cB_{1-c}, the procedure begins with (i) guessed partially averaged charge and magnetisation densities for both A and B sites. (ii) Effective potentials are set up. (iii) The problem of an electron propagating through a random array of these potentials is solved for each spin polarisation, via the C.P.A. (iv) The partially averaged densities are recalculated and (v) steps (ii) to (iv) repeated until consistency is achieved. (vi) Finally the converged quantities are used to calculate the Free Energy and other equilibrium properties.

Such self-consistent field SCF-KKR-CPA calculations can be routinely carried out for both ferro- and paramagnetic alloys usually with good agreement with experimental data. For example, Pinski et al. [17] calculated magnetic moments for a range of b.c.c. $Fe-V$ and both b.c.c. and f.c.c. $Fe-Ni$ alloys and presented them on a Slater-Pauling curve with a comparison with experimental measurements.

As is well known, Stoner theory, applied to pure metals provides a picture of near rigidly exchange split bands [9]. This picture, however, is not appropriate for alloys. The Stoner instability requires a large paramagnetic density of states and in a disordered paramagnetic alloy, electrons of either spin polarisation will be equally affected by the chemical disorder. In the spin polarised magnetic state it is possible for the effects of disorder to be confined to electrons of one 'spin'. (Later the importance of this for compositional ordering will be shown.) In this way the rigidly split bands of Stoner theory are lost [17].

The $Fe-V$ calculations for a range of compositions indicate that the minority spin electrons 'see' little difference between Fe and V sites and it is the majority spin electrons that are most strongly affected by the disorder. The Fermi energy is pinned in the bonding-antibonding valley of the minority density of states fixing a constant number of minority spin electrons for a range of iron-rich alloys. This produces a mechanism for the linear dependence of the average moment on the electron per atom ratio. (This has also been noted by Williams et al.[18] for ordered alloys.) An understanding of the more complicated dependence of the moments on Fe and V sites requires a more detailed analysis of the electronic structure. The moments on the V sites are formed anti-parallel to the larger ones on the Fe sites (a similar situation is found in $Fe-Mn$ alloys). When the concentration of Fe is less than 30% the alloys are found to be paramagnetic also in agreement with experiment.

The $Ni-Fe$ calculations indicate the opposite picture. Here it is the majority spin electrons which are scarcely affected by the chemical disorder in sharp distinction from the behaviour of the minority spin electrons. In the f.c.c. alloys the average magnetisation per site increases linearly with increase in Fe concentration. As more iron is added the partially averaged moment on a Fe site decreases whilst that on the nickel sites increases. In the Invar composition range, as the iron concentration approaches 70%, the size of moment is very sensitive to the lattice spacing. For a lattice parameter of 6.76a.u. an average moment of $2\mu_B$ exists per site whereas on a lattice spacing with a 4% smaller spacing no moment can be sustained [17]. A small increase in Fe concentration in this range also causes a marked moment collapse. With volume decrease or a larger iron concentration the Fermi level is pushed into the majority d-bands precipitating the moment disappearance. A similar analysis is valid for interpreting calculations on pure f.c.c. Fe on a range of lattice spacings[19] and is useful in understanding the magnetic properties of many f.c.c. iron alloys.

As a summary of this section, it can be stated that at low temperatures (well below T_c, it is now accepted that the magnetic properties of metallic systems can be understood and predicted from the study of the behaviour of an electron moving through a lattice of effective potentials. These are set up self-consistently by all the electrons and the magnetic components are all aligned along one 'global' direction. Such SDF calculations have been carried out with success for pure transition metals [9], ordered alloys [20] and , within the framework of the KKR-CPA [11,13,17], for disordered alloys. It is now therefore possible to investigate in a parameter-free way the wide range of magnetic properties exhibited by the transition metal alloys and the influence of magnetism upon other physical characteristics. The following discussion is concerned with the interrelation of these magnetic properties and compositional ordering.

3 Compositional order and magnetic structure of transition metal alloys

A 'first principles' theory of magnetic and compositional phase transitions in alloys should ideally treat both magnetic and compositional thermal fluctuations on an equal footing but here we will take the first step towards such a theory and look at the effect of the magnetic state upon compositional ordering. This is basically the generalisation to magnetic alloys of the theory that Gyorffy and Stocks developed for paramagnetic alloys [21] and which has been recently extended by us [22].

We begin by writing down formally the Grand Potential for a system of electrons in a particular configuration of nuclei using the SDF theory. On a site i in the underlying lattice, the variable $\xi_i = 1$ if the site is occupied by a A atom, $= 0$ if occupied by a B atom, i.e. it is an Ising-like variable. A configuration is denoted $\{\xi_i\}$ and the Grand Potential $\Omega\{\xi_i\}$. Averaging over the compositional fluctuations with measure

$$P\{\xi_i\} = \exp(-\beta\Omega\{\xi_i\})/\prod_i \sum_{\xi_i=0,1} \exp(-\beta\Omega\{\xi_i\}) \tag{25}$$

gives an expression for the Free Energy

$$F = -(1/\beta)\ln\prod_i \sum_{\xi_i=0,1} \exp(-\beta\Omega\{\xi_i\}) \tag{26}$$

where $\beta = (k_B T)^{-1}$. $\Omega\{\xi_i\}$ is thus playing the role of a highly complicated concentration fluctuation Hamiltonian. By expanding about a suitable reference Hamiltonian, Ω_0, using the Feynman-Peierls Inequality [23]

$$F \leq F_0 + \langle\Omega - \Omega_0\rangle^0 = \breve{F} \tag{27}$$

where

$$\langle X\rangle^0 = \prod_i \sum_{\xi_i=0,1} X\{\xi_i\}\exp(-\beta\Omega_0\{\xi_i\})/\prod_i \sum_{\xi_i=0,1} \exp(-\beta\Omega_0\{\xi_i\}) \tag{28}$$

A mean field theory is produced with the choice

$$\Omega_0 = \sum_i f_i(\xi_i) \tag{29}$$

in which the best functions f_i are found following a functional minimisation of $\breve{F}$ and

$$f_i(\xi') = \langle\Omega\rangle^0_{\xi_i=\xi'} \tag{30}$$

where the restricted average is explicitly

$$\langle\Omega\rangle^0_{\xi_i=\xi'} = \prod_{j\neq i} \sum_{\xi_j=0,1} \Omega(\xi_1,\xi_2,\cdots,\xi_i=\xi',\cdots)\exp(-\beta f_j(\xi_j))/\prod_{j\neq i} \sum_{\xi_j=0,1} \exp(-\beta f_j(\xi_j)) \tag{31}$$

i.e. the average of the Grand Potential is taken over all configurations with the restriction that on the ith site the occupation is fixed, $\xi_i = \xi'$. These partial averages are accessible from the SCF-KKR-CPA framework sketched in the last section and the mean field picture is consistent with that of the coherent potential approximation. The probability of finding an A atom on a particular site, i, at a temperature T, is

$$c_i = \exp(-\beta f_i(1))/(\exp(-\beta f_i(1)) + \exp(-\beta f_i(0)) \tag{32}$$

Note that, in principle, the probability of occupation can vary from site to site but it is only the case of a homogeneous probability distribution, $c_i = c$, that can be tackled in practice with our KKR-CPA method.

Using response theory and the fluctuation dissipation theorem, it is possible to write down the various correlation functions appropriate to the system and hence investigate ordering. Firstly we will consider the response of the homogeneous system to the application of an inhomogeneous external field,

$$V^{ext.} = \sum_i (v_i - \eta) \tag{33}$$

which couples to the occupation variables $\{\xi_i\}$ such that a term

$$\sum_i (v_i - \eta)\xi_i = \sum_i u_i \xi_i \tag{34}$$

is added to the Grand Potential $\Omega\{\xi_i\}$ (η is the chemical potential difference such that the number of A and B sites remains constant). This applied field, different on every site, induces changes in the probabilities of occupation on every site $\{\delta c_i\}$. Furthermore, of crucial importance for this discussion, it also causes changes to the moments from site to site $\{\delta\mu_i^{A(B)}\}$ as well as re-arranging the charge. It is possible to formulate a theory through the formal framework of an inhomogeneous C.P.A. [21]. Although the resulting equations, shown below, are not soluble in practice,

$$c_i \tau_{inh.}^{A,ii} + (1 - c_i)\tau_{inh.}^{B,ii} = \tau_{inh.}^{c,ii} \tag{35}$$

$$\tau_{inh.}^{c,ii} = t_{c,i,inh.} + \sum_{k \neq i} t_{c,i,inh.} G^0(\mathbf{R}_i - \mathbf{R}_k; \epsilon)\tau_{inh.}^{c,ki} \tag{36}$$

it is possible to find an approximate solution by perturbing about the homogeneous case. Clearly the single site C.P.A. t-matrix $t_{c,i,inh.}$ and the site diagonal C.P.A. scattering path operator $\tau_{inh.}^{c,ii}$ are different on every site, but to lowest order in the perturbation, they can be written

$$t_{c,i,inh.}^{-1} = t_c^{-1} + \delta t_{c,i}^{-1} \tag{37}$$

$$\tau_{inh.}^{c,ii} = \tau^{c,00} + \delta\tau^{c,ii} = \tau^{c,00} - \sum_j \tau^{c,ij}\delta t_{c,j}^{-1}\tau^{c,ji} \tag{38}$$

These equations with the homogeneous solution $t_c, \tau^{c,ij}$ can now be solved to find $\{\delta t_{c,i}^{-1}\}$.

The induced change in probability of occupation on the ith site is written to low order in the applied field

$$\delta c_i \simeq -\beta c(1 - c)(\langle\delta\Omega\rangle_{iA} - \langle\delta\Omega\rangle_{iB}) \tag{39}$$

and from the expressions for the induced changes in the charge and magnetisation densities on each site, closed expressions for the various correlation functions can be obtained [22,23] in terms of quantities based on the electronic structure of the homogeneously disordered alloy. A key result is then that the compositional correlation function, related to $\alpha(\mathbf{q}) = 2\beta c(1-c)/(1-\beta c(1-c)S^{(2)}(\mathbf{q}))$ by

$$\alpha(\mathbf{q}) = (1/N)\sum_j \beta(\langle\xi_i\xi_j\rangle - \langle\xi_i\rangle\langle\xi_j\rangle)\exp(i\mathbf{q}\cdot(\mathbf{R}_i - \mathbf{R}_j)) \tag{40}$$

The quantity, $S^{(2)}$, is essentially the spin polarised version of the expression derived by Gyorffy and Stocks (21) plus charge rearrangement effects discussed in [22] and those changes to the magnetic moments as the chemical environment alters. If there were no spins and the system were described by a pairwise interacting lattice gas model in the random phase approximation then $S^{(2)}$ would represent the Krivoglaz-Moss-Clapp formula [24] and the real space version of $S^{(2)}(\mathbf{q})$, $S^{(2)}_{ij}$ would be the interchange energy

$$V_{ij} = V^{AA}_{ij} + V^{BB}_{ij} - 2V^{AB}_{ij} \tag{41}$$

with obvious notation. When the effect of magnetic moments interacting via phenomenological exchange interactions ,J_{ij}, in a moments-aligned case, is included then

$$S^{(2)}_{ij} = V_{ij} + J^{AA}_{ij} + J^{BB}_{ij} - 2J^{AB}_{ij} \tag{42}$$

Thus $S^{(2)}_{ij}$ can be interpreted as an effective pairwise ordering energy enhanced by the presence of rigid moments.

The magneto-compositional function $\Upsilon(\mathbf{q})$ is specified

$$\begin{aligned}\Upsilon(\mathbf{q}) &= (1/N)\sum_j(\delta\mu_i/\delta u_j)\exp(i\mathbf{q}\cdot(\mathbf{R}_i-\mathbf{R}_j)) \\ &= (1/N)\sum_{jk}(\delta\mu_i/\delta c_k)(\delta c_k/\delta u_j)\exp(i\mathbf{q}\cdot(\mathbf{R}_i-\mathbf{R}_j)) \\ &= \gamma(\mathbf{q})\alpha(\mathbf{q})\end{aligned} \tag{43}$$

Another important result is that the magneto-compositional 'cross' correlation function is related to $\Upsilon(\mathbf{q})$ by

$$\Upsilon(\mathbf{q}) = (1/N)\sum_j \beta(\langle\mu_i\xi_j\rangle - \langle\mu_i\rangle\langle\xi_j\rangle)\exp(i\mathbf{q}\cdot(\mathbf{R}_i - \mathbf{R}_j)) \tag{44}$$

which itself is written as a product of $\alpha(\mathbf{q})$ and $\gamma(\mathbf{q})$ as shown above. $\gamma(\mathbf{q})$ is expressed in terms of quantities γ_A and γ_B (associated with A and B sites respectively) and the lattice Fourier transforms of these have a particularly revealing physical interpretation.

$$(1/V_{BZ})\int d\mathbf{q}\gamma_{A(B)}(\mathbf{q})\exp(-i\mathbf{q}\cdot(\mathbf{R}_i - \mathbf{R}_j)) = \gamma^{ij}_{A(B)} = \delta\mu^{A(B)}_i/\delta c_j \tag{45}$$

$\gamma^{ij}_{A(B)}$ describes the change in moment on a site i in the lattice if it is occupied by an $A(B)$ atom if the probability of occupation is altered on another site j. It measures the chemical environment effect on the moments' sizes. Thus

$$\gamma^{ij}_{A(B)}\Delta c_j = \gamma^{ij}_{A(B)}(1-c) \tag{46}$$

describes the moment change on the ith site occupied by an $A(B)$ atom if the site j in the random alloy is now *definitely* occupied by an A atom. Similarly,

$$\gamma^{ij}_{A(B)}\Delta c_j = \gamma^{ij}_{A(B)}(-c) \tag{47}$$

describes the change if site j is now *definitely* occupied by a B atom. Consequently once these quantities $\gamma^{ij}_{A(B)}$ have been calculated, it is possible, in a perturbative fashion, to investigate the magnetic structure of alloys of varying composition. Hence it provides a complementary approach to that supplied by explicit cluster calculations. For example the magnetic properties of modulated or clustering alloys can be examined and alloys designed theoretically to have specified magnetic properties [23].

In a similar vein, the calculation of the response to an external inhomogeneous magnetic field applied along the spontaneous magnetisation direction,

$$H^{ext.} = \sum_i h_i^{ext.} \tag{48}$$

delivers quantities $\chi(\mathbf{q})$, $\Upsilon(\mathbf{q})$ which are themselves related to the magnetic and magneto-compositional 'cross' correlation functions respectively.

$$\chi(\mathbf{q}) = (1/N)\sum_j \beta(\langle\mu_i\mu_j\rangle - \langle\mu_i\rangle\langle\mu_j\rangle)\exp(i\mathbf{q}\cdot(\mathbf{R}_i - \mathbf{R}_j)) \tag{49}$$

$$\chi(\mathbf{q}) = c\bar{\chi}^A(\mathbf{q}) + (1-c)\bar{\chi}^B(\mathbf{q}) + (\gamma(\mathbf{q}))^2\alpha(\mathbf{q}) \tag{50}$$

These theoretical quantities, correlation functions, determined fully by the underlying electronic structure of the homogeneous disordered alloy are closely related to information obtained from X-ray and neutron scattering, nuclear magnetic resonance and Mössbauer spectroscopic measurements. In particular, the cross-sections obtained from diffuse polarised neutron scattering data can be written

$$(d\sigma/d\omega)_\zeta = d\sigma^N/d\omega + \zeta(d\sigma^{NM}/d\omega) + d\sigma^M/d\omega \tag{51}$$

where $\zeta = +1(-1)$ if the neutrons are polarised (anti-) parallel to the magnetisation. The nuclear component $d\sigma^N/d\omega$ is proportional to the compositional correlation function (closely related to the Warren-Cowley short range order parameters) i.e. $\alpha(\mathbf{q})$.

$$d\sigma^N/d\omega = (\Delta b)^2\alpha(\mathbf{q}) \tag{52}$$

where Δb is the difference between the two nuclear scattering amplitudes. Similarly the magnetic component is proportional to $\chi(\mathbf{q})$. Finally $d\sigma^{NM}/d\omega$ describes the magnetic-compositional 'cross' correlation function, proportional to $\Upsilon(\mathbf{q})$. It is clear that the sum of the polarised cross-sections measures a combination of $\alpha(\mathbf{q})$ and $\chi(\mathbf{q})$ whilst the difference gives $\Upsilon(\mathbf{q})$. In order to separate $\alpha(\mathbf{q})$ and $\chi(\mathbf{q})$ experimentally the technique of isotope substitution can be used. Once $\alpha(\mathbf{q})$ and $\Upsilon(\mathbf{q})$ are measured, $\gamma(\mathbf{q})$,i.e. the moments' dependence on their chemical environment can be extracted. By considering a linear superposition of perturbations and phenomenological parameters, Marshall [25] gives formulae of the same form as the above equations and the work of Cable and co-workers and others [26,27] has continued from this and attempted to describe the experimental neutron scattering cross-sections in terms of a parameter for the range of perturbations and others to account for the chemical environment effect, a measure of the exchange interactions and also the type of lattice. It has been the aim of this section to outline a parameter-free theory which can be tested against these experiments. In concluding this section we now provide one such illustration.

Iron-vanadium is a rather suitable system upon which to carry out a comprehensive theoretical and experimental study. Ferromagnetic b.c.c. solid solutions form over a wide range of concentrations. The Fe and V atoms are of a similar size but are quite different in their neutron scattering amplitude ($b_{Fe} - b_V = 0.99$) and the diffuse scattering is consequently relatively intense with little distortion from static lattice displacement effects [26]. Extensive Mossbauer and NMR studies [27] of iron-vanadium magnetic multilayers also make this system an excellent candidate for an application of our scheme for the theoretical study and design of such structures. Cable et al. [26] carried out diffuse neutron scattering measurements on a single crystal of a quenched ferromagnetic $Fe_{86.5}V_{13.5}$ alloy. By applying a saturating magnetic field parallel to the scattering vector they were able, in an unpolarised experiment, to extract the nuclear cross-section for three directions [1,0,0],[1,1,0] and [1,1,1]. We compared the results of our 'first principles' calculation of $\alpha(\mathbf{q})$ with these data using the same values of the scattering amplitude difference, the background incoherent and multiple scattering cross-sections (0.141 b.sr^{-1}.at^{-1}.) and the same quench temperature (1270K) as quoted in the experimental report. Moreover the average magnetic moment of $1.78\mu_B$ ($2.18\mu_B$ associated with each Fe site and $-0.81\mu_B$ with each V site) compared favourably with the experimental estimate of $1.8\mu_B$ and the observation that the vanadium moment is anti-parallel with that of iron. Our calculations successfully reproduced all the main features of the experiment. The dominant peaks at $\mathbf{q} = (1,0,0)$ and $(1,1,1)$, indicative of a $\beta - CuZn$ type ordering tendency, were evident in both theory and experiment as was the peak at (0.5,0.5,0) for the [1,1,0] direction. Particularly striking was the theoretical reproduction of the double peak structure around (0.5,0.5,0.5) in the [1,1,1] direction. This long ranged atom pair correlation was proposed as a specific challenge to theory by the experimentalists [26].

We made a fit our calculations to real space quantities using

$$S^{(2)}(\mathbf{q}) = \sum_n \sum_{i\epsilon n} \exp(i\mathbf{q} \cdot \mathbf{R}_i) S_n^{(2)} \tag{53}$$

in which the index n runs over the shells. In accordance with Cable et al.'s findings for the experimental data we needed many parameters in order to fit accurately our **q** space results illustrating the long ranged nature of the correlations. The traditional interpretation of $\alpha(\mathbf{q})$ in terms of Warren-Cowley short order parameters is not useful in this case. Our theoretical work successfully described the experimental measurements and we could unravel the calculations and isolate the electronic mechanisms driving the ordering - the ultimate aim of an alloy theory. It turned out that the prominent **q**-dependent features were driven by the majority spin electronic structure. For this alloy our calculations for the electronic density of states showed that the minority spin electrons 'see' little difference between the iron and vanadium sites whereas the majority spin electrons are strongly affected by the disorder. The importance of the exchange splitting of the electronic structure for compositional order is apparent. It is the half-filling of these split majority spin states that accounts for the overall ordering tendency of this alloy. A double peak [1,1,1] structure was found to derive from parallel segments for the majority spin Fermi surface [23].

4 Metallic Magnetism at Finite Temperatures.

One of the oldest problems in magnetism is its nature in transition metals. A picture of itinerant electrons interacting with spin fluctuations, which are themselves the consequence of the collective behaviour of the electrons, has proved useful[30]. This resolves the dichotomy of itinerant and localised spin models noted by Herring [6]. The directional attributes of these spin fluctuations also vary on a time scale much longer than the electronic intersite transit time. This approximation gives rise to a 'local moment' picture [30]. For example, in b.c.c. Fe, a mechanism derived from Hund's rule coupling between d-holes[31] sets up local moments in the paramagnetic state whose orientations thermally fluctuate whereas the actual sizes remain roughly constant ($\approx 2\mu_B$).

Recently [32]),two of us presented a theoretical framework for the correlations between fluctuations of these sorts in metals. The electronic structure basis to the theory is treated in a realistic 'first principles' fashion. For the first time, the idea of an Onsager cavity field[33] was used to describe the interplay between the fluctuations and the electronic structure in a parameter free way. The method of construction of the reaction field fixes the fluctuation dissipation theorem and delivers the spherical model results in the appropriate limit[33]. We applied the theory to spin fluctuations in the paramagnetic state and found a description of a range of magnetic behaviour. In one limit it can describe the type of metallic system in which well defined moments can be envisaged, whereas in sharp contrast it can also model systems in which the spin fluctuations have a larger extent. In this limit the theory is reminiscent of the high temperature, static version of those designed for weak ferromagnetic materials [34]. As a first application of the theory we studied the wave-vector dependent,static,paramagnetic spin susceptibility $\chi(\mathbf{q},T)$ of b.c.c.Fe and f.c.c.Ni.

We first specify matters with a brief outline of the Onsager arguments for the classical Heisenberg model,$H = -\sum_{ij} J_{ij}\hat{e}_i \cdot \hat{e}_j$ ($\hat{e}_i$ is a unit vector), along the lines of ref.(33). A small magnetic field $\{\mathbf{h}_i\}$ is applied to the paramagnetic state inducing small magnetic moments $\{\mathbf{m}_i\}$ where $\mathbf{m}_i = \langle \hat{e}_i \rangle$. A mean field theoretical description gives $\mathbf{m}_i \approx (\beta/3)(\sum_j J_{ij}\mathbf{m}_j + \mathbf{h}_i)$,$(\beta = 1/k_B T)$. Onsager's important idea is contained in the comment that the local field acting on a magnetic moment on a particular site i must not include the contribution which comes from that part of the magnetisation on surrounding sites which arises from that magnetisation on the site i, i.e.$\mathbf{m}_i \approx (\beta/3)(\sum_j J_{ij}(\mathbf{m}_j - \delta\mathbf{m}_j^{(i)}) + \mathbf{h}_i)$. We call $\sum_j J_{ij}\mathbf{m}_j + \mathbf{h}_i$ the 'Weiss' field,$\mathbf{h}_i^W$,$\sum_j J_{ij}(\mathbf{m}_j - \delta\mathbf{m}_j^{(i)}) + \mathbf{h}_i$ the 'cavity' field,$\mathbf{h}_i^C$ and $\sum_j J_{ij}\delta\mathbf{m}_j^{(i)}$ as the 'reaction' field $\mathbf{h}_i^R$. We expect the last quantity to be proportional to the magnetisation on the site i,$\mathbf{h}_i^R = \lambda_i \mathbf{m}_i$. We can define a susceptibility $\chi_{ij} = (\beta/3)\sum_l (J_{il}\chi_{lj} - \lambda_i \chi_{ij} - \delta_{ij})$ which, in reciprocal wave-vector space, becomes $\chi(\mathbf{q}) = (\beta/3)/(1 - \beta(J(\mathbf{q}) - \lambda)/3)$. So far we have not provided a prescription for λ,it is merely a phenomenological parameter. We could use it to fix the fluctuation dissipation theorem $\chi_{ii} = (\beta/3)(\langle \hat{e}_i^2 \rangle - \langle \hat{e}_i \rangle^2)$ which equals $\beta/3$ in the paramagnetic phase. Instead we shall set it by the following transparent construction.

$\delta m_j^{(i)}$ is the part of the magnetisation on a site j which is set up by the magnetisation on site i. For small m_i,$\delta m_j^{(i)} = \chi_{ji} h_i^C$, in terms of the cavity field. Now m_i is immediately given by $m_i = \chi_{ii} h_i^C$ and hence $\delta m_j^{(i)} = \chi_{ji} m_i / \chi_{ii}$. It follows that $\chi_{ij} = (\beta/3)(\sum_l (J_{il}\chi_{lj} - J_{il}\chi_{li}\chi_{ii}^{-1}\chi_{ij} + \delta_{ij}))$ and we need to solve the pair of coupled equations $\chi(\mathbf{q}) = (\beta/3)/(1 - \beta(J(\mathbf{q}) - \lambda)/3)$,$\lambda = \chi_{ii}^{-1} \int d\mathbf{q} J(\mathbf{q})\chi(\mathbf{q})$. The result of the solution of these equations is that $\chi_{ii} = \beta/3$, the fluctuation dissipation theorem is satisfied although it was not imposed explicitly at the outset. We now use these ideas to spin fluctuations in itinerant electron paramagnetic system but point out here that the approach is rather general.

We start with the assumption that, on a time scale τ long compared to that of a typical spin fluctuation, the spins of the electrons are sufficiently correlated to leave the magnetisation averaged over a time τ at a site non zero. The orientations of these local magnetisations $\{\hat{e}_i\}$ vary slowly whereas their actual magnitudes vary on the fast electronic time scale and are dependent on the particular orientational configurations, i.e.$\mu_k = \mu_k\{\hat{e}_i\}$. An adaptation of Spin Density Functional theory is used for the Grand Potential $\Omega\{\hat{e}_i\}$ of a system of electrons constrained to produce local magnetisations along a prescribed set of directions. For a given set $\{\hat{e}_i\}$, the local SDF approximation produces a set of local moments $\{\mu_i\}$. The second step is to assume that in the reduced phase space (labelled by $\{\hat{e}_i\}$) the long time averages are found from averaging over the ensemble of these configurations with measure $P\{\hat{e}_i\} = \exp -\beta\Omega\{\hat{e}_i\} / \prod_i \int d\hat{e}_i \exp -\beta\Omega\{\hat{e}_i\}$ $(\beta = (k_B T)^{-1})$ and the Free Energy is $F = -k_B T \log \prod_i \int d\hat{e}_i \exp -\beta\Omega\{\hat{e}_i\}$. In a way $\Omega\{\hat{e}_i\}$ is a complicated 'spin' Hamiltonian.

In the past [34] a mean field theory was formulated to make this theory tractable. Calculations have been carried out using the SCF-KKR-CPA method and have found the sizes of the local moments $\langle \mu_i\{\hat{e}_j\} \rangle_{\hat{e}_i} = \mu_i(\hat{e}_i) = \bar{\mu}$ for various systems in their paramagnetic states [11] including one of $1.91\mu_B$ for b.c.c. Fe. As for disordered alloys, the Coherent Potential Approximation (CPA) fixes an effective medium to treat the motion of an electron on the average through the orientational configurations [34] and sticks neatly to the tenor of a mean field theory. In contrast, nickel is unable to support a finite local moment on the average since the magnitude of the moment associated with a particular orientation is strongly dependent on its environment. The mean field theory then reduces to conventional Stoner theory with all the shortfalls which that entails.

We improved the above theory [32] by including a description of the Onsager cavity field. Once again a small magnetic field $\{\mathbf{h}_i\}$ is applied to the paramagnetic system. The induced magnetisation on a site i $\mathbf{M}_i = \int d\hat{e}_i \mu_i(\hat{e}_i)\hat{e}_i P(\hat{e}_i)$ can be written to lowest order in the applied field as a sum of two parts,$\mathbf{u}_i = \int d\hat{e}_i \delta\mu_i(\hat{e}_i)\hat{e}_i/4\pi$ and $\bar{\mu}\mathbf{m}_i = \bar{\mu}\int d\hat{e}_i \hat{e}_i \delta P(\hat{e}_i)$ and consequently the paramagnetic susceptibility $\chi_{ij} = \chi^u_{ij} + \chi^m_{ij}$. The first term $\mathbf{u}_i$ describes how the magnitudes of the local moments respond whereas the second $\bar{\mu}\mathbf{m}_i = \bar{\mu}\langle\hat{e}_i\rangle$ describes how they tend to align with the field. If we apply Onsager's reaction field idea to the response of the C.P.A. effective medium itself and subtract those changes $\{\delta\mathbf{u}^{(i)}_l\}$,$\{\bar{\mu}\delta\mathbf{m}^{(i)}_l\}$ that are derived from the induced magnetisation at site i we find to lowest order $\bar{\mu}\mathbf{m}_i = (\beta/3)\sum_{l\neq i}(S^{mm}_{il}(\mathbf{m}_l - \delta\mathbf{m}^{(i)}_l) + \bar{\mu}S^{mu}_{il}(\mathbf{u}_l - \delta\mathbf{u}^{(i)}_l)) + (\beta/3)\sum_l(\bar{\mu}^2\delta_{il} + \bar{\mu}\Sigma_{il}(1-\delta_{il}))\mathbf{h}_l$. In [10] it was found that $S^{mm}_{il} = \delta^2\bar{\Omega}/\delta\mathbf{m}_i\delta\mathbf{m}_l$ and $S^{mu}_{il} = \delta^2\bar{\Omega}/\delta\mathbf{m}_i\delta\mathbf{u}_l$ where $\bar{\Omega}$ is the Grand Potential averaged over all configurations within the C.P.A. The SCF-KKR-CPA method also provides an expression for $\mu_i(\hat{e}_i)$ the magnitude of the moment on a site i which is orientated along a unit vector $\hat{e}_i$ and the orientations on all the other sites are averaged over [34]. In terms of a partially averaged single electron Green function $\mu_i(\hat{e}_i) = -Im/\pi \int d\varepsilon f(\varepsilon - \nu)\int d\hat{e}_i tr\sigma\cdot\hat{e}_i\langle\tilde{G}(\mathbf{r}_i,\mathbf{r}_i;\varepsilon)\rangle_{\hat{e}_i}$ ($f(\varepsilon-\nu)$ is the Fermi factor, ν the chemical potential) and with the Onsager correction we can find an explicit expression for the form $\mathbf{u}_i = (1/4\pi)\int d\hat{e}_i\delta\mu_i(\hat{e}_i,\{\mathbf{m}_l - \delta\mathbf{m}^{(i)}_l\},\{\mathbf{u}_l - \delta\mathbf{u}^{(i)}_l\}) = \sum_{l\neq i}(\gamma^{um}_{il}(\mathbf{m}_l - \delta\mathbf{m}^{(i)}_l) + \sum_l\gamma^{uu}_{il}(\mathbf{u}_l - \delta\mathbf{u}^{(i)}_l) + \sum_l\chi^0_{il}\mathbf{h}_l$. The quantities S^{mm}_{ij},S^{mu}_{ij},γ^{um}_{ij},$gamma^{uu}_{ij}$,Σ_{ij} and χ^0_{il} are all available from the basic SCF-KKR-CPA calculations of the paramagnetic state and are all temperature dependent from single particle excitations.

Following our earlier discussion we proposed $\bar{\mu}\delta\mathbf{m}^{(i)}_l = \chi^m_{li}\mathbf{h}^C_i = \chi^m_{li}\chi^{-1}_{ii}\mathbf{M}_i$ and $\delta\mathbf{u}^{(i)}_l = \chi^u_{li}\chi^{-1}_{ii}\mathbf{M}_i$ in terms of the total induced magnetisation $\mathbf{M}_i$ on a site i. Finally in wave-vector space the set of coupled equations are found

$$\chi^m(\mathbf{q}) = (\beta/3)(S^{mm}(\mathbf{q})\chi^m(\mathbf{q}) + \bar{\mu}S^{mu}(\mathbf{q})\chi^u(\mathbf{q}) - \Lambda_1(\chi^m(\mathbf{q}) + \chi^u(\mathbf{q})) + (\beta/3)(\bar{\mu}^2 + \bar{\mu}\Sigma(\mathbf{q})) \tag{54}$$

$$\chi^u(\mathbf{q}) = \gamma^{um}(\mathbf{q})\chi^m(\mathbf{q}) + \gamma^{uu}(\mathbf{q})\chi^u(\mathbf{q}) - \Lambda_2(\chi^m(\mathbf{q}) + \chi^u(\mathbf{q})) + \chi^0(\mathbf{q}) \tag{55}$$

with $\Lambda_1 = \chi^{-1}_{ii}\int d\mathbf{q}(S^{mm}(\mathbf{q})\chi^m(\mathbf{q}) + \bar{\mu}S^{mu}(\mathbf{q})\chi^u(\mathbf{q}))$ and $\Lambda_2 = \chi^{-1}_{ii}\int d\mathbf{q}(\gamma^{um}(\mathbf{q})\chi^m(\mathbf{q}) + \gamma^{uu}(\mathbf{q})\chi^u(\mathbf{q}))$. On integrating eq.1 and 2 over $\mathbf{q}$ we find $\chi_{ii} = \beta\bar{\mu}^2/3 + \chi^0_{ii}$ from which a high temperature of the magnetic correlations can be extracted,$\langle M^2_i\rangle \approx \bar{\mu}^2 + 3\chi^0_{ii}/\beta$.

For a rigid local moment system in which itinerancy effects are small S^{mu},γ^{um},γ^{uu} and χ^0 vanish and we obtain the classical Heisenberg results with 'spins' of length

$\bar{\mu}$. As shown by ref.[33], the approach is equivalent to the spherical model. On the other hand, in a system where no 'local moment' is set up on the average in the paramagnetic phase,$\bar{\mu} = 0$ and S^{mm},S^{mu} and γ^{um} also vanish and $\chi^{m}(\mathbf{q}) = 0$,$\chi(\mathbf{q}) = \chi^{u}(\mathbf{q}) = \chi^{0}(\mathbf{q})/(1 - \gamma^{uu}(\mathbf{q}) + \Lambda_2)$ with $\Lambda_2 = (\chi^{0}_{ii})^{-1} \int d\mathbf{q}\gamma^{uu}(\mathbf{q})\chi^{u}(\mathbf{q})$. γ^{uu} is a product of a Stoner exchange-correlation term and a Pauli susceptibility χ^{0}. The theory now has the form of the static,high temperature limit of those set up to describe weak itinerant ferromagnets [35]. The interactions between spin fluctuations are dealt with self consistently through Λ_2. Recent work has adapted and extended these ideas to compositional order in alloys [22].

Comparison between theory and experiment [36] of the static uniform susceptibility for b.c.c iron (the experimental values are for $Fe(5.7\%Si)$) was made [32]. The Curie temperatures T_c of $1015K$ and $1040K$ were in very good agreement (the 'mean field' estimate was 1280K [31]). For temperatures in excess of $T_c + 300K$, both fit Curie-Weiss (C-W) laws $\chi = C/(T - \theta)$, each with $\theta \approx 1100K$. The effective moments,μ_{eff} ($C = \mu^2_{eff}/3$), of $1.96\mu_B$ and $3.13\mu_B$ respectively are not in such good agreement although the theoretical value is slightly enhanced over the 'local' moment value of $1.91\mu_B$. On analysing the $\mathbf{q}$-dependence of the susceptibility we find that at $1.25T_c$, for $|\mathbf{q}| \leq 1\dot{A}^{-1}$, $\chi(\mathbf{q})$ varies as $\chi(0)/(1 + q^2/\kappa^2)$ with the inverse correlation length $\kappa \approx 0.37\dot{A}^{-1}$. (This is similar to analogous results for a nearest neighbor Heisenberg model with the same T_c.) This value compares favorably with $\kappa \approx 0.4\dot{A}^{-1}$ quoted by Shirane et al. [37] who interpreted their inelastic neutron scattering experiments in terms of simple paramagnetic scattering. In summary we find that a Heisenberg model provides a qualitative description of b.c.c. iron.

Our earlier 'mean field' theory [34] had not been able to give a satisfactory treatment for nickel. The new developments described above provide a reasonable account of experimental data. No 'local' moments,$\bar{\mu} = 0$,are found and $\chi(\mathbf{q})$ is determined solely by $\chi^{u}(\mathbf{q}) = \chi_0(\mathbf{q})/(1 - \gamma^{uu}(\mathbf{q}) + \Lambda_2)$, $\Lambda_2 = (\chi^{0}_{ii})^{-1} \int d\mathbf{q}\gamma^{uu}(\mathbf{q})\chi^{u}(\mathbf{q})$. Both the theoretical and experimental results show approximate C-W behaviour. Although at 450K and some 200K lower than the experimental value, T_c is substantially reduced from the Stoner value of 3000K. Effective moments of $1.21\mu_B$ from theory and $1.6\mu_B$ are also in fair agreement. Evidently there is little connection with the low temperature saturated magnetisation per atom of 0.6 μ_B as accorded by a Rhodes-Wohlfarth plot [34]. For $|\mathbf{q}| \leq 1\dot{A}^{-1}$ the wave-vector dependent susceptibility $\chi(\mathbf{q}) \approx \chi(0)/(1 + q^2/\kappa^2)$ where κ^2 has roughly the same temperature dependence as $\chi(0)^{-1}$, a feature which has also been noted from experiment [34]. At $T = 1.25T_c$,$\kappa \approx .28\dot{A}^{-1}$ which agrees well with the inverse spin correlation length of $\approx .22\dot{A}^{-1}$ extracted from the simple paramagnetic scattering analysis of Shirane et al.[37] of their inelastic neutron scattering data. In summary then we find that iron is related superficially to a Heisenberg model but nickel can be analysed in terms of traditional Stoner theory although spin fluctuations have drastically renormalised the exchange interaction and lowered T_c.

In this article the spin polarised electronic structure of both magnetic metals and alloys has been focussed upon which has included a 'first principles' framework for

Onsager cavity fields in itinerant electron systems. This has been shown to provide the basis for a theory for a 'first principles' description of the mechanism driving the magnetic phase transition in pure metals at finite temperatures and in alloys it demonstrates how compositional ordering depends upon the state of magnetic order. Recent work has combined these treatments of magnetic and compositional fluctuations and has obtained a parameter-free microscopic theory of magnetic and compositional ordering in alloys. Applications of this work is in progress. A picture in which these magnetic and compositional fluctuations are subtly interrelated via the underlying electronic structure will emerge and the results of this theory will be able, step by step, to be compared with measurements from a wide range of experiments. The prospect of understanding a wide range of magnetic properties of alloys is in sight.

5 References

1. S.Chikazurin and C.D.Graham,*Magnetism and Metallurgy* eds.A.E.Berkowitz and E.Kneller, **II**,Academic Press, (1969),p.577.

2. R.Kuentzler,*Physics of Transition Metals 1980* eds. P.Rhodes, Institute of Physics Conference Series no.55, (1980),p.397.

3. W.Kohn and P.Vashishta,*Physics of Solids and Liquids* eds. B.I.Lundqvist and N.March, Plenum,(1982).

4. E.C.Stoner,*Proc.Roy.Soc.***A139**,(1939),339.

5. E.P.Wohlfarth,*Rev. Mod. Phys.*,**25**,(1953),211.

6. C.Herring,*Magnetism* **IV** eds. G.T.Rado and H.Suhl,Academic Press,(1966).

7. A.K.Rajagopal,*Adv.Chem.Phys.* **41**,(1980),59.

8. O.Gunnarsson, and B.I.Lundqvist,it Phys.Rev. **B13**,(1976),4274.

9. V.L.Moruzzi,J.F.Janak,and A.R.Williams,it Calculated Electronic Properties of Metals,Pergamon,(1978).

10. G.M.Stocks and H.Winter,*Electronic Structure of Complex Systems* eds. P.Pharisea W.M.Temmerman, NATO ASI series **B113**,Plenum, (1984).

11. D.D.Johnson,F.J.Pinski and G.M.Stocks,*J.Appl.Phys.*, **57**,(1985),3018.

12. P.Soven,*Phys.Rev.***156**,(1967),809.

13. B.L.Gyorffy and G.M.Stocks, *Electrons in Finite and Infinite Structures* eds. P.Phariseau and L.Scheire,NATO ASI series **B24**,Plenum,(1977). and *Electrons in Disordered Metals and at Metallic Surfaces* eds. P.Phariseau,B.L.Gyorffy and L.Scheire,NATO ASI series **B42**,Plenum,(1979).

14. J.S.Faulkner and G.M.Stocks,*Phys.Rev.* **B21**,(1980),3222.

15. B.L.Gyorffy and M.J.Stott,**Band Structure Spectroscopy of Metals and Alloys** eds. D.J.Fabian and L.M.Watson,Academic Press,(1973).

16. P.J.Durham,B.L.Gyorffy,and A.J.Pindor,*J.Phys.***F10**,(1980),661.

17. D.D.Johnson,F.J.Pinski and J.B.Staunton,*J.Appl.Phys.***61**,(1987),3715.

18. A.R.Williams,V.L.Moruzzi,A.P.Malozemoff and K.Terakura,IEE Trans. Magn. MAG-19,(1983).

19. C.S.Wang,B.M.Klein and H.Krakauer,*Phys.Rev.Lett.***54**,(1985),1852.

20. A.R.Williams,R.Zeller,V.L.Moruzzi,C.D.Gelatt and J.Kubler, *J.Appl.Phys.***53**,(1982),2019.

21. B.L.Gyorffy and G.M.Stocks,*Phys.Rev.Lett.***50**,(1983),374. B.L.Gyorffy,G.M.Stocks,B.Ginatempo,D.D.Johnson,D.M.Nicholson,F.J.Pinski, J.B.Staunton and H.Winter,*Phil.Trans.R.Soc.Lond.A*,**334**,(1991),515.

22. J.B.Staunton,D.D.Johnson and F.J.Pinski,(1993) to be submitted to Phys.Rev.B.

23. J.B.Staunton,D.D.Johnson and F.J.Pinski,*Phys.Rev.Lett.***65**,(1990),1259.

24. M.A.Krivoglaz,*Theory of X-ray and Thermal Neutron Scattering by Real Crystals*,Plenum,(1969). and D.de Fontaine, *Solid State Physics***34** eds. Ehrenreich,Seitz and Turnbull,(1979).

25. W.Marshall,*J.Phys.***C1**,(1968),88.

26. J.W.Cable and R.A.Medina,*Phys.Rev.***B13**,(1976),4868.

27. B.D.Rainford,*J.Phys.Colloq.***43**,C-7,(1982),33.

28. J.W.Cable,H.R.Child, and Y.Nakai,*International Conference on Neutron Scattering*,Grenoble,France,July 12-15,(1988).

29. I.Mirebeau,M.C.Cadeville,G.Parette and I.A.Campbell,*J.Phys.* **F12**,(1982),25. and N.K.Jaggi,L.H.Schwartz,H.K.Wang and J.B.Ketterson,*J.Mag.Magn.Mat.***49**,(1985),1.

30. *Electron Correlation and Magnetism in Narrow Band Systems* ed. T.Moriya,Berlin:Springer,(1981).

31. D.M.Edwards,*J Mag.Magn.Mat.***45**,(1984),151.

32. J.B.Staunton and B.L.Gyorffy,*Phys.Rev.Lett.***69**,(1992),371.

33. R.Brout and H.Thomas,*Physics***3**,(1967),317.

34. B.L.Gyorffy,A.J.Pindor,J.Staunton,G.M.Stocks, and H.Winter, *J.Phys.***F15**,(1985),1337.
J.Staunton,B.L.Gyorffy,G.M.Stocks and J.Wadsworth, *J.Phys.***F16**,(1986),1761.

35. T.Moriya and A.Kawabata,*J.Phys.Soc.Jpn.***34**,(1973),639.
K.K.Murata and S.Doniach,*Phys.Rev.Lett.***29**,(1972),285.
G.G.Lonzarich and L.Taillefer,(1985),J.Phys.C**18**,4339.
T.Moriya and Y.Takahashi,*Ann.Rev.Mat.Sci.***14**,(1984),1.
M.Cyrot,J.Mag.Magn.Mat.**45**,(1984),9.

36. S.Arajs and D.S.Miller,*J.Appl.Phys.***31**,(1960),986.
W.Sucksmith and R.R.Pearce,*Proc.Roy.Soc.*bf A167,(1938),189.

37. G.Shirane,P.Boni and J.P.Wicksted,*Phys.Rev.***B33**,(1986),1881.
Y.J.Uemura,G.Shirane,O.Steinsvoll and J.P.Wicksted, *Phys.Rev.Lett.***51**,(1985),232

38. R.Rhodes and E.P.Wohlfarth, *Proc.Roy.Soc.***273**,(1963),247.

FIRST PRINCIPLES INVESTIGATION OF EPITAXIAL INTERFACES USING LMTO-SUPERCELL APPROACH

G. P. DAS
Solid State Physics Division,
Bhabha Atomic Research Centre,
Bombay 400085, India

ABSTRACT

Interfaces with high structural perfection can be grown with modern epitaxial techniques. Such epitaxial heterojunctions are the building blocks of the artificial multilayers like quantum well structures and superlattices, which have novel and tunable properties. Various theoretical attempts have been made to get a microscopic understanding of the electronic properties of these interfaces. In this talk, we shall use the self-consistent LMTO-supercell approach based on local density approximation (LDA), for handling epitaxial metal-semiconductor interfaces and calculating its Schottky barrier height (SBH). We deal with a specific example of $NiSi_2/Si(111)$ interface, whose SBH depends on its growth orientation viz. whether it is A- or B-type. A detailed analysis of the interface specific electronic properties, followed by the so-called 'dipole approach' for SBH determination will be presented. Finally we shall discuss the success and failure of LDA for handling the problem of Schottky barrier.

1. Introduction.

An interface between two crystalline solids is called epitaxial when there exists a well defined structural and orientational relationship between the two single crystals on either side of the contact plane. Both or either of the constituents may be metals, semiconductors, insulators or (ceramics). Such hetero-epitaxial systems have novel properties and are becoming increasingly important for advanced electronic, optoelectronic and materials science applications. Multiple epitaxial interfaces with alternate thin layers of two solids give rise to artificially modulated multilayer structures like superlattices and quantum wells. A large volume of literature has already been published in this field in recent past [1–4]. Single hetero-epitaxial interfaces may be classified into two groups [5]. If the structures of the two solids constituting the interface are the same with a marginally different lattice constants, then only the compositions differ e.g. a semiconductor heterojunction like $GaAs/Al_xGa_{1-x}As$. If both the structure and composition differ between the two sides, as in the case of a metal-semiconductor (M-S) interface like $NiSi_2/Si$ or an insulator-semiconductor (I-S) interface like CaF_2/Si, then the lattice matching criterion between the two half crystals is desirable but not an essential requirement for the formation of an epitaxial interface. Although the chemistry and metallurgy of these interfaces have been widely studied, the electronic properties are still not properly understood. Many of the interface-specific electronic properties like the band offset in a semiconductor heterojunction or the Schottky barrier height in a metal-semiconductor interface have a strong dependence on the exact interface structure. Polycrystallinity, stoichiometry and other defects complicate the microscopic understanding of the formation mechanism of band offset or Schottky barrier height. Density functional calculations have therefore been attempted only on atomically abrupt epitaxial interfaces [6], in order to throw some insight into the underlying physics of these problems.

There are two broad lines of attack to these classes of epitaxial interfaces as well as superlattices [7] viz. (a) Supercell method and (b) Green's function method. In both methods, first principles as well as empirical implementations are possible and have been performed with various degrees of sophistications. The most appealing theoretical method is probably the one based on a self-consistent Green's function technique [8,9]. But the supercell method is simpler and more straightforward, though computationally heavier. In fact, the supercell technique, implemented in conjunction with some band structure methods such as the linear combination of muffin-tin orbital (LMTO), the linear augmented plane wave (LAPW) or pseudopotential, is now being extensively used to investigate single epitaxial hetero-interfaces as well as superlattices [10,11]. In this contribution, I shall confine myself to the LMTO-supercell approach and illustrate its applicability to calculate the electronic structure and LDA Schottky barrier height in a representative M-S interface viz. $NiSi_2/Si(111)$ [12].

We shall start in section 2 by summarizing the existing theoretical models of Schottky barrier and emphasizing the need to go to a full self-consistent electronic structure calculation. In section 3 we shall describe the structure of $NiSi_2/Si(111)$ interface. Section 4 will be devoted to some necessary details of the LMTO-supercell approach which has been used in our calculations. The results on electronic structures of A- and B-type $NiSi_2/Si(111)$ interfaces will be presented in section 5, while the calculation of SBH will be covered in details in section 6. We shall conclude this article with a discussion on how good is LDA for determining SBH and its orientation dependence.

2. Theoretical models of Schottky barrier height.

In a M-S junction, the semiconductor band adjusts itself to the metal Fermi level by means of charge flow and a rectifying potential barrier is created, which is known as the Schottky barrier (SB). For an n-type semiconductor, the barrier height ϕ_{B_n} is conventionally defined as the relative position of the Fermi level with respect to the conduction band (CB) minimum. For a p-type semiconductor, the SBH ϕ_{B_p} is defined as the difference between the Fermi level and the valence band (VB) maximum. ϕ_{B_n} and ϕ_{B_p} together should add up to the value of the bulk band gap (E_g) of the semiconductor. If ϕ_{B_p} is vanishingly small, one gets an ohmic junction without any rectification. It is to be noted that the interface dipole, which builds up at the junction due to charge transfer, extends usually over atomic distances (approximately $10\AA$) from the interface. So, for the purpose of estimating the SBH, the band bending effect (usually extending to $\geq 500\AA$) may be neglected, and the *'local'* bands may be assumed to be flat.

According to the original model proposed by Schottky [13], it was assumed that there is no electron transfer across the junction, implying that there is no modification of electron states due to interface formation. The barrier height ϕ_{B_n} is then just the difference between the metal work function ϕ_m and the semiconductor electron affinity χ_s i.e.

$$\phi_{B_n} = \phi_m - \chi_s \tag{1}$$

and hence

$$\phi_{B_p} = E_g - (\phi_m - \chi_s) \tag{2}$$

Although (1) is usually attributed to Schottky, it was also stated implicitly by Mott [14] and hence it is sometime referred to as the Schottky-Mott model. According to this model, for junctions formed between different metals and the same semiconductor, the barrier height and therefore the degree of rectification increases linearly with metal work function;

$$\Delta\phi_B = s\ \Delta\phi_m \tag{3}$$

with $s = 1$ in the Schottky model. Experiments however indicate $s \simeq 0.1$ in most cases [15], suggesting that the SBH for a particular semiconductor remains more or less constant and independent of (a) the work function of the metal deposited, (b) the degree of doping of the semiconductor, (c) the crystallographic orientation of the semiconductor surface etc. The failure of this simplistic Schottky-Mott model is attributed to the fact that it describes the barrier formation in terms of quantities that are characteristics for the two separate bulk components, viz. the metal and the semiconductor.

This prompted Bardeen [16] to underscore the importance of *surface states* in pinning the Fermi level. These states, lying at a well defined energy position E_s with respect to the CB bottom, will be partially filled, thus pinning the Fermi level of the clean surface. When a metal comes in contact with the semiconductor, a charge flow between the metal and these surface states will occur until a dipole Δ (whose magnitude compensates the difference between ϕ_m and χ_s) will be formed at the interface. In Bardeen model, therefore

$$\phi_{B_n} = E_s \tag{4}$$

Then came the metal-induced gap states (MIGS) model of Heine [17] who pointed out that when the metal conduction band overlaps with the semiconductor energy gap, states will be formed from the tailing of the metal wave functions into the semiconductor gap. Subsequently Louie and Cohen [18] quantified the MIGS concept by performing a realistic LDA calculation on Al/Si(111) interface and actually showing the onset of MIGS in the local density of states (DOS). Tejedor et al [19] further isolated the essential ingradient of Bardeen-Heine picture of Fermi level pinning by introducing the concept of *charge neutrality level* (CNL). The CNL is nothing but an energy E_n within the semiconductor band gap, where the MIGS wave function crosses over from being largely VB derived to being CB derived. Thus the role of E_n of the semiconductor at the interface is equivalent to that of Fermi level E_F of the metal. Later Tersoff [20] refined this idea and proposed a very simple technique for calculating the *'neutrality point'* from the bulk semiconductor band structures without any adjustable parameter. The Fermi level is pinned at E_n for both n- and p-doped semiconductors, and the SBH is given by

$$\phi_{B_n} = E_n + \gamma_{i_s} \left(\phi_m - \chi_s - E_n\right) \tag{5}$$

where γ_{i_s} is a screening factor related to the semiconductor dielectric function. Tersoff's prediction of SBH's differ from experimental data by $\sim \pm 0.2eV$. A schematic history of all these model theories of Schottky barrier is summarized in table 1.

Table 1. Various proposed models, empirical theories and self-consistent calculations of Schottky barrier.

	Year	Author(s)	Models and calculations
.	1939	Schottky-Mott	Work function diff.
.	1947	Bardeen	Surface state
.	1965	Heine	Metal induced gap state
.	1976	Louie & Cohen	MIGS [in Al/Si(111)]
.	1977	Tejedor et al	Charge Neutrality Level
.	1984	Tersoff	CNL + MIGS
.	1974	Phillips	Chemical bond model
.	1978	Brillson	Chemical reaction
.	1981	Freeouf & Woodall	Effective Work Function
.	1979	Spicer	Defect pinning
.	1982	Dow & Allen	Defect pinning
.	1983	Mönch	Defect pinning
.	1987	Walukiewicz	Defect pinning
.	1988	Ludeke et al	Defect pinning
.	1987	Bisi & Ossicini	LMTO-supercell
.	1989	Das et al	LMTO-supercell
.	1990	Fujitani & Asano	LMTO-supercell
.	1990	Schilfgaarde & Newman	LMTO-supercell
.	1990	Pickett & Erwin	LCAO & LAPW-supercell
.	1991	Ciraci et al	Nonlocal pseudopotential

There is another school of thought which advocates the *extrinsic* aspects of interface formation to be the driving force for predicting SBH. The most prominent amongst these are the interface chemical bond models [21,22], the effective work function model [23] and various defect models[24–28] (see table 1). The defect models, for example, essentially suggest that the deposition of a metal on a semiconductor induces defect levels at the interface (donor level for n-type and acceptor level for p-type), which pin the metal Fermi energy. Its proponents site in its favor, a number of experimental evidences, whose status report can be obtained in the literature [29–31]. Lately there have also been attempts [32–35] to reconcile these two extreme viewpoints and to establish some kind of duality between *MIGS pinning* and *defect pinning*. While there is no reason to disbelieve the non trivial role played by defects in SB formation, it does not explain the well established pinning behaviour in defect free interfaces, as has been succinctly put forward by Tersoff in ref. [36].

One important lacuna in all the abovementioned model theories of SBH is the assumption that the dipole formation at the interface can be derived from the bulk electronic properties of the constituents alone. Consequently these models are not strictly *interface specific* in the sense that they contain no information about the atomic structure of the interface, the growth direction of the interface etc. In order to study these interface specific features, it is necessary to do a *full microscopic electronic structure calculation*, taking into account the actual atomic arrangement at the interface. Following the success of LDA in predicting the band offsets in semiconductor heterojunctions [10,11], similar calculations have been attempted on some ideal M-S interfaces and most of these (see table 1) are based on LMTO supercell approach[12,37–39]. Recently LAPW [40] and pseudopotential [41] calculations have also been performed on some idealized M-S interfaces. The calculational procedures though inherently accurate, yield mixed success. The atomic structure of M-S interfaces, being relatively inaccessible spectroscopically, are not so well known as those of lattice matched semiconductor heterojunctions. We have therefore chosen for our theoretical investigation [12], one of the most ideal and experimentally well studied system viz. $NiSi_2/Si(111)$ which is described in the following section.

3. Structure of $NiSi_2/Si(111)$ interface.

Epitaxial silicide - Si interfaces have shown very high structural perfection and are relatively 'clean' i.e. without any external impurity at the interface [42]. So these interfaces have been extensively studied using various macroscopic and microscopic probes. Two chief obstacles to understand SB formation, viz. the obscurity of the atomic structure of the M-S interface and presence of defect states, can be removed in a well defined silicide/silicon interface. In particular, $NiSi_2$ and $CoSi_2$ are the two most well known cubic silicides with lattice constants close to that of Si, which can be grown epitaxially on $Si(111)$ with a strict continuation of the lattice symmetry. Being abrupt both in structure and composition, these are ideal systems from the point of view of first principles calculation. Furthermore, there is unique interest in $NiSi_2/Si(111)$ interface, which has been grown in two distinct

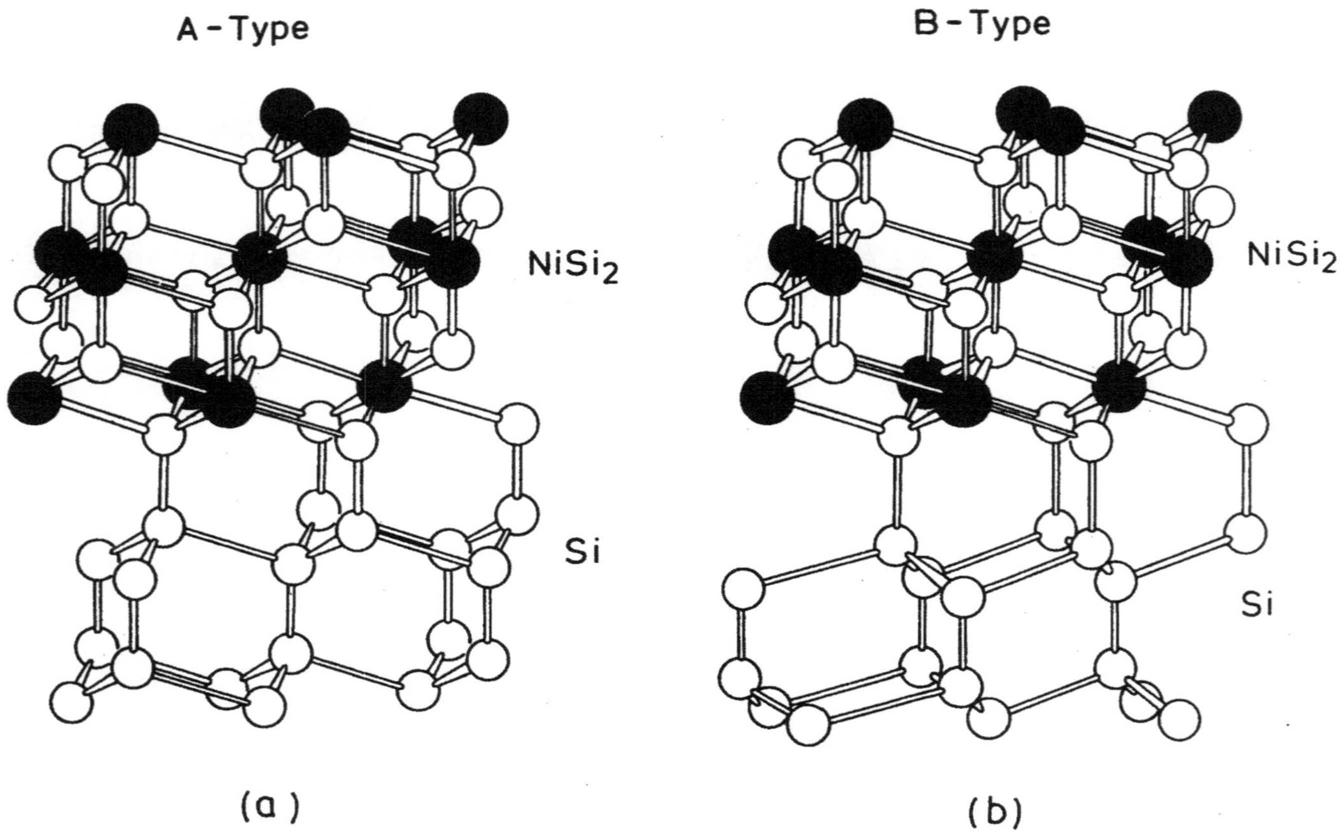
A - Type
B - Type
NiSi2
Si
NiSi2
Si
(a)
(b)

structures, called type-A and type-B, by carefully controlling the rate of deposition of Ni and using what is known a 'template technique' [43]. Type-A silicide has the same orientation as the Si substrate, while type-B is rotated 180° about the Si(111) axis (fig. 1). Tung [44] has recently reviewed on the growth and characterization of $NiSi_2/Si$ interfaces.

The most striking observation first made by Tung [45] in these interfaces is that the B-type orientation has considerably higher (n-type) SBH compared to that of A-type; the magnitudes are respectively 0.79eV and 0.64eV, which have later been reproduced by other groups. Now the origin of this 0.14ev difference in SBH has been an enigma, even though various possible explanations have been put forward on the basis of heuristic arguments. SBH is the single most important electronic parameter characterizing a M-S interface and to understand its formation mechanism, one must know the exact atomic structures of the interface. A large number of interface sensitive experimental probes have been used to investigate the atomic structures of A- and B-type $NiSi_2/Si(111)$ systems (see ref. 6 for a brief overview). The high resolution cross-section TEM images of $NiSi_2/Si(111)$ interfaces show exceptional uniformity and sharpness. X-ray standing wave (XSW) and Rutherford Back Scattering (RBS) studies suggest that the Ni atom at the interface is 7-fold coordinated, with the silicide layer terminated by a silicon-layer. Later cluster calculations [46] as well as total energy calculations [47] confirm that the $NiSi_2/Si(111)$ interface is indeed 7-fold coordinated, in contrast to the closely related $CoSi_2/Si(111)$ interface, where the interfacial Co atom is 8-fold coordinated. For $NiSi_2/Si(111)$ interface, a slight contraction in the bond length across the interface has been independently observed by RBS-cum-ion-channeling, XSW, and more recently by X-ray interference experiments.

4. Details of calculation.

We have used self-consistent LMTO method [48–52] with Barth-Hedin form of the LDA exchange-correlation potential. The foundation of this method and its applications have been dealt with in great details in this school [53] by Andersen and by Skriver; so no attempt will be made here to discuss the methodology. LMTO calculations have been performed in conjunction with supercell geometry[12,54–56], and mostly using the atomic sphere approximation (ASA), although a few full potential (FP) results will also be discussed. The results reported here are obtained using the tight-binding representation (TB-LMTO) [56,57] of the basis in both levels of sophistications (i.e. ASA and FP). We have included s-, p- and d-partial waves in the basis set on all the spheres.

Both A- and B-type supercells can be described by a hexagonal (centric) basis (space group $R\bar{3}m$), with the c-axis along (111) direction being the 3-fold rotation axis. The small (0.46%) lattice mismatch between Si and $NiSi_2$ is neglected in our calculation, and Si bulk lattice constant has been used for the entire supercell. Futhermore, if we neglect the interface relaxation and retain the Si-Si bond length across the interface to be same as that of bulk Si (i.e. 2.35Å), the interface distance d_{IF} turns out to be 3.52Å. The supercells which are repeated with Bloch periodicity along the z-axis, contain two identical interfaces buried between n triple layers of $NiSi_2$ and m double layers of Si, so that each formula unit becomes $(NiSi_2)_n(Si_2)_m$, designated as (n+m) supercell. In order to ensure close packing, it is necessary to include two empty spheres per unit cell of Si and one E-sphere per unit cell of $NiSi_2$. In the artificially constructed supercells, one must choose the numbers, positions and sizes of the atomic spheres appropriately so that the overlap between any two spheres, defined as $[(s_1 + s_2 - d) * 100 / s_1]$ remains within the permissible limit (~ 30 %) . We use the so-called atomic Hartree potential plot [54,55] details) to determine the relative sphere sizes at the interface. A-interface has an overall bcc packing i.e. ...ABC||ABC..., with all atomic spheres (including E-spheres) having equal radius. In case of B-interface, the packing changes at the interface to ...ABC||ACB... i.e. there is a stacking fault. Including the E-spheres, the formula unit corresponding to our (n+m) supercell is $(ESiNiSi)_n E(ESiSiE)_m$. The size of the supercell must be chosen to be large enough such that the central $NiSi_2$ and Si layers are bulk-like in the sense that the potentials, apart from a spatially constant contribution, agree with those of bulk $NiSi_2$ and Si respectively. Cell size convergence test is especially crucial for SBH determination, and can be achieved through the so-called *'Frozen potential'* calculations of E_F versus E_v profiles [58]. By 'trial and error' test on different (n+m) type-A supercells, it was found [12,54–56] that the (8+6) cell is optimum for the electronic structure and SBH study of $NiSi_2/Si(111)$ system. For this supercell, the total number of atoms in the unit cell (including E-spheres) is 57, that means 513 orbitals in total for spd-basis.

The Brillouin zone (BZ) integrations in k-space are performed by means of the tetrahedron method [59], implemented in its latest version, which avoids misweighting [60] and also corrects for errors caused by the linear approximation of the bands inside each tetrahedron [56]. The k-points are chosen on an equispaced mesh. For supercells, the c-axis dimension in real space is so large that the corresponding BZ is like a thin tablet. The reciprocal lattice vectors in the (111)-plane are therefore divided into 8 parts, and one (or at most two) division are used along the c-axis. A specific mesh is denoted, for example, as (8,8,2), which corresponds to 10 k-points in the irreducible BZ standing on the ΓMK plane.

5. Results on electronic structure.

Layer-projected densities of states (LDOS) of A- and B-type (8+6) supercells (fig. 2) show how the basic features of the electronic structure of a layer changes while going from the interfacial layer to the bulk-like layer. As the exact transition from a Si-like layer to a $NiSi_2$-like layer at the interface is rather unclear, we have chosen this line as midway between the last Si of the $NiSi_2$ triple layer at the interface and the last Si of the interfacial Si_2 double layer. Only half of the total number of layers is included in fig. 2, the other half being equivalent by symmetry of the

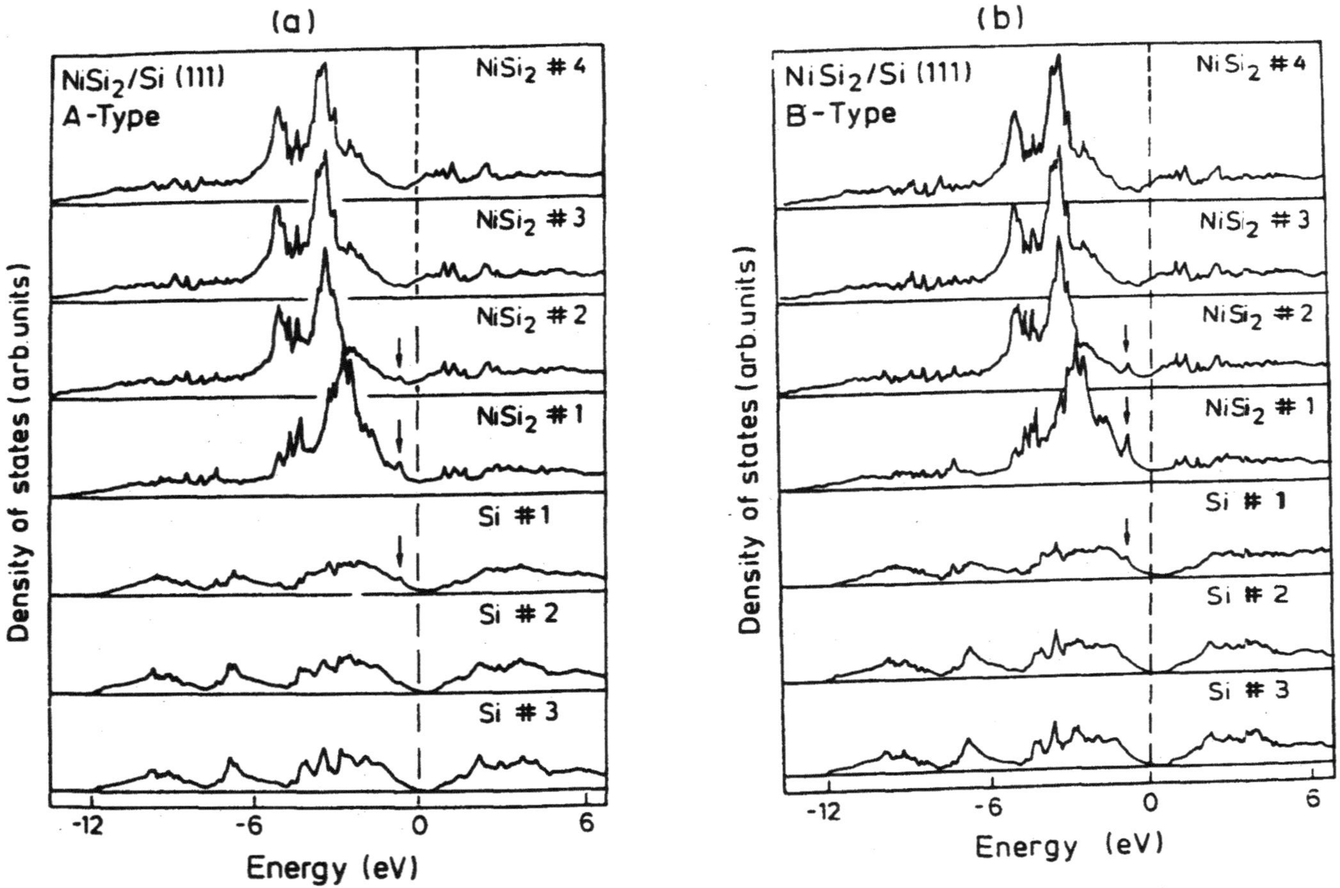
(a)
NiSi$_2$/Si (111)
A-Type
NiSi$_2$ # 4
NiSi$_2$ # 3
NiSi$_2$ # 2
NiSi$_2$ # 1
Si #1
Si # 2
Si # 3
Density of states (arb.units)
-12
-6
0
6
Energy (eV)
(b)
NiSi$_2$/Si (111)
B-Type
NiSi$_2$ # 4
NiSi$_2$ # 3
NiSi$_2$ # 2
NiSi$_2$ # 1
Si # 1
Si # 2
Si # 3
Density of states (arb.units)
-12
-6
0
6
Energy (eV)

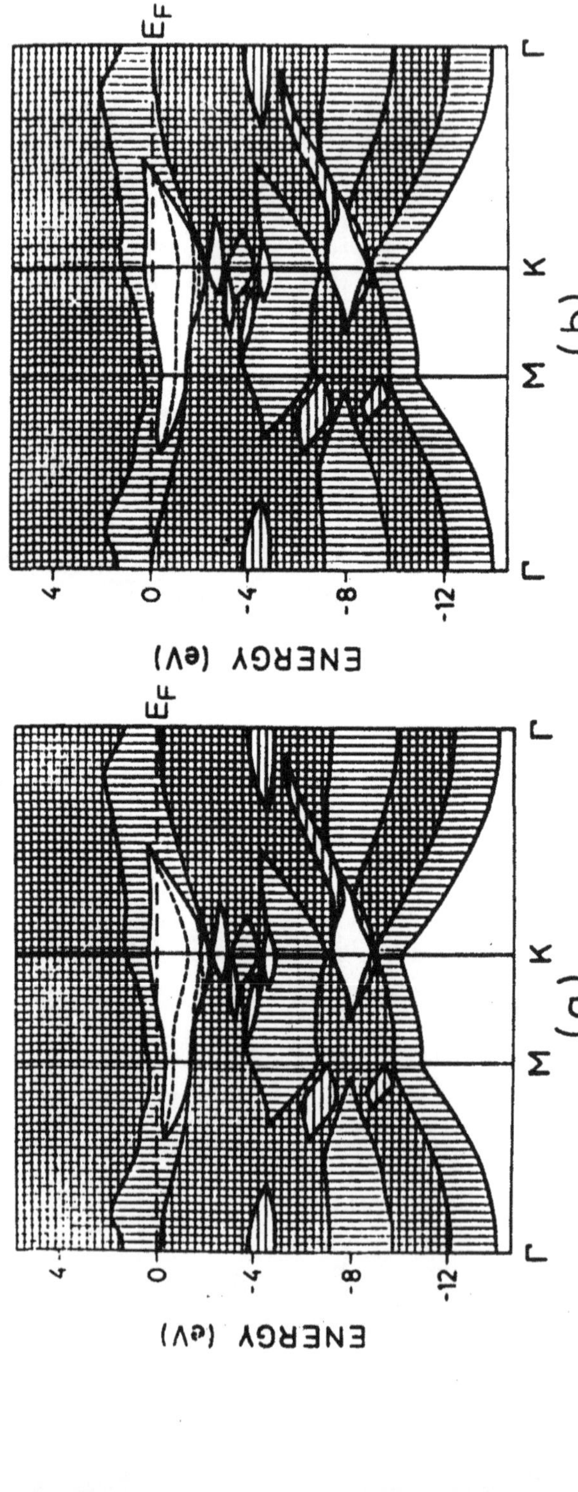
ENERGY (eV)
4
0
-4
-8
-12
Γ
M
K
Γ
E_F
(a)
(b)

supercell. There are three notable features in the LDOS of A-type interface. *Firstly* the bulk-like 3rd Si_2 layer (i.e. ESiSiE) clearly exhibits the LDA band gap $\sim 0.6eV$ and the familiar three-humped structure characteristic of bulk Si [61]. In contrast, the first and to some extent the second Si_2 LDOS show metallic character due to the presence of MIGS. *Secondly*, the Ni-$d(t_{2g})$ like bonding peak of the fourth bulk-like $NiSi_2$ layer is located about 5.3eV below E_F, the non-bonding Ni-$d(e_g)$ peak is located 3.7eV below E_F, while the anti-bonding peak having mostly Si character is very broad, and is centered ~2eV above E_F. All these features resemble the DOS of bulk $NiSi_2$ [62]. As we move to the first $NiSi_2$ layer at the interface, we observe an upward energy shift of the non-bonding peak by ~1eV w.r.t. the bulk-like layer; this is due to the shift of the electrostatic (or Madelung) potential while crossing the interface. The exact values of the 'dipoles' produced at the A- and B-interfaces have, in fact, been found to be 132mRy and 142mRy respectively, the resulting 10mRy (i.e. 0.14eV) difference being embarrassingly close to the observed SBH difference! *Thirdly*, a 'tiny' peak (shown by arrow in fig. 2) is seen in the LDOS of the interfacial layers ~0.7eV below E_F, whose intensity gradually vanishes as we proceed towards the bulk-like layers on either side. This is due to a localized interface state. All these three features exist in the LDOS of B-interface; only the intensity of the interface state peak is somewhat sharper.

Two dimensional band structure of the supercell, projected onto the interface plane (i.e. the ΓMK plane of the hexagonal BZ) clearly shows the band dispersions (fig. 3). Because of the fact that there are large number of bands (corresponding to the number of planes in the supercell) folded inside the first BZ, the 2D band structure looks like a 'bunch of spaghetti'. In order to bring out the essential features, we have simply indicated in fig. 3, the band edges of the bulk Si and bulk $NiSi_2$ bands. By aligning the two sets of bands (vertically hatched for $NiSi_2$ and horizontally hatched for Si) with respect to the electrostatic dipole created due to interfacing, we get the so-called common gap and the band edges along $\Gamma MK\Gamma$ contour. What is of special interest in the present context is the relative position of the VB top with respect to the Fermi level of the supercell. Even though the energy scale of this figure is rather crude, it can be seen that for the A-interface the Fermi level lies slightly above the VB top, while for the B-interface the two levels almost coincide, suggesting that $(E_F - E_v)$ has a higher value for A-type compared to the B-type interface. Also seen in the picture, a doubly degenerate interface band cutting right across the centre of the common gap near the symmetry line MK, and intersecting the Fermi surface between ΓM and ΓK. These interface band dispersion is more flat for B-type than in the A-type, resulting in a more pronounced peak of the van Hove singularity in LDOS of the B-type (fig. 2b). The bands are partially filled in both cases (approximately 1.3 electrons per interface and two dimensional cell), but the occupancies are slightly different as will be seen below. These interface bands should be observable in an angle resolved photoemission experiment performed *in situ* as a function of layer thickness. There have been very few cases [63] in which such localized interface states have been identified and

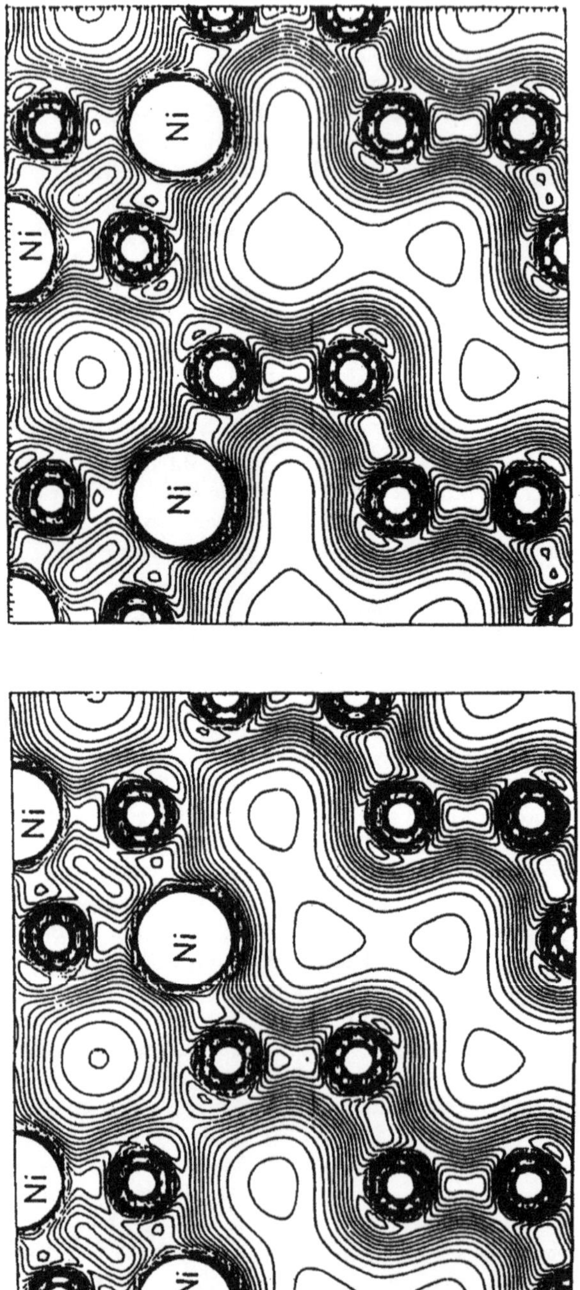
Ni
(b)
(a)

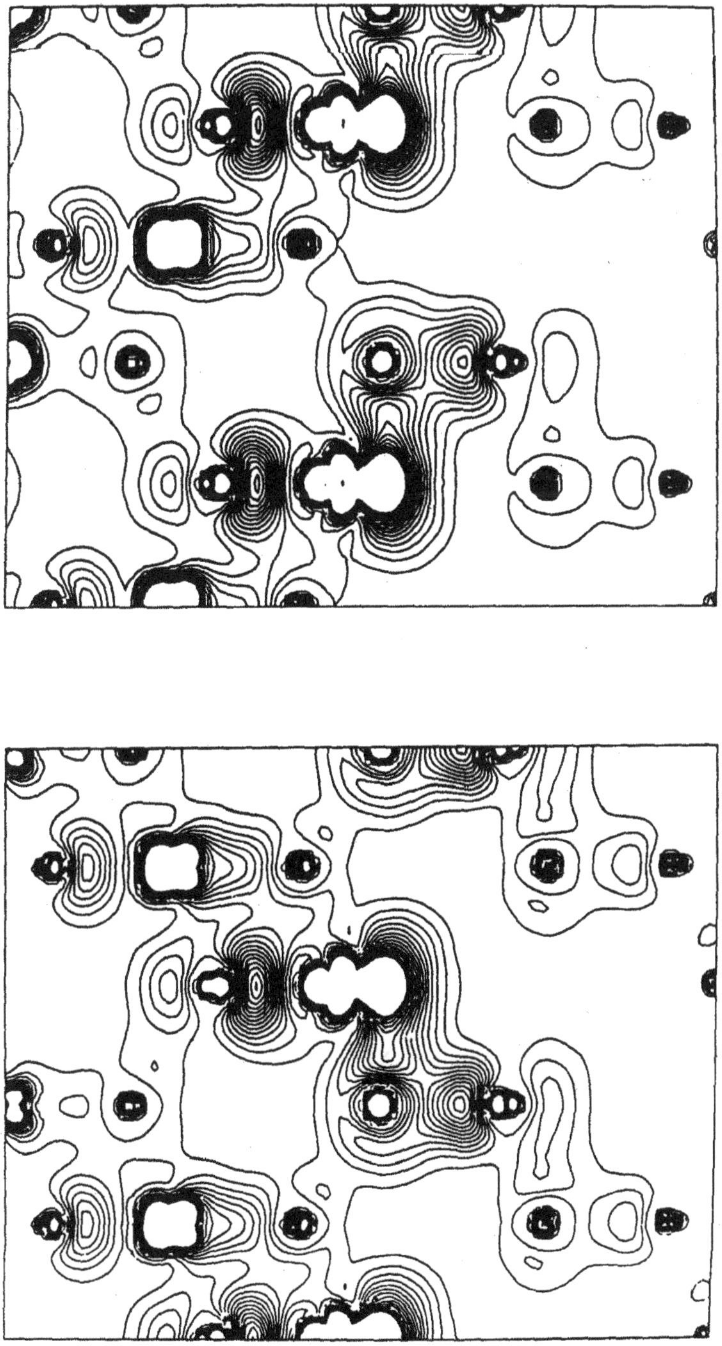

(b)

(a)

their band dispersions measured experimentally.

The valence electron charge density contours in the $(1\bar{1}0)$ plane (fig. 4) are obtained by solving the wave equation for potentials self-consistent in the ASA. The ASA charge density has been established [64] to be almost as accurate as those obtained from the best state of the art full potential or first principles pseudopotential calculation. For both A- and B-type interfaces, the $sp^3 - sp^3$ bonds on the Si side and the $d(t_{2g}) - sp^3$ bonds on the $NiSi_2$ side resemble very much the corresponding bulk bonds [12]. Furthermore, a $Si - Si$ bond is formed across both the interfaces. The subtle differences can be observed by means of a closer look into the charge distribution *only* of the interface bands (fig. 5). Common to all states in the interface band is an anti-bonding between the interfacial Ni atom and the Si atom lying along the z-direction inside the disilicide. On the opposite side, the 7-fold coordinated Ni atom misses the Si atom, and this anti-bonding is dangling. Comparing fig. 5a and fig. 5b one can see that more electrons percolate from $NiSi_2$ to Si side in A-type, as compared to the B-type interface, thus accounting for the slight difference in occupancies (viz. 1.333 electrons in A-type and 1.309 electrons in B-type, per interface and two dimensional cell). This is reminiscent of the recent theoretical calculation of ballistic electron emission (BEEM) spectrum [65] of these two interfaces, which shows that the transmission probability of electrons across the M-S junction is higher for A-type than for B-type.

6. SBH calculation.

In absence of a full fledged density functional theory of SBH, we shall outline how to evaluate this quantity from the Kohn-Sham LDA band structure [66]. The details of this approach is published elsewhere [55]. Here we give only the essential features of this approach. It is well known that unlike the quasiparticle spectrum ϵ_i, the KS eigenvalue spectrum E_i^{DFT} has no rigorous physical meaning except for only one energy viz. the highest occupied state [67–69]. The two quantities relevant for SBH calculation are the self-consistent values of E_F and E_v coming out of the supercell calculation. Although individually E_F is the highest occupied state of a metal and E_v that of a semiconductor, the picture changes in a coupled M-S junction. One can no more look at E_v, for example, as the highest occupied state of the coupled system.

Let us consider [12,70] a thick slab with an interface between a metal (to the left say) and an intrinsic semiconductor (to the right say), with spatially flat band (i.e. no band bending). It can be rigorously shown [70] for such a slab geometry, that the difference $\epsilon_F - \epsilon_v$ between the exact quasiparticle energies can be related to the corresponding difference $E_F - E_v$ between the DFT eigenvalues via the relation

$$\epsilon_F - \epsilon_v = E_F - E_v + \Delta v_{xc} \tag{6}$$

where

$$\Delta v_{xc} = v_{xc}(\infty) - v_{xc}(-\infty) \tag{7}$$

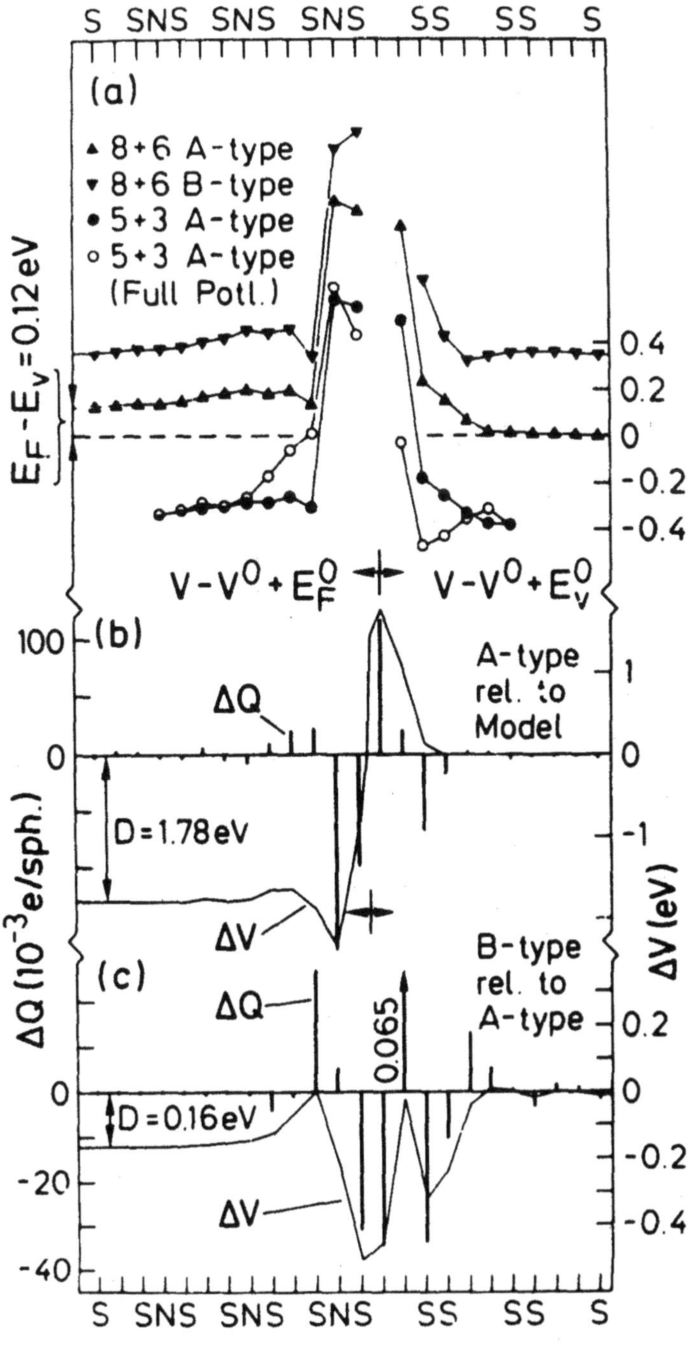
S SNS SNS SNS SS SS S
(a)
8+6 A-type
8+6 B-type
5+3 A-type
5+3 A-type
(Full Potl.)
$E_F - E_v = 0.12 eV$
0.4
0.2
0
-0.2
-0.4
$V - V^0 + E_F^0$
$V - V^0 + E_v^0$
(b)
100
0
ΔQ
A-type
rel. to
Model
1
0
-1
D=1.78eV
ΔV
ΔQ (10^{-3}e/sph.)
ΔV (eV)
(c)
B-type
rel. to
A-type
ΔQ
0.065
0.2
0
D=0.16eV
-0.2
-20
ΔV
-0.4
-40
S SNS SNS SNS SS SS S

This is because, in the vacuum outside the metal, the asymptotic behaviour of the density (and hence the electrostatic potential ϕ) is characterized by the work function $\phi(-\infty) - \epsilon_F$ in many-body theory, and by $\phi(-\infty) + v_{xc}(-\infty) - E_F$ in DFT. Hence $\epsilon_F = E_F - v_{xc}(-\infty)$ [12]. Similar considerations for the decay of the density outside the semiconductor yields $\epsilon_v = E_F - v_{xc}(\infty)$. These two relations combined yield the result (6). Here "∞" means a distance which is *large* compared to the atomic spacing but *small* compared to the thickness of the semiconductor. Since the formation of SB takes place on a length scale of the order of atomic distances (which is much less than the length scale for band bending), it is possible to exploit the abovementioned slab geometry to derive the formula (6). Clearly, for distances much larger than the slab width, i.e. at ∞ and $-\infty$, v_{xc} vanishes in the LDA i.e. Δv_{xc}=0, implying $\phi_{B_p} = E_F - E_v$. In the exact DFT, however, Δv_{xc} is *not zero*. It has been shown by a model calculation [55], that Δv_{xc} may indeed be a nontrivial additional contribution to ϕ_{B_p}. As of today, one can at best calculate the SBH within LDA and that is precisely what has been attempted by us [12,54–56], as well as a number of other groups [37–40].

After performing a self-consistent supercell calculations using either ASA or the more accurate Full potential calculation, one can use the so-called *'reference potential method'* [6,55] (which is in some sense a generalization of the frozen potential method) to do the layer-by-layer extraction of the representation energy viz. E_F for the metal and E_v for the semiconductor. The variation of such E_F/E_v profile can be traced out as a function of the supercell layers, as shown in fig. 6a for various cell sizes. As mentioned earlier, the m=8 and n=6 size cell is the optimum choice for E_F and E_v to become 'flat' as one approaches the respective bulk-like layers. The difference $E_F - E_v$ corresponding to the central layers on either side yields the SBH ϕ_{B_p}. For extracting SBH, other methods have also been reported in the literature (see ref. 6 and 55 for details). The most intuitive as well as accurate method, according to us, is the so-called *dipole approach*, which been used to extract SBH in M-S interfaces [12] as well as band offset in semiconductor heterojunction [11]. We shall discuss it briefly in this section.

The SBH ϕ_{B_p} is related to a reference $\phi^0_{B_p}$ and an interface induced dipole D formally defined as

$$D = 8\pi \int_{-\infty}^{\infty} z\Delta\rho(z)dz \tag{8}$$

where $\Delta\rho(z)$ is the change in electron density caused by the formation of the interface, averaged over a 2D cell at the position 'z' measured from the interface 'plane'. The choice of this 'plane' being ambiguous, D does not have a direct physical meaning. In spite of this, it is possible to construct a procedure where the ambiguities in D cancel and one is left with the physical quantity of interest viz. $E_F - E_v$.

Choice of the *'reference'* is crucial, if not controversial. We choose a reference density $\rho^o(z)$, consistent with certain zeroth order barrier height ϕ^o, such that

$$\Delta\rho(z) = \rho(z) - \rho^o(z) \tag{9}$$

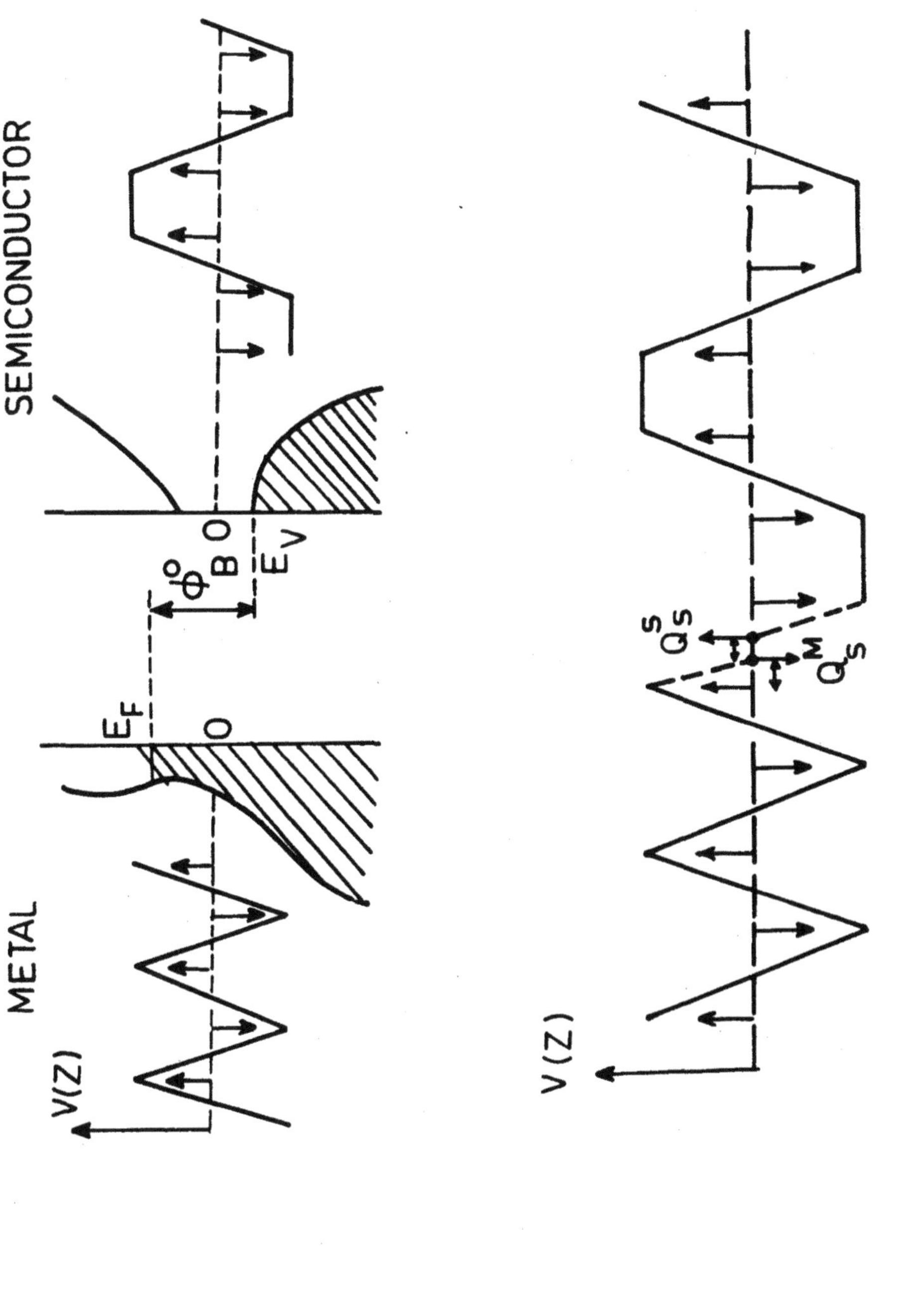
METAL
SEMICONDUCTOR
V(Z)
E_F
0
Φ_B^0
0
E_V
V(Z)
Q_s^S
Q_s^M

and

$$\phi = \phi^o - D \tag{10}$$

Within ASA, there is a common reference potential by which energy bands obtained from separate bulk calculations can be aligned. This provides a convenient definition of ϕ^0 as the *'natural barrier height'* i.e. the difference between the self-consistent bulk values in the LMTO-ASA energy scale

$$\phi^o = E_F^o - E_v^o \tag{11}$$

Now $\rho^o(z)$ does not consist solely of the bulk charge densities of the two solids, because *'cutting'* the solids in an arbitrary way and *'gluing'* the two half solids together with the frozen charge densities will automatically lead to a macroscopic electric field (and hence a *'dipole'*) at the interface. In order to avoid this, and force D^0 (i.e. dipole of the reference model) to be zero, we must add a compensating surface charge density to each of the two half crystals. The so-called *'fictitious surface charge density'* can be chosen such that the electric field emerging in the vacuum from either half crystal is compensated to zero. For the point-charge densities and potentials of the ASA, one may take the compensating surface density of the reference model to be a planar-averaged point charge distribution in the vacuum, positioned at a distance *'z'* such that the surface point-charge dipole is zero and the initial line-up corresponds to zero dipole.

This *reference model* with the abovementioned *'cut-and-glue'* prescription and the resulting *'fictitious surface charges'*, for the specific case of $NiSi_2/Si(111)$ interface, is shown schematically in fig. 7. Our self-consistent ASA calculations for the bulk constituents yield for $NiSi_2$ $E_F^0 = 0.0654Ry$ and for Si $E_v^0 = -0.0761Ry$, resulting in $\phi_{B_p}^0 = 0.1415Ry = 1.92eV$. The planar averaged electrostatic potential V(z) for a given interface orientation is a piecewise linear function whose slope changes by the amount

$$\Delta V'(z) = -\frac{8\pi}{A}\sum Q(z) \tag{12}$$

at the plane $z = z_i$ which contain the point charges. In the present case of point charges on a bcc lattice, the one-dimensional Poisson's equation reduces to

$$-V(z_{i-1} + 2V(z_i) - V(z_{i+1}) = \frac{8\pi c}{A} Q(z_i) \tag{13}$$

with area of the plane $A = 12.768\mathring{A}^2$ and spacing between the planes $c = 0.784\mathring{A}$. The self-consistent bulk occupancies Q_i^o, i.e. the excess number of valence electrons within the atomic spheres **(a)** of Si are $Q^o(Si) = -Q^o(E) = -0.786$, and **(b)** of $NiSi_2$ are $Q^o(Ni) = 0.421$, $Q^o(Si) = -0.832$, $Q^o(E) = 1.243$. The corresponding bulk point charge potentials $V^o(Z_i)$ can be calculated using relation (13); **(a)** for Si, $V^o(Si) = -V^o(E) = -4.365$ and **(b)** for $NiSi_2$, $V^o(Ni) = 0.028$, $Q^o(Si) = -2.310$, $Q^o(E) = 4.529$. Thus the average Q_i^o and $V^o(z_i)$ over a period in the z-direction is zero. Following the prescription described above, we can also find the magnitude and location of the fictitious point charges to each of the truncated half-crystals. On $NiSi_2$ side, its magnitude is $\frac{1}{2}Q^{\circ}_{E} = 0.62$ electrons, located at a distance $\frac{c}{2}(-Q^{\circ}_{Si}/Q^{\circ}_{E}) = 0.335c$ outside the last Si-plane. On Si side, its magnitude is $Q^{\circ}_{Si} = -0.786$ electrons located at a distance of $0.5c$ outside the last E-plane. Once the reference density and potential are known, the $\Delta Q(z)$ and $\Delta V(z)$ profiles may be obtained and are shown in fig. 6b. The dipole calculated from the formula

$$D = \frac{8\pi c}{A} \sum_{i} i\, \Delta Q(z_i) \tag{14}$$

yields for the A-interface, $D = -1.80eV$ and hence using (10), $\phi_{B_p} = 0.12eV$.

Now, the numerical accuracy of this estimate should be checked in view of the number of approximations we have made in our calculations viz. (a) finite cell-size, (b) limited number of **k**-points, (c) neglect of the small ($\sim 0.5\%$) lattice mismatch as well as interface relaxation ($\sim 4\%$ contraction), (d) neglect of relativistic spin-orbit term, (e) use of Barth-Hedin exchange-correlation potentials, (d) LDA band-gap underestimation, (f) use of ASA and hence neglect of the intra-atomic polarization effects. After meticulously performing all these tests [6,55] we found that their combined effect on A-type $NiSi_2/Si(111)$ interface is an increase of SBH by a net amount of only $\sim 0.02eV$. Thus our calculated value of ϕ_{B_p} for the A-type interface (after taking into account all the possible correction term) is 0.14eV, which is $\sim 0.3eV$ two low compared to experiment value (table 2).

Table 2. Values of $\phi_{B_p} = E_F - E_v$ (in eV) for A- and B-type $NiSi_2/Si(111)$ interfaces obtained from different LMTO calculations.

.	Bisi and Ossicini	Das et al	Fujitani and Asano		Experiment (Tung)
.	(1987)	(1989)	(1988)	(1990)	(1984)
.	(5+6)	(8+6)	(12+11)	(12+11)	
A-type	-0.0	0.14	0.39	0.36	0.47
B-type	0.3	-0.0	0.33	0.19	0.32

Treating B-type interface exactly on the same footing i.e. using (8+6) supercell, (8,8,2) **k**-mesh etc. it is found[12] that the charges $\Delta Q(z)$ per sphere relative to that of the A-type, are only hundredths of an electron (fig. 6c). The potential difference $\Delta V(z)$ add up to generate a dipole difference of 0.14eV i.e. the B-type SBH is *lower* than A-type SBH by 0.14eV, in agreement with experiment. But the absolute value of B-type SBH is ~ 0, which means that even though our results show the correct A-B difference, the absolute values are $\sim 0.3eV$ off from the experiment.

Let us now compare our results with the other LMTO-supercell calculations reported in the literature. Bisi and Ossicini[37] used (5+3) supercells and found ϕ_{B_p} to be higher for B-type, as opposed to experimental trend. This is presumably due to (a) an inefficient choice of the atomic spheres in the supercell and (b) a rather crude way of extracting SBH from LDOS. The results reported by Fujitani and Asano[38] using larger supercells, differ from their own earlier calculation[71]. The earlier calculations were reported to be nonrelativistic, have used the $NiSi_2$ lattice constant (5.406Å) for the entire supercell and have deployed the *wave function matching method* for SBH extraction. The later calculations [38] were made scalar relativistic, with Si lattice constant (5.43Å) for the entire supercell and using the *frozen potential approach* [54] for SBH estimation; and most important of all is that they improved their choice of interfacial E-sphere radii to match with that of ours [12]. These modifications seem to jack up their A-B difference in SBH from 0.06eV to 0.17eV, and the later value remains more or less same for the various supercell sizes used viz. (8+9) and (11+12). Thus, as far as the difference in SBH between A-type and B-type $NiSi_2/Si(111)$ interfaces is concerned, the LDA calculations of ours as well as of Fujitani and Asano seem to support Tung's experimental finding [45]. Furthermore, by performing detailed calculations for the A-interface with and without the interface state, we argued [12,55] that the partially occupied interface bands are responsible for this (A-B) difference. The interface states are found to be screened more efficiently in the A-type than in the B-type. This may be related to the fact that the lines of [111]- and [$\bar{1}\bar{1}$1]-bonds running in the [110] direction continue uninterrupted across the A-interface (fig. 4a), but breaks at the B-interface (fig. 4b).

7. Summary and discussion.

SBH $\phi_{B_p} = E_F - E_v$ is the single most important electronic quantity in a M-S interface, just like the band gap $E_g = E_c - E_v$ in a semiconductor. Our attempt in this article is to illustrate with the $NiSi_2/Si(111)$ example, the success as well as limitations of LDA for estimating this important quantity. The reason why we have chosen this particular interface is that it is one of the most ideal, lattice matched and atomically abrupt interface which can be produced; also it shows the unique orientation dependence of SBH. The experimentally observed difference of 0.14eV in the SBH of A- and B-type $NiSi_2/Si(111)$ interface is exactly reproduced by the LMTO supercell calculations. Also we have attributed its origin to the different screening of the partially filled interface bands which are of semi-dangling bond character. The states decay on both sides and are localized at the interface, showing the MIGS characteristics. A recent non-local pseudopotential calculation [41] on $Al/Ge(001)$ interface also shows high density of MIGS pinning the Fermi level, without involving any extrinsic states such as defect or impurity states.

The disturbing feature common to all the LDA-supercell calculations is that the absolute values of ϕ_{B_p} are underestimated by an amount which is comparable (if not higher) in magnitude than the (A-B) difference mentioned above. After detailed analysis, we found [12,55] that the discrepancy of $\sim 0.3eV$ remains, even after accounting for all the possible errors in LDA calculation; it is therefore attributed to a highly nonlocal correction Δv_{xc} to the LDA-SBH, as discussed in section 6. This is a very important conclusion and should be substantiated by further LDA calculations on other M-S interface systems. Following our work, two groups have performed similar LDA supercell calculations, one on *'ideal metal'/GaAs(110)* interface [39] and the other on Ni/diamond interface [40]. For details of their calculations, the reader is referred to the original papers and to the discussion given in ref. 6.

Table 3. Calculated and experimental SBH's $\phi_{B_p} = E_F - E_v$ (in eV) for several epitaxial M-S junctions (see text for details). The photothresholds of the semiconductors and the work functions of the metals used are also shown. $\phi^o_{B_p}$ is the bulk offset, used as the reference barrier height, to which the self-consistent dipole D is added in order to get the ϕ_{B_p}.

Semicon-ductor	Photo-threshold (eV)	Metal	Work fn. (eV)	Offset $\phi^o_{B_p}$ (eV)	SBH ϕ_{B_p} (eV)	Exptl. ϕ_{B_p} (eV)
Si(111)	4.85	NiSi2 (A-type)	-	1.92	0.14	0.52
.		NiSi2 (B-type)	-	1.92	0.0	0.38
GaAs(111)	5.45± 0.30	Fe	4.5	0.52	0.34	0.75
.		Cr	4.5	1.84	0.36	0.78
.		Ag	4.26	-0.38	0.39	0.56
.		Au	5.1	0.04	0.29	0.53
.		Ga	4.2	0.81	0.94	-
.		Al	4.28	0.91	1.05	0.63
.		Cd	4.22	2.07	1.07	-
Diamond (001)	5.5	Ni top-site	5.15	-	<0.1	1.3*
.		Ni T4-site			0.9	2.2

The results summarized in table 3 clearly show that (p-type) SBH's for all the M-S interfaces are more or less consistently underestimated from the corresponding experimental values by a few tenths of an eV. We are therefore inclined to ascribe this discrepancy to the use of LDA itself [12,55]. The non-local contribution Δv_{xc} should therefore be added to the $E_F - E_v$ calculated from LDA, in order to get the *true* SBH. Godby et al have renently commented [72] that the interface with the vacuum is irrelevant for Schottky barrier and that v_{xc} goes to zero at infinity. However, using a simple model calculation based on a one-dimensional Hubbard-like chain, it has been rigorously shown [55] that Δv_{xc} is crucial to determine the SBH. As of today, it is not possible to do an *exact* estimation of Δv_{xc}. But it may interesting to examine how such a potential may affect the eigenvalue difference $E_F - E_v$. If we assume that the finit value of Δv_{xc} is mainly caused by a variation of the exchange-correlation potential in the region of the metal-semiconductor interface, we can mimic its effect on the eigenvalue difference by adding a potential step at the interface in the self-consistent calculations. Shifting the potentials on the semiconductor side by ΔV_0, one finds that the self-consistent value of $E_F - E_v$ increases by an amount $\Delta V = \Delta V_0/\varepsilon_{eff}$,

where ε_{eff} is the effective (static) dielectric constant of the supercell. Taking $\Delta V_0 = 0.4eV$ (which is ~ SBH discrepancy), we find for A-type $NiSi_2/Si(111)$ interface, $\Delta V = 0.02eV$ [55], implying $\varepsilon_{eff} \simeq 20$ i.e. the potential step is screened very heavily within layers close to the interface. Schilfgaarde and Newman [39] found ε_{eff} as high as 120 for interfaces involving transition metals like Cr and Fe, and ~ 30 for the rest of the metals studied. All these ε_{eff} values are significantly larger than the dielectric constants of the typical semiconductors. Thus due to such efficient screening, the applied potential difference is not expected to modify the barrier height by changing $E_F - E_v$, but it enters with its full *'unscreened'* weight via Δv_{xc}. Further investigation in this direction is required in order to settle the enigma of Schottky barrier.

8. Acknowledgements.

I am grateful to Dr. P. Blöchl, Prof. O.K. Andersen, Prof. N.E. Christensen and Dr. O. Gunnarsson, whose excellent collaboration has resulted in this work. It is a pleasure to thank Prof. O.K. Andersen and his group in Max Planck Institute, Stuttgart for kind hospitality and support during the course of this work. Finally I would like to thank the International Centre for Theoretical Physics, Trieste, Italy for providing me this opportunity to attend this excellent workshop.

9. References

1. F. Capasso and G. Margaritondo (eds.), *'Heterojunction band discontinuities : physics and device applications'*, (North Holland, Amsterdam, 1987).

2. R.A. Abram and M. Jaros,in *Band structure engineering in semiconductor microstructures* (Plenum, New York, 1989) NATO ASI Sr. B Vol. 189.

3. P. Dhez and C. Weisbuch, *Physics, fabrication and applications of multilayered structure* (Plenum, New York, 1990) NATO ASI Sr. B Vol. 182.

4. E.G. Bauer, B.W. Dodson, D.J. Ehrlich, L.C. Feldman, C.P. Flynn, M.W. Geis,

J.P. Harbison, R.J. Matyi, P.S. Peercy, P.M. Petroff, J.M. Phillips, G.B. Stringfellow and A. Zangwill, J. Mater. Res. **5**, (1990), 852.

5. A. Ourmazd, R. Hull and R.T. Tung, in *Electronic structure and properties of semiconductors*, ed. W. Schröter (VCH, Weinheim, 1991).

6. G.P. Das, Pramana - J. Phys. **38**, (1992), 545.

7. D.L. Smith and C. Mailhiot, Rev. Mod. Phys. **62**, (1990), 173.

8. W.R.L. Lambrecht and O.K. Andersen, Sur. Sci **178**, (1986), 256.

9. H.L. Skriver and N.M. Rosengaard, Phys. Rev. **B43**, (1991), 9538

10. N.E. Christensen, Phys. Rev. **B37**, (1988), 4528.

11. W.R.L. Lambrecht, B. Segall and O.K. Andersen, Phys. Rev. **B41**, (1988), 2813.

12. G.P. Das, P. Blöchl, O.K. Andersen, N.E. Christensen and O. Gunnarsson, **63**, (1989), 1168.

13. W. Schottky, Z. Phys. **113**, (1939), 367.

14. N.F. Mott, Proc. Roy. Soc. (London), **A171**, (1939), 27.

15. E.H. Rhoderick and R.H. Williams, *Metal semiconductor contacts*, 2nd edition (Clarendon, Oxford, 1988).

16. J. Bardeen, Phys. Rev. **71**, (1947), 717.

17. V. Heine, Phys. Rev. **138**, (1965), A 1689.

18. S.G. Louie and M.L. Cohen, Phys. Rev. **B15**, (1976), 2461.

19. C. Tejedor, F. Flores and L. Louis, J. Phys. **C10**, (1977), 2163.

20. J. Tersoff, Phys. Rev. Lett. **52**, (1984), 465.

21. J.C. Phillips, J. Vac. Sci. Technol. **11**, (1974), 947.

22. L.J. Brillson, J. Vac. Sci. Technol. **15**, (1978), 1378.

23. J.L. Freeouf and J.M. Woodall, Appl. Phys. Lett. **39**, (1981), 727.

24. W.E. Spicer, P.W. Chye, P.R. Sketh, C.Y. Su and I. Lindau, J. Vac Sci. Technol.**16**, (1979), 1427. [Note : An improved defect model has later been proposed by Spicer et al, J. Vac. Sci. Technol. **B6**, (1988), 1245].

25. J.D. Dow and R. Allen, J. Vac. Sci. Technol. **20**, (1982), 659.

26. W. Mönch, Surf. Sci. **132**, (1983), 92.

27. W. Walukiewicz, J. Vac. Sci. Technol. **B5**, (1987), 1062.

28. R. Ludeke, G. Jezequel and A. Taleb-Ibrahimi, Phys. Rev. Lett. **61**, (1988), 601.

29. J.L. Freeouf, Surf. Sci. **132**, (1983), 233.

30. O.F. Sankey, R.E. Allen, S. Ren and J.D. Dow, J. Vac. Sci. Technol. **B3**, (1985), 1162.

31. I. Lindau and T. Kendelewicz, CRC Crit. Rev. Solid State Sci. **13**, (1986), 27.

32. W. Mönch, Phys. Rev. Lett. **58**, (1988), 1260.

33. T.T. Chiang, C.J. Spindt, W.E. Spicer, I. Lindau and R. Browning, J. Vac. Sci. Technol. **B6**, (1988), 1409.

34. R. Cao, K. Miyano, T. Kendelewicz, I. Lindau and W.E. Spicer, Phys. Rev. **B39**, (1989), 11146.

35. J.L. Freeouf, J.M. Woodall, L.J. Brillson and R.E. Viturro, Appl. Phys. Lett. **56**, (1990), 69.

36. J. Tersoff, in *Metallization and metal-semiconductor interfaces*, ed. I.P. Batra (Plenum, New York, 1989), NATO ASI Sr. B 195, p.281.

37. O. Bisi and S. Ossicini, Surf. Sci. **189/190**, (1987), 285; a modified calculation (with correct choice of interfacial empty spheres) was presented later by these authors in the MRS fall metting 1987.

38. H. Fujitani and S. Asano, Phys. Rev. **B40**, (1990), 1696.

39. M. van Schilfgaarde and N. Newman, Phys. Rev. Lett. **65**, (1990), 2728.

40. W.E. Pickett and S.C. Erwin, Phys. Rev. **B41**, (1990), 9756.

41. S. Ciraci, A. Baratoff and I.P. Batra, Phys. Rev. **B43**, (1991), 7046.

42. J. Derrien, J. Chevrier, A. Younsi, V. Le Thanh, J.P. Dussanley and N. Cherief, Phys. Scr. **T35**, (1991), 251. Jacobson, Thin Solid Films, **93** (1982), 77.

43. R.T. Tung, J.M. Gibson and J.M. Poate, Phys. Rev. Lett. **50**, (1983), 429.

44. R.T. Tung, in *Atomic level properties of interface materials*, ed. D. Wolf and S. Yip (Chapman & Hall, London, 1992).

45. R.T. Tung, Phys. Rev. Lett. **52**, (1984), 462

46. P.J. van den Hoek, W. Ravenek and E.J. Baerends, Phys. Rev. Lett. **60**, (1988), 1743.

47. D.R. Hamann, Phys. Rev. Lett. **60**, (1988), 313.

48. O.K. Andersen, Phys. Rev. **B12**, (1975), 3060.

49. H.L. Skriver, *The LMTO Method* (Springer, Heidelberg, 1984).

50. O.K. Andersen, in *The electronic structure of complex systems*, ed. W. Temmerman and P. Phariseau (Plenum, New York, 1984) p.1.

51. O.K. Andersen, O. Jepsen and D. Glötzel, in *Highlights in condensed matter theory*, ed. F. Bassani, F. Fumi and M.P. Tosi (North Holland, Amstardem, 1985), p.59.

52. O.K. Andersen, O. Jepsen and M. Sob, in *Electronic band structure and its applications*, ed. M. Yussouff (Springer Verlag, Berlin, 1987), Lecture Notes in Physics Vol. 281, p.1. [**Note :** A revised version of this article is published as lecture notes for CECAM Workshop on Interatomic Forces, 1987].

53. Present volume, see article by O.K. Andersen and coworkers.

54. G.P. Das, P. Blöchl, N.E. Christensen and O.K. Andersen, in *Metallization and metal-semiconductor interfaces*, ed. I.P. Batra (Plenum, New York, 1989), NATO ASI Sr. B Vol. 195 p.215.

55. P. Blöchl, G.P. Das, O.K. Andersen, N.E. Christensen and O. Gunnarsson, to be published.

56. P. Blöchl, *Total energies, forces and metal-semiconductor interfaces*, Ph.D. dissertation, University of Stuttgart (1989).

57. O.K. Andersen and O. Jepsen, Phys. Rev. Lett. **53**, (1984), 2571.

58. O.K. Andersen and N.E. Christensen, unpublished.

59. O. Jepsen and O.K. Andersen, Solid State Comm. **91**, (1971), 1763.

60. O. Jepsen and O.K. Andersen, Phys. Rev. **B29**, (1984), 5965.

61. D. Glötzel, B. Segall and O.K. Andersen, Solid State Comm. **36**, (1980), 403.

62. W.R.L. Lambrecht, N.E. Christensen and P. Blöchl, Phys. Rev. **B36**, (1987), 2493.

63. A.B. McLean and F.J. Himpsel, Phys. Rev. **B39**, (1989) 1457.

64. O.K. Andersen, Z. Powlowska and O. Jepsen, Phys. Rev. **34**, (1986), 5253.

65. M.D. Stiles and D.R. Hamann, Phys. Rev. Lett. **66**, (1991), 3179.

66. O. Gunnarsson, O.K. Andersen, P. Blöchl, G.P. Das and N.E. Christensen, Bull. Am. Phys. Soc. **35**, (1990), 666.

67. L.J. Sham and W. Kohn, Phys. Rev. **145**, (1966), 561.

68. R.W. Godby, M. Schlüter and L.J. Sham, Phys. Rev. Lett. **56**, (1986), 2415.

69. R.O. Jones and O. Gunnarsson, Rev. Mod. Phys. **61**, (1989), 689.

70. G.P. Das, P. Blöchl, O.K. Andersen, N.E. Christensen and O. Gunnarsson, Phys. Rev. Lett. **65**, (1990), 2084.

71. H. Fujitani and S. Asano, J. Phys. Soc. Japan **57**, (1988), 2253.

72. R.W. Godby, M. Schlüter and L.J. Sham, Phys. Rev. Lett. **65**, (1990), 2083.

Fig. 1 'Ball and stick' model of the structures of $NiSi_2/Si(111)$ interfaces (a) untwinned A-type and (b) twinned B-type. Filled circles represent Ni atoms and open circles Si atom.

Fig. 2 Layer projected density of states (LDOS) for (a) A-type and (b) B-type interfaces, as obtained from self-consistent LMTO-ASA calculations on (8+6) supercells with (8,8,2) bf k-mesh. The labelling of the layers are as follows : the (ESiNiSi) at the interface is the 1st $NiSi_2$ layer, while the (ESiSiE) at the interface is the 1st Si_2 layer. The 2nd, 3rd etc. layers are correspondingly defined away from the interface. The arrows indicate the van Hove singularities of the interface bands.

Fig. 3 Two-dimensional band structures of (a) A-type and (b) B-type $NiSi_2$ / $Si(111)$ interface. The interface states indicated by dashed curves are localized in the region MK of the Brillouin zone and then become resonant states cutting across the Fermi level.

Fig. 4 Valence electron charge density in $(1\bar{1}0)$ plane for (a) A-type (b) B-type $NiSi_2/Si(111)$. The lowest contour value, found in Si and $NiSi_2$ interstices, is $0.005 electrons/(a.u.)^3$. The contour step is also 0.005.

Fig. 5 Valence electron charge density only of the interface states for (a) A-type and (b) B-type $NiSi_2/Si(111)$. The lower portion is Si and the upper portion is $NiSi_2$. The lowest contour value as well as the contour step are $0.0005 electrons/(a.u.)^3$.

Fig. 6 Potentials and sphere charges as functions of z ($\hat{z}$ is perpendicular to the interface). Layers of Si and Ni atoms are labeled 'S' and 'N', whereas layers of interstices (E) are left unlabeled. The (8,8,2) **k**-mesh was used. The quantities displayed are (a) E_F/E_v profiles, obtained from ASA and FP calculations on (5+3) and (8+6) supercells (b) charges and potentials calculated for the (8+6) supercell relative to those of the reference model of fig. 7 (as explained in the text). The fictitious surface charges of the reference model are not shown (c) charges and potentials calculated for the B-type minus those of the A-type.

Fig. 7 Schematic representation of the reference model required for SBH calculation of $NiSi_2/Si$ interfaces using the *'dipole approach'*. The upper part shows separately for the metal and the semiconductor, the sphere averaged point charges and the corresponding potentials obtained by solving the one-dimensional Poisson's equation. $\phi^o_{B_p}$ is the zeroth order (or reference) offset given by the difference between the bulk Fermi level and the bulk VB top (see equation 11). The lower part of the figure shows the fictitious surface charge Q^M_s placed at a distance z^M from the truncated metal surface and another fictitious surface charge Q_s^S placed at a distance z^S from the truncated semiconductor surface, such that the 'cut-and-glue' surface defined as $ESiNiSi \parallel EESiSi$ has no net electric field (see text for details).

AB-INITIO MOLECULAR-DYNAMICS: THE CAR-PARRINELLO METHOD

GIORGIO PASTORE
Dipartimento di Fisica Teorica, Università di Trieste, Strada Costiera 11
Trieste, I-34014, Italy

ABSTRACT

The theory of the Car-Parrinello method to perform molecular dynamics calculations with interionic forces derived from first-principle total energy calculations is reviewed. Emphasize is put on the possibility of using the recently developed theory of the method to improve its computational efficiency. Some recent proposals of extensions of the original algorithm are briefly discussed.

1. Introduction

Few years ago, Car and Parrinello [1] unified the density functional theory (DFT) and the molecular dynamics (MD) technique into a new computational method which, for the first time, allowed to perform first-principle computer simulations of realistic systems. In such a method, status-of-the-art total energy calculations are married with classical MD to deal with dynamical problems and, at the same time, it becomes possible to get rid of most of the uncontrolled approximations unavoidable with empirical interaction laws.

It is difficult to maintain an updated list of the systems studied with the "Car-Parrinello" (CP) method. The continuously increasing number of papers witness the spreading interest for such a computational technique. The basic algorithms have been described quite in detail in the literature (ref. [2] and [3]) while technical improvements as well as potentially alternative methods have been proposed. The theoretical principles on which the method hinges as well as an analysis of its limits have been investigated in a recent paper [4] (hereafter ref. [4] will be referred to as PSB).

In the present paper, I will review the theory of the CP method and I shall discuss in more details a few practical issues related to the choice of the parameters. In section 2 the main results of the analysis of PSB will be recalled. Some complements to the discussion in PSB will also be included. In section 3 some recent proposals aiming to improve the original algorithm will be qualitatively examined. Some open problems of the theory of AIMD and directions of possible improvements will be briefly discussed.

2. Theory of the Car-Parrinello method

2.1 The CP dynamical system

Let us summarize how the CP method is used to perform ab-initio MD, i.e.

an MD simulation where the interionic forces are derived from a suitable approximation to the exact density functional of the electronic ground state. We have explicitly focused on the usual implementation of DFT at the level of the local density approximation (LDA) within the Kohn-Sham (KS) formalism [5]. However, most of our conclusions remain valid in connection with other approximate DFT (and with the Hartree-Fock functional). The only crucial requirement is that the approximate electronic ground state must be obtained from a variational principle.

In the usual LDA-KS scheme, single particle orbitals $\psi_i(r)$ are introduced to generate the electronic density ($\rho(r) = \sum_i f_i \, |\psi_i(r)|^2$, where the f_i's are the occupation numbers). The energy of the electrons in the presence of ions at positions $\{R_\alpha\}$ is given by the total energy functional $E\,[\{\psi_i(r)\ \},\{R_\alpha\}]$. The forces on the ions should be evaluated according to the Hellmann-Feynman theorem by taking the gradient of the energy with respect to the $\mathbf{R}_\alpha$ at the value of $\psi_i(r)$ minimizing E. We call the resulting ionic evolution the Born-Oppenheimer (BO) dynamics.

A straightforward implementation of a BO dynamics algorithm is a cycle of elementary iterations where the evaluation of the ionic forces, the updating of the ionic positions and a new minimization of the functional E are sequentially executed. However, such a scheme is faced with the necessity of a full minimization at each time step accurate enough to allow an acceptable integration of the ionic equations of motion. This requirement can be computationally very expensive if the number of variational parameters is large and sufficient precision is required[6] [3]. It is true that with a strict BO dynamics relatively large time steps can be used, however, often this advantage does not compensate for the heavier calculations.

The main idea behind the CP method is that the adiabatic evolution of the KS orbitals can be approximated by a suitably chosen dynamics such that the electronic degrees of freedom are significantly faster than the ionic ones. In such a case the classical dynamical system will experience a (classical) adiabatic decoupling between the fast and the slow degrees of freedom. Such a decoupling will maintain the total system in a metastable state where the wavefunctions perform fast and small oscillations around their instantaneous ground state. In this way, when averaging the forces on the ions over the short period of the fast variables, the resulting ionic dynamics will be arbitrary close to the BO surface. In a way, reversing the BO approximation, the CP method uses an artificial fast dynamics of the KS orbitals to follow the instantaneous ground state without explicit minimizations. The exact nature of the wavefunction dynamics is largely arbitrary and actually can be tailored according to specific needs.

Since the main task of the ab-initio MD is to perform a sampling of the phase space according to a well definite statistical ensemble, a conservative dynamics is highly recommended. In the following, I will focus only on algorithms belonging to this class.

A convenient way of introducing a second-order system is writing a time-independent lagrangian. The reasons for choosing a second-order dynamical system will be discussed in the following. For the moment, let us notice that even within a second order conservative dynamics, we still have some freedom in determining the exact lagrangian. The original CP lagrangian for the ionic coordinates and the KS wavefunctions is

$$\mathcal{L}_{CP} = \mu \sum_i \int \left|\dot{\psi}_i(r)\right|^2 dr + \frac{1}{2} \sum_\alpha M_\alpha \dot{R}_\alpha^2 - E\left[\{\psi_i(r)\}, \{R_\alpha\}\right] + \sum_{k,l} \Lambda_{kl}\left(\int \psi_k(r)\psi_l(r) dr - \delta_{kl}\right) \quad (1)$$

where μ is a parameter affecting the time-scale of the KS orbitals ($\psi_i(r)$) evolution, M_α are the ionic masses and the last term contains the Lagrange multipliers taking care of the orthonormalization constraint. The resulting Euler equations give newtonian equations of motion which, by discretizing the $\psi_i(r)$ on a grid, can be integrated numerically with the usual MD techniques.

2.2 Classical adiabatic behaviour

In general (i.e. for arbitrary initial conditions) the dynamics generated by $\mathcal{L}_{CP}$ is very different from the BO dynamics. However, provided the starting electronic configuration is close to the ground state, the effective coupling to the ionic degrees of freedom is weak enough, and the time scale of the electronic and the ionic motions are well separated, the classical adiabatic approximation does hold and the trajectory in the phase space of the electronic variables stays arbitrarily close to the BO surface for very long times. As a consequence, also the *average* forces on the ions will be arbitrarily close to the *exact* Hellmann-Feynman forces. Then, we can assume that the physically relevant dynamics of the ions will remain close to the true BO dynamics. This argument may be made more precise by recalling the exact statement of the classical adiabatic theorem for a perturbed harmonic oscillator and applying it to the classical normal modes of the KS wavefunctions.

In a regime of small oscillations, the ratio between the energy and the frequency of the i-th mode is an adiabatic invariant. Formally this can be expressed as the condition that, if ν is proportional to the rate of change of a slow perturbation,

$$\left|\frac{E_i(T)}{\omega_i(T)} - \frac{E_i(0)}{\omega_i(0)}\right| < \nu \quad (2)$$

for every time $T < \nu^{-\tau}$. The exact value of τ is strongly system dependent but in no case less than 1/2. Actually, in some cases it may be significantly larger than 1. What formula (2) says is that the total energy of each normal mode divided by the frequency of the mode remains arbitrary close to its initial value for arbitrary large times when the typical time of variation of the perturbation becomes large. In other words, it is always possible to reduce the change in E_i/ω_i over a given time T just making larger the separation of the fast and slow time scales. Notice however, that the adiabatic invariance of the ratio E_i/ω_i does not imply immediately that the fast variables will stay arbitrary close to the minimum. Only if the frequency ω_i does not change significantly from the original value we can derive the approximate equality between $E_i(T)$ and $E_i(0)$. In such a case, the fact that the potential energy is less or equal than E will prove that the fast variables have not moved very far from the neighborhood of the instantaneous minimum. In this sense the value of $K_f = \mu \sum_i \int \left|\dot{\psi}_i(r)\right|^2 dr$ (the classical kinetic energy of the KS states) can be taken as a qualitative measure of the distance from the instantaneous minimum.

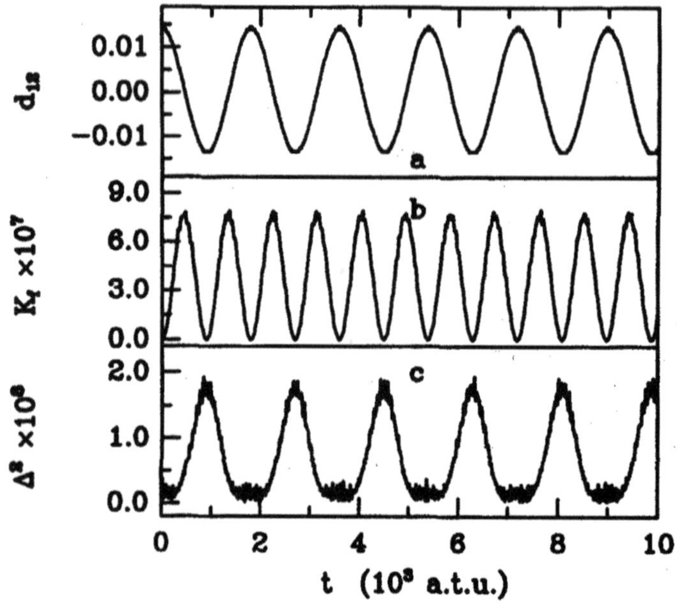

Figure 1: Part a: time variation of the distance from the equilibrium position of two silicon atoms in the diamond structure (distances in a.u.). Part b: wavefunction classical kinetic energy (K_f, in hartree). Part c: distance from the BO surface during the ionic evolution (see text).

However, K_f is not a direct measure of the distance from the BO surface for two reasons. First of all we have to take into account the trivial part of K_f coming from the strictly adiabatic part of the motion, i.e. the kinetic energy that even pure BO wavefunctions will have as a consequence of the underlying ionic dynamics. This contribution is clearly not related with any departure from BO. Within a plane waves expansion scheme it is not simple to separate this contribution. However, following Bloechl and Parrinello [8] it is possible to give a reasonable estimate of it. The second reason is that since the adiabatic invariant is the ratio E_i/ω_i, the identification between high values of K_f and strong deviations from the adiabaticity are possible only in the cases where the frequencies of the normal modes do not show big variations. As illustration of this point, in parts a,b and c of figure 1, I have plotted the displacement from the equilibrium position, the wavefunction kinetic energy and the distance from the BO surface as measured by the norm $\Delta^2 = \|\sum_i(-H\psi_i + \sum_j H_{ij}\psi_j)\|$ for the same "toy" system modelling bulk silicon already studied in PSB. As it is evident, the largest deviation from the BO energy surface takes place at each second turning point of the optical ionic mode and not at the maximum of the electronic kinetic energy. The reason is that, in correspondence with each second turning point, there is the largest deviation of the electronic eigenfrequencies from their initial value. Only after another half oscillation, the system comes back to a configuration such that $\omega(T) = \omega(0)$ and the distance from the instantaneous minimum again becomes very small. We can conclude that the condition of a limited variation of the eigenfrequencies (and as we will see in the next subsection of the KS eigenvalues) is important to allow a meaningful use of K_f as a probe of the departure from the BO surface.

2.3 Classical frequencies and KS eigenvalues

In the experience so far accumulated on the CP method, a strong correlation has been noticed between the rate of departure from the BO surface and the eigenvalue spectrum of the hamiltonian. For example, solid Si or C, which exhibit relatively large energy gap, were found to strictly obey the classical adiabatic approximation. On the contrary, in their liquid phases, where no steady gap is present, a fast transfer of energy from the ionic to the electronic degrees of freedom was observed [9] [10]. This implies a connection between dynamical matrix of the electronic part of the classical Lagrangian and the eigenvalues of the quantum KS hamiltonian. Actually, it is easy to show that the classical dynamical matrix of the electronic degrees of freedom (the hessian of the KS energy functional) is related to the linear density response function of the inhomogeneous electrons. However, it is difficult to extract from such a relation a precise information about the role played by the one particle eigenvalues of the KS hamiltonian on the classical motion.

PSB showed that, by neglecting the non-linearity of the effective hamiltonian, the normal modes frequencies associated to the small amplitude CP dynamics of the wavefunctions should be given by the formulae:

$$\omega_{ij}^{(1)} = (f_j(\epsilon_i^* - \epsilon_j)/\mu)^{1/2} \tag{3}$$

and

$$\omega_{ij}^{(2)} = ((f_j - f_i)(\epsilon_i - \epsilon_j)/2\mu)^{1/2} \tag{4}$$

where ϵ_i^* indicates the eigenvalue of the i-th unoccupied and ϵ_j the j-th occupied level and f_i are the occupation numbers. As one could expect, in the case of the actual KS Hamiltonian, the nonlinearities quantitatively modify this result, although in many cases a semiquantitative agreement holds between the ideal linear case and the real normal-mode spectrum.

Direct comparisons of the normal mode spectrum and the spectrum resulting from equations (3) and (4) for a few selected cases showed a semiquantitative agreement particularly on the position and bandwidth of the eigenvalue spectrum.

On the ground of these equations we can rationalize the findings of the real calculations using the CP algorithm. For an initial state slightly different from the ground state, small "electronic" oscillations of the fourier coefficients of the KS wavefunctions will start at the frequencies $\omega_{ij}^{(1)}$ (if, as usual, all the filled states have the same occupation). For large values of the "gaps" $(\epsilon_i^* - \epsilon_j)$ and for a suitable fixed μ the smallest electronic frequency is large with respect to the typical ionic frequencies and we can expect no overlap between them thus having a negligible energy transfer between ionic and electronic components. As soon as the gap decreases, the frequency of the electronic mode will decrease becoming closer to the ionic frequencies and eventually overlapping with them. In this regime the transfer of energy will be progressively more efficient and the time to achieve the equipartition of energy shorter. This transfer of energy has two main consequences: i) the kinetic energy of the ions (and their temperature) decreases steadily; ii) the electronic degrees of freedom will execute oscillations of larger amplitude and the electronic system will spend most of its time far from the equilibrium position. This implies sizable errors affecting the forces on the ions.

In addition to this case, where the break-down of the adiabaticity comes from a small or null gap, an accidental crossing between an occupied and an unoccupied level represents a catastrophic event: the change of sign of $(\epsilon_i^* - \epsilon_j)$ in equation (3) means that the equilibrium position becomes unstable. In this case, the evolution of the wavefunction is not able to remain very close to the instantaneous ground state.

Notice that, since in practical situations it is unlikely that more than a few levels will cross at the same time, the analysis of a few level model like that done in PSB should have a quite broad validity. A thorough classification of the possible cases of level crossing should also be possible by means of the formal machinery of singularity theory.

2.4 The choice of the parameters

The fictitious mass μ plays a role similar but not equivalent to that of the energy gap. The electronic frequency controlling the classical adiabatic behavior of the system is the lowest one. The highest frequency is responsible for the maximum possible time step which still allows a proper integration of the differential equations of motion for the electronic degrees of freedom. For a given adiabatic decoupling, i.e. for fixed lowest frequency, the largest possible time step will be obtained by minimizing the bandwidth of the normal mode spectrum.

In this context we can discuss possible alternative schemes of wavefunction dynamics based on a first order (Schroedinger-like) dynamics. Although more physically appealing, such dynamical systems have a bandwidth of the normal modes spectrum increasing linearly with the highest excited level. This should be contrasted with the square root increasing of the second order CP dynamics. Then, in this respect, the choice of a second order dynamics is computationally more efficient.

We see that in the practical choice of the value for μ, we need to achieve a tradeoff between the best possible adiabatic decoupling (if there is a finite gap in the electronic spectrum) and the practical requirement of having the maximum possible time step. A variation of μ implies a homogeneous shift of the whole spectrum. This fact limits the possibility of improving the classical adiabatic decoupling of the electronic and ionic subsystems by decreasing μ. More refined schemes where

μ becomes state- (and maybe $\mathbf{G}$-) dependent in order to minimize the band-width of the vibrational spectrum are presently under study. For example, equation (5) would suggest that masses

$$\mu_i = \frac{\mu_0}{(\epsilon_i - \epsilon_0)} \tag{5}$$

(ϵ_0 and μ_0 are two constants and the condition $\epsilon_i > \epsilon_0$ holds for each i) could give a reduced band-width of the vibrational spectrum. However, preliminar numerical results with such a simple formula reveal that only a moderate reduction of the bandwidth can be obtained. I would conclude that the $\mathbf{G}$ dependence should be taken into account as well.

Finally, I note that, in general, the coupled ionic-electronic motion will cause a renormalization of the eigenfrequencies of the decoupled subsystems. The ionic frequencies will be slightly smaller than those obtained on the BO surface. However, for the values of μ which ensure a good adiabatic behaviour the relative shift is usually negligible. Only if particularly large values of μ are used or whenever the coupling between the two subsystems is large, it could be advisable to change the value of the ionic masses in such a way that the "dressed" ions have the same inertial properties of the physical ones. For that, the mass renormalization factor can easily be found by directly comparing any simple inertial property of an ion evaluated with the CP method with the same property obtained from a true BO dynamics.

3. Recent progress

Recently, some important modifications of the original method have been proposed in the literature. I will not examine in detail all of them but I will focus on those that (very subjectively) appear as the most interesting and promising.

Bloechl and Parrinello [8] have proposed a practical solution of the problem of broken adiabaticity in closing gap systems. Their idea is to further extend the CP dynamical system by means of the Nosé method [11] [12] to put two deterministic thermostats separately on the ionic and electronic subsystems. In this way, via a suitable feedback, the electrons cannot acquire a large kinetic energy and their escape from the neighborhood of the instantaneous minimum is greatly slowed down. Moreover, the ions are also constrained to move consistently with an average kinetic energy in a way that guarantees the correct sampling of the canonical phase space.

The limited experience collected up to now with this algorithm seems to confirm the ability of the feedback to avoid a fast departure from the BO surface in systems with small gaps or crossing levels. Apparently the method works just preventing the electronic states to acquire a too large kinetic energy and then substantially reducing their possibilities of climbing high potential energy hills. At a level crossing, in the absence of a thermostat, the highest KS state would experience a strong mixing between the two states that crossed for long times after the event. The electronic thermostat provides a deterministic damping mechanism that rapidly restores the small oscillations around the ground state. Further analysis of the algorithm to study the behaviour of the averaged forces on the ions is in progress.

Some other extensions of the original algorithm have been presented aiming to allow a dynamical variation of the occupation numbers f_i [13] [14]. Reasons for such extensions can be found either in the aim of dealing with finite temperature effects in the valence electrons, or as a possible way to cure from the beginning the instability associated to the level crossing in the CP dynamics. The simplest implementation of a variable-occupation scheme requires to deal with the f_i as additional independent dynamical variables [14] and skipping from the Energy functional to the Mermin's free energy functional [15]. Then the forces on the occupations can be easily evaluated. Different techniques may be exploited to ensure the constraint of having occupations limited between 0 and 2 (in a LDA approach). However, up to now no convincing results have been presented for dynamics. The main problem is in devising a suitable lagrangian such that no dramatic softening of modes takes place during the evolution of the system.

Another variation sometimes discussed in the literature is the use a first order Schroedinger-like unitary evolution for the wavefunctions. A real unitary evolution, consistent with some time-dependent versions of DFT, is possible but is more affected than the usual CP dynamics by the bandwidth of the wave-function oscillation spectrum. Recently Teilhaber [16] has used a similar approach but introducing a scaling parameter equivalent to modify the value of the electron mass to allow larger time steps. Even if in some cases the resulting time steps may be comparable with that of the second order CP dynamics the scaling behaviour of the time step with the total number of plane waves (i.e. with the dimensionality of the Hilbert space) is certainly worse. However, there is a potential interest for improved forms of such a scheme, since the unitary evolution could offer a simple way to avoid the orthonormalization step of the usual algorithm with its bad scaling with the system size.

4. Conclusions

The recent analysis of the CP method performed in PSB [4], has been able to account for most of the known phenomenology of the numerical simulations with the original quasi-microcanonical ionic dynamics. I think that some interesting directions for future investigations would come from the study of the CP method from the point of view of the classical dynamical systems. In particular, it would be interesting to assess the dependence of the non-adiabatic energy transfer rate on the parameter μ and on the KS spectrum as well as to clarify the role of possible internal resonances of the electronic degrees of freedom. Notice that the theory of the classical adiabatic approximation is not yet complete for systems with more than 2 degrees of freedom [7]. In this respect, the CP method could be considered as an interesting applied case study and, in turn, it could profit from advances in the theory of classical dynamical systems to achieve a larger efficiency.

It is also clear from the previous analysis that the systems the most suitable for the original quasi-microcanonical algorithm are insulators or semiconductors with a quite stable gap. Apparently, the deviations from adiabaticity observed in systems with a varying gap or undergoing level crossings can be cured by the introduction of two thermostats. From the existing experience with the two thermostat method, this is sufficient to keep the electronic subsystem close enough to the BO in long runs. However, some additional investigations on the effect of such a method on the ionic dynamics is in order. In particular in some cases, for example in the study of localized atomic dynamics (impurity diffusion mechanism, vacancy migration etc.), it would be useful to have a true microcanonical evolution.

The efficient implementation of a Car-Parrinello microcanonical evolution for metallic systems remains a challenge for the theory. Some progress in this direction could come from the extension to finite temperature density functionals. Indeed, there are indications that in a scheme with dynamically varying occupation numbers the relative importance of the dynamical instability caused by an accidental level crossings should be reduced. However, one cannot exclude that more drastic changes of the original CP algorithm are required to solve this problem or even that these cases could require to implement a direct BO MD by using the efficient minimization methods developed in the last few years.

Finally, it should not be neglected the possibility of exploiting non-second order dynamical systems to approximate the BO evolution.

4. References

1. R. Car and M. Parrinello, *Phys. Rev. Lett.* **55** (1985) 2471.
2. R. Car and M. Parrinello in *Simple Molecular Systems at Very High Density*, ed. A. Polian, P. Loubeyre and N. Boccara,p.455 (Plenum, New York, 1989).
3. D. K. Remler and P. A. Madden, Mol. Phys. **70** (1990) 921.
4. G. Pastore, E. Smargiassi, F. Buda *Phys. Rev.* **A44** (1991) 6334.
5. W. Kohn and L. J. Sham, *Phys. Rev.* **140** (1965) A1133.
6. I. Štich, *Magister Philosophiae Thesis* (S.I.S.S.A, Trieste, 1987).
7. P. Lochak and C. Meunier, *Multiphase Averaging for Classical Systems* (Springer, Berlin, 1989).
8. P. Bloechl and M. Parrinello, *Phys. Rev.* **B 45** (1992) 9413.
9. G. Galli, R. M. Martin, R. Car, M. Parrinello, *Phys. Rev. Lett.* **63** (1989) 988.
10. I. Štich, R. Car and M. Parrinello, *Phys. Rev. Lett.* **63** (1989) 2240.
11. S. Nosé, *Mol. Phys.* **52** (1984) 255.
12. S. Nosé, *Progr. Theor. Phys. Suppl.* **103** (1991) 1.
13. G. Qian, M. Weinert, G. W. Fernando, J.W. Davenport *Phys. Rev.* **B 40** (1989) 7985.
14. Y. Yamamoto and T. Fujiwara *Phys. Rev.* **B 46** (1992) 13596.
15. N. D. Mermin, *Phys. Rev.* **137** (1965) A1441.
16. J. Theilhaber *Phys. Rev.* **B 46** (1992) 12990.

ATOMIC AND ELECTRONIC STRUCTURE OF CLUSTERS FROM CAR-PARRINELLO METHOD

VIJAY KUMAR
International Centre for Theoretical Physics
34100 Trieste, Italy

and

Materials Science Division, Indira Gandhi Centre for Atomic Research, Kalpakkam - 603 102, India[1]

ABSTRACT

With the development of *ab-initio* molecular dynamics method, it has now become possible to study the static and dynamical properties of clusters containing upto a few tens of atoms. Here I present a review of the method within the framework of the density functional theory and pseudopotential approach to represent the electron-ion interaction and discuss some of its applications to clusters. Particular attention is focussed on the structure and bonding properties of clusters as a function of their size. Applications to clusters of alkali metals and Al, non-metal - metal transition in divalent metal clusters, molecular clusters of carbon and Sb are discussed in detail. Some results are also presented on mixed clusters.

1. Introduction

During the past decade much experimental and theoretical progress has been made in our understanding of the physical and chemical properties of several elemental, binary and compound clusters[1−5]. Interest in these studies has basically arisen from the technological importance of clusters e.g. in catalysis, photographic films, magnetic recording, etc. and from the quest to understand the evolution of materials properties as a function of the size of an aggregate. Very recently it has also become possible[6] to prepare a new form of solid carbon from large carbon clusters which are now referred to as fullerenes. These fullerenes are caged structures having 12 pentagons and a varying number of hexagons of carbon atoms. The most beautiful and interesting among these fullerenes is the C_{60} molecule which has the truncated icosahedral structure. These molecules can be crystallized in a $Pa\bar{3}$ structure at low temperatures. The exciting discovery of superconductivity at relatively high temperatures in K[7](18 K), Rb[8](28 K) and Cs and Rb[9](33 K) doped solid C_{60} has provided another dimension to cluster research and efforts are now also being made[10−12] to find other clusters/molecules which could be used as building blocks for making new materials. It has also been possible to encapsulate atoms and molecules within fullerenes[13] and to prepare other forms of carbon such a buckytubes[14] and

[1]Permanent address

bucky onions[15] etc . These developments have therefore broaden the scope of the studies of finite aggregates and opened up new avenues in materials research. Also it is hoped that the progress in cluster research will enhance our understanding of complex structures and pave the way for new molecular architecture.

One of the important factors which governs the properties of clusters is their structure which is in general very different from a bulk fragment. This is due to the fact that in a small cluster most of the atoms are on the surface and therefore have reduced coordination. This leads to a reconstruction of the bulk fragment so that the *free energy* of the cluster becomes minimum. In addition there are quantum effects which lead to an oscillatory behaviour of e.g. the binding energy, ionization potential etc. as the cluster size grows. Such changes in the atomic and electronic structure can affect significantly the bonding and other physical and chemical properties of clusters. Some interesting examples are small clusters of divalent and tetravalent elements. Si clusters have been found to have closed packed structures[16] as compared to strongly directional bonding in the bulk whereas small carbon clusters have structures ranging from chains, rings, fullerenes, tubes, onions and possibly some other forms also. Dimers of divalent metals such as mercury and magnesium are very weakly bonded due to ns^2 closed shell atomic electronic structure and a large promotional energy to the p state but these are good metals in bulk. Therefore for such elements a transition to bulk chemical bonding should occur as the cluster size grows. However, in most cases it is still not clear when such a transition would occur. This may range from a few tens to a few hundred or thousand atoms depending upon the species involved. For mercury, experiments[17] indicate this transition to occur for a moderate size of the clusters ($70 \geq N \geq 13$).

Clusters also differ in an important way from bulk materials when alloyed. In the bulk, alloying is not possible if the atomic size difference is larger than about 15%. However, for clusters there is no such restriction on the relative size of atoms. Thus binary clusters such as Cu-Os could be formed[18]. This provides another interesting way of improving the reactivity and selectivity of clusters by effectively changing the *local* electronic structure. Several studies on the reactivity of clusters illustrate the importance of the variation of the electronic structure with cluster size. Iron clusters have been found to show[19] several orders of magnitude change in the reaction rate of deuterium as the cluster size changes from a few to few tens of atoms. This correlates well with the variation in the electron binding energy of the clusters as shown in Fig. 1. The reactivity increases as the electron binding energy decreases. The latter generally decreases with an oscillatory behaviour to the value of the work function as the cluster size grows. Similarly Al clusters show marked variation in the reactivity with oxygen as a function of their size[20]. The magic clusters Al_{13}^- and Al_{23}^- are found to be unreactive with oxygen whereas the non-magic clusters do react.

The magnetic properties of small clusters are also in general very different from the bulk due to the discrete nature of the electronic energy spectrum. Clusters of elements which are non-magnetic in bulk can possess magnetic moments while in the case of the magnetic elements, the moments can be significantly different from

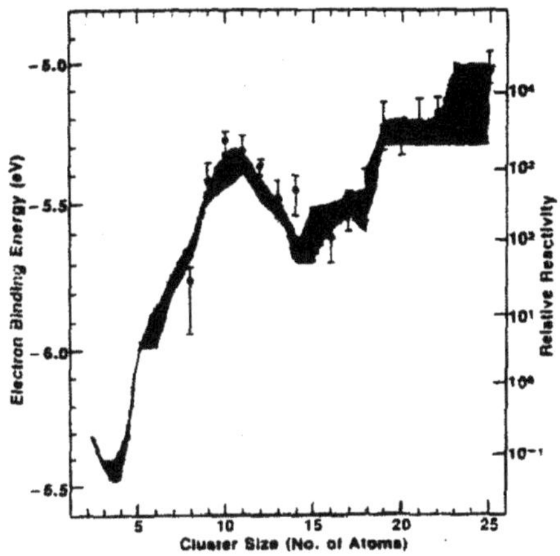

Figure 1: Comparison of electron binding energy and reactivity of Fe_N clusters for dissociative chemisorption of D_2 and H_2. The shaded portion reflects the uncertainty in ionization threshold measurements while the vertical bars indicate uncertainty in reactivity results. (After Whetten et al[19])

their bulk value. Also the different sites in a cluster will in general have different magnetic moments. There is little progress in this direction so far. Similarly due to changes in the charge density distribution, the optical properties of clusters are expected to be different from the behaviour on semi-infinite surfaces. In principle clusters form a much broader class of materials because different clusters of even the *same* material may behave quite differently and in the case of more than one component, the atomic size of the constituents is not the limiting factor. Such studies can therefore help to prepare better catalysts, photographic films, magnetic tapes, energy storage devices, etc. by suitable choices of material and particle size and can open doors for making new classes of materials such as fullerites, the crystals obtained from fullerenes.

Experimentally while direct information on the structure of clusters is difficult, it has been possible to study the ionization energy, magnetic and electric moments, density of states, polarizability, photoabsorption cross section and effects of multiple charging etc. of mass selected clusters. Experiments on mass selected clusters containing upto several thousand atoms have become possible[21]. However, most theoretical studies based on ab-initio methods have been possible on clusters containing a few atoms only since it is difficult to know the lowest energy structures due to the very many atomic and electronic configurations that are possible for a given size of the cluster. Calculations based on a jellium model[22] are simpler and can be done for large clusters. These have been very successful in providing a qualitative understanding of the variations in the stability, ionization potentials, polarizability etc. of clusters of simple metals as a function of their size[4,5]. However, in several cases of interest such as clusters of semiconductors, transition metals and alloys, a jellium model may not be suitable. Also for a quantitative understanding it is necessary to know the atomic structure and the changes in the electronic structure as a function of the size and geometries and their temperature dependence. This is also likely to enhance our understanding of the dynamical properties, structural transformation and melting etc. of clusters and will be helpful in understanding chemical reactions on clusters.

Theoretical studies (see e.g. Refs. 1-5) of the atomic and electronic structure of clusters of various elements have been numerous. These include studies based on the use of some interatomic potentials as well as ab-initio calculations. While studies based on some interatomic potentials can be done for clusters having upto several thousand atoms and have given some useful information, one has to be cautious about the validity of such results (in particular for small clusters) as potentials generated from some known structures of a material may not be appropriate for other structures of the same material. One can hope to get a good description of clusters of rare gases using pair- potentials whereas interactomic potentials based upon effective medium theories or tight binding methods may particularly be useful for large clusters of metals and semiconductors. Ab-initio calculations are much more difficult and have been possible for clusters having at best upto a few tens of atoms. A majority of these calculations have been done for a fixed atomic distribution in the cluster. However, in several cases the atomic structure has been relaxed and efforts have been made to find structures with lowest energies. These include calculations based on configuration interaction as well as density functional methods. However, most of these have been restricted to clusters with a few atoms only. The most promising development which has given theoretical study of clusters a new perspective, has been the ab-initio molecular dynamics (MD) approach developed by Car and Parrinello[23] (CP). In this appraoch the many body interatomic potential is calculated from ab-initio calculations based upon the density functional theory (see articles by von Barth and M P Das in this volume) which has proved very successful in the study of the structural and electronic properties of a variety of materials. The ab-initio molecular dynamics method not only allows a search of the lowest energy structures within the framework of the simulated annealing (SA) but also it can be used to study the static and dynamical properties of clusters having upto a few tens of atoms. So far its applications have been to clusters of $s-p$ bonded elements. However, recent developments of dealing with the d-orbitals within this scheme (see article by Laasonen in this volume) will make the study of technologically more important clusters with transition metal elements feasible. Some of the systems that have already been studied successfully include clusters[24,25] of Si[17], Se[26], S[27], Na[28,29], P[30], Be[31], Mg[32], Sb[33,34], Sn[35], Al[36,37], Ga[37,38] , C[39–41], B[42], Ge[24] and mixed clusters of GaAs, GaP and AlAs[43], NaK[28], Na-Mg[44], Al-Si, Al-C and B-C[45]. Good agreement has been obtained with experimental data wherever available. Here I will discuss the CP method and present results of some of these applications. There are a few other reviews[46–49] on this method which have appeared recently and contain some details of this method and applications to other systems such as liquid and amorphous materials, surface structure, diffusion, defects etc. Some details can also be found in the articles by Pastore and Laasonen in this volume.

In the following section I present the calculational procedure. Section 3 contains a review of most of the results on clusters obtained so far using the ab-initio molecular dynamics procedure. Section 4 contains an outlook for future work.

2. ab-initio Molecular dynamics method

In classical molecular dynamics approach an empirical interatomic potential, often a pairwise interaction, is used to calculate the energy of a system. Such empirical potentials are constructed from a fit to some known properties of that material. However, except for rare gas solids and strongly ionic materials, in most other systems it is difficult to justify a pairwise interaction from first principles. Further, during MD simulation the ionic positions can change very significantly from their starting configuration and structural transformation may occur for which the starting potential may not be valid. In cases such as clusters where the structure of the system is *a priori* not known, it would be difficult even to find a good interatomic potential. Car and Parrinello[23] combined one of the most successful theories of total energy calculation of an aggregate, namely the density functional method[50,51], with the molecular dynamics approach to develop an efficient combined electron-ion minimization procedure. Hereafter we shall refer the methods based on the idea of the density functional molecular dynamics approach to as the ab-initio molecular dynamics methods. In the following we first discuss briefly the Kohn-Sham equations[52] which are central to the solution of the electronic structure problem within the density functional formalism and then describe the standard and the MD approaches for solving these equations.

2.1 Kohn-Sham equations

The density functional theory as developed by Hohenberg and Kohn[50] and Kohn and Sham[51] offers a formal exact treatment of solving the ground state electronic properties of a system. However, it requires the knowledge of a functional for exchange and correlation which is not known for most systems. An approximate treatment of the exchange and correlation energy based on the local density or local spin density approximation (LDA or LSD) has been successfully used in a wide variety of systems ranging from atoms and molecules to surfaces and bulk materials and will also be the basis of the ab-initio MD to be discussed here.

For systems such as clusters one of the principal tasks of an ab-initio calculation is to obtain the lowest energy isomers by minimizing the total energy E with respect to the ionic and electronic degrees of freedom. Within the density functional theory this can be achieved by solving the Kohn-Sham equations (for details see the article by M. P. DAs),

$$[-\frac{1}{2}\nabla^2 + V_{ext}(\mathbf{r}) + V_H(\mathbf{r}) + V_{xc}(\mathbf{r})]\psi_i(\mathbf{r}) = \epsilon_i\psi_i(\mathbf{r}), \tag{1}$$

where V_{ext}, V_H and V_{xc} are the external (electron - ion), Hartree and exchange-correlation potentials respectively (atomic units $e = \hbar = m_e = 1$ are used.). The electron-ion potential can be an all electron or a pseudopotential. The other two contributions to the potential are

$$V_H(\mathbf{r}) = \int d\mathbf{r}' n_e(\mathbf{r}')/|\mathbf{r} - \mathbf{r}'| \tag{2}$$

and

$$V_{xc}(\mathbf{r}) = \frac{dE_{xc}}{dn_e(\mathbf{r})}, \tag{3}$$

where the electronic charge density $n_e(\mathbf{r})$ is given by,

$$n_e(\mathbf{r}) = \sum_i^{occ} f(i)|\psi_i(\mathbf{r})|^2, \tag{4}$$

$f(i)$, being the occupation number for the state i. The ground state total energy E for a given configuration of the ions $\{\mathbf{R}_I\}$ is

$$E[\{\psi_i\},\{\mathbf{R}_I\}] = \sum_i^{occ} f(i)\int d\mathbf{r}\psi_i^*(\mathbf{r})(-\frac{1}{2}\nabla^2)\psi_i(\mathbf{r}) + \int d\mathbf{r}V_{ext}(\mathbf{r})n_e(\mathbf{r})$$

$$+\frac{1}{2}\int d\mathbf{r}d\mathbf{r}'\frac{n_e(\mathbf{r})n_e(\mathbf{r}')}{|\mathbf{r}-\mathbf{r}'|} + E_{xc}[n_e] + \frac{1}{2}\sum_{I\neq J}\frac{Z_I Z_J}{|\mathbf{R}_I - \mathbf{R}_J|}. \tag{5}$$

Here Z_I is the charge on the Ith ion at the position $\mathbf{R}_I$. $E_{xc}[n_e]$ is the exchange-correlation energy. This has been calculated within the LDA/LSD in almost all the applications of the ab-initio MD,

$$E_{xc}[n_e] = \int d\mathbf{r} n_e(\mathbf{r})\epsilon_{xc}[n_e(\mathbf{r})], \tag{6}$$

where $\epsilon_{xc}[n_e(\mathbf{r})]$ is the exchange-correlation energy per particle for a homogeneous electron gas with density $n_e(\mathbf{r})$. In the case of spin polarized calculation, this energy should correspond to the density with the same spin polarization as in the actual system. There are several approximate forms available for this and therefore E_{xc} can be easily calculated in real space once the density is known. This approximation yields reliable binding energy trends and equilibrium structures for a variety of molecules[52] and solids[53]. However, recently there have been interesting developments where non-local exchange-correlation functionals have been successfully used which have provided improved agreement with experiments for several systems (see von Barth in this volume). Their inclusion in the ab-initio MD scheme is also likely to give improved results but since most applications have been done within LDA/LSD we shall restrict our discussion to LDA/LSD only.

2.1.1 The standard approach

In the standard approach, for a given configuration of the ions, one has to guess a starting wavefunction, construct the Hamiltonian matrix and then solve (1) iteratively untill self-consistency is achieved. One can then use the steepest descent technique to relax the ionic positions by calculating the forces on ions,

$$\mathbf{F}_I = -\frac{\delta E}{\delta \mathbf{R}_I} \tag{7}$$

following the Hellman-Feynman theorem[54]. This would lead to a *local* minimum in the energy surface. However, for finding global minimum one has to then take a new configuration and repeat this procedure until a configuration is obtained which has the lowest energy. This is a challenging task and it would be nearly impossible to explore all the electronic and ionic configurations that may be possible for a given number of atoms in a cluster except in the case where this number is restricted to a few atoms. Though much progress has been made by the steepest descent root, the energy surface is in general very complicated for most systems and in such situations other strategies based on simulated annealing become more efficient.

2.1.2 The molecular dynamics approach

The development of the ab-initio molecular dynamics method[23] has provided an efficient way for the optimization of the energy functional with respect to the electronic and ionic degrees of freedom *simultaneouly*. This method relies on the assumptions that the ions can be regarded as classical particles and that the motion of the electrons and the ions can be treated within the Born-Oppenheimer(BO) approximation. The ab-initio MD exploits the fact that in the BO approximation the electrons follow the ions *instantaneously* as they move and therefore remain very close to the ground state of the corresponding ionic configuration. This eliminates the need to find the electronic ground state for each ionic configuration and therefore reduces the computaional effort very significantly.

In the MD approach the electronic $\{\psi_i\}$ and the ionic $\{\mathbf{R}_I\}$ degrees of freedom in the energy functional $E[\{\psi_i\},\{\mathbf{R}_I\}]$ are taken to be time dependent and a Lagrangian is introduced :

$$L = \sum_i^{occ} f(i)\mu_i \int d\mathbf{r}|\dot{\psi}_i(\mathbf{r})|^2 + (1/2)\sum_I M_I \dot{\mathbf{R}}_I^2 - E[\{\psi_i\},\{\mathbf{R}_I\}], \tag{8}$$

which generates a dynamics for ψ_i and $\mathbf{R}_I$ through the Newton's equations of motion:

$$\mu_i \ddot{\psi}_i(\mathbf{r},t) = -H\psi_i(\mathbf{r},t) + \sum_j \Lambda_{ij}\psi_j(\mathbf{r},t) \tag{9a}$$

and

$$M_I \ddot{\mathbf{R}}_I = -\frac{\delta E}{\delta \mathbf{R}_I(t)}, \tag{9b}$$

where μ_i's are the fictitious masses for the electronic degrees of freedom which are generally taken to be independent of the state i. Λ_{ij} are the Lagrangian multipliers for the orthonormalization of the single particle orbitals ψ_i,

$$\int \psi_i{}^*(\mathbf{r})\psi_j(\mathbf{r})d\mathbf{r} = \delta_{ij}. \tag{10}$$

The ion dynamics described by Eq.(9) is real whereas the dynamics of the electronic degrees of freedom is fictitious and should be regarded as a means of solving the Kohn-Sham equations. For a fixed ionic configuration when the acceleration of the orbitals becomes zero, then Eq.(9a) reduces to the standard problem and the eigenvalues of the Λ matrix are the eigen values of the Kohn-Sham equations. Once the orbitals are converged, one can then use the set of Eqs.(9) to follow the coupled electronic-ionic motion. As we shall discuss later, the calculation of the forces on ions can be easily done using the Hellmann-Feynman theorem. In principle for an exact calculation of the Hellmann-Feynman forces the electrons should be in the ground state at any instant of time during the simulation. This can be done but it is computationally very expensive. In the CP method, however, the dynamical simulation can be performed accurately (within a certain bound) when the deviation from the BO surface is small. This is due to a cancellation of errors in the calculation of the Hellmann-Feynman forces as the electron- ion dynamics proceeds. This cancellation follows from the fact that plasma frequencies are generally much higher than the ionic frequencies. Therefore during a period of the ionic motion the orbitals would oscillate several times around the ions. The forces exerted on the ions by the electrons due to deviation from the BO surface would oscillate around the correct value and would be nearly cancelled when averaged over several oscillations. Therefore for a successful CP dynamics the smallest plasma frequency should be larger than the largest ionic frequency. Further, the energy gap between the highest occupied and the first excited state in the electronic energy spectrum should be larger than the thermal energy associated with the ionic motion (see also the chapter by Pastore in this volume). This is typically the case for semiconductors and insulators. In such cases, for a given temperature there is very little transfer of energy between the ions and the electrons for a significant and in some cases even for the whole observation time of the simulation and the electrons remain very close to the BO surface. However, in metals the energy gap between the highest occupied and the lowest unoccupied states is zero and therefore it becomes a problem to keep the electrons moving adiabatically. As the simulation proceeds the kinetic energy associated with the electronic degrees of freedom increases at the expense of the kinetic energy of the ions. In such situations one should ,in principle, use[55] the fractional occupation of various states according to the Fermi-Dirac distribution function at any finite temperature. An easy way to overcome this problem is to bring the electrons close to the instantaneous ground state after a certain number of time steps and then proceed with the simulation again. This would generally lead to a cooling of the ions. To avoid this problem Nose[56] thermostat has been used to keep the ionic temperature at a desired value. This, however, does not eliminate the problem of energy transfer to electrons. Blöchl and Parrinello[57] have used a Nose thermostat for the electrons also so that the electrons remain close to their ground state. However, such constraints increase the computational effort significantly. For clusters as the electronic energy spectrum is discrete, the condition of large gap between the highest occupied molecular orbital (HOMO) and the lowest unoccupied molecular orbital (LUMO) is satisfied in some cases (in particular for the magic clusters) whereas for other clusters and more so for larger clusters of metals, deviation from adiabatic behaviour could be significant to warrant frequent electron minimization or adopt other strategies. The formulation presented here is general and can be applied to solids, liquids or finite systems.

2.2 Implimentation of molecular dynamics using pseudopotentials

In the foregoing discussion we did not assume any particular form of ionic potential. Though there have been interesting developments of doing all electron[58] MD simulations, most studies have been done using pseudopotentials[59] which allow the use of a plane wave basis set. A plane wave basis is convinient particularly when the ions have to be moved as the plane waves are not localized on a particular site. Then the wavefunction in (1) can be written as

$$\psi_i^{\mathbf{k}}(\mathbf{r}) = \sum_{\mathbf{G}} C_{\mathbf{G}}^{i\mathbf{k}} exp[i(\mathbf{k}+\mathbf{G}).\mathbf{r}]. \tag{11}$$

The sum over $\mathbf{G}$ is usually truncated to include plane waves upto a certain kinetic energy which determines the accuracy of the calculation. Further, the use of plane waves assumes imposition of periodic boundary conditions which are also used in classical MD simulations in bulk systems. In the case of clusters, the MD cell is taken to be sufficiently large so that the interaction between the cluster and its periodic images is negligible. $\mathbf{k}$ is a wave vector in the Brillouin zone. For achieving self-consistency one generally uses a set of a few special $\mathbf{k}$ points[60]. However, in the case of clusters, as the cell dimensions are large, one can use just a single $\mathbf{k}$ point , the Γ point, in (11) to sample the Brillouin zone of the MD supercell and neglect the band dispersion. This has an additional advantage that the single particle orbitals become real and therefore the computational effort reduces significantly. Henceforth, we shall drop the $\mathbf{k}$ index.

The total potential from the ions can be written as

$$V_{ext}(\mathbf{r}) = \sum_I v^I_{ps}(\mathbf{r} - \mathbf{R}_I), \tag{12}$$

where $v^I_{ps}(\mathbf{r})$ is the pseudopotential for the Ith ion. There are several ionic pseudopotentials available in the literature[61,62]. However, in most applications norm-conserving pseudopotentials of Bachelet *et al*[61] have been used as these have been tabulated. In general, we can write the ionic pseudopotential as,

$$v^I_{ps}(\mathbf{r}) = \sum_{l=0}^{\infty} v^I_l(r)\hat{P}_l. \tag{13}$$

$\hat{P}_l$ is the angular momentum projection operator which projects out the lth angular momentum component of a function. $v^I_l(r)$ is the pseudopotential which operates on the lth angular momentum component of the wave function. If one assumes $v^I_l = v^I_{l_{max}}$ for $l \geq l_{max}$ then the infinite sum in (13) can be rewritten as

$$v^I_{ps}(\mathbf{r}) = v^I_{l_{max}}(r) + \sum_{l=0}^{l_{max}-1} \Delta v^I_l(r)\hat{P}_l, \tag{14}$$

where

$$\Delta v^I_l(r) = v^I_l(r) - v^I_{l_{max}}(r). \tag{15}$$

$v^I_{l_{max}}$ is local and therefore Δv^I_l becomes the non-local part of the ionic pseudopotential. Thus we can write the total ionic potential as a sum of a local contribution, $V_{loc}(\mathbf{r})$ and a non-local contribution, $\Delta V_{nl}(\mathbf{r})$.

In the MD approach for solving the Kohn-Sham equations the expansion coefficients become time dependent and therefore the equation of motion for ψ_i can be recast into equation of motion for $C_{\mathbf{G}}$. Substituting Eq.(11) in (9a) and multiplying with $exp(-i\mathbf{G}'.\mathbf{r})$ and integrating over $\mathbf{r}$ gives

$$\mu_i \ddot{C}^i_{\mathbf{G}} = \sum_{\mathbf{G}'} [1/2|\mathbf{G}|^2 \delta_{\mathbf{G}\mathbf{G}'} + V_{loc}(\mathbf{G}-\mathbf{G}') + \Delta V_{nl}(\mathbf{G},\mathbf{G}')$$
$$+ V_H(\mathbf{G}-\mathbf{G}') + V_{xc}(\mathbf{G}-\mathbf{G}')]C^i_{\mathbf{G}'} + \sum_j \Lambda_{ij} C^j_{\mathbf{G}}. \tag{16}$$

These equations are then numerically integrated using the standard MD procedures such as Verlet algorithm or the conjugate gradient technique. In the Verlet algorithm, the second derivative is written as a second order difference equation:

$$C^i_{\mathbf{G}}(t+\Delta t) = 2C^i_{\mathbf{G}}(t) - C^i_{\mathbf{G}}(t-\Delta t) + \Delta t^2 \ddot{C}^i_{\mathbf{G}}(t). \tag{17}$$

It introduces an error of order $(\Delta t)^4$ into the integration of the equations of motion. A more sofisticated finite difference equation could also be used but this would require storing the coefficients for a larger number of time steps and since these are large arrays, the storage can become a problem.

2.2.1 Electron minimization

The simulation proceeds with an initial choice of the ionic distribution and of the coefficients in the expansion of the wave functions which are orthonormalized. The latter can be used to construct the charge density $n_e(\mathbf{r})$ and therefore the exchange correlation potential and energy. The Hartree contribution to potential is easily calculated by performing a fast Fourier transform (FFT) of $n_e(\mathbf{r})$ and it is given by

$$V_H(\mathbf{G}) = 4\pi n_e(\mathbf{G})/G^2. \tag{18}$$

The local contribution to the total ionic potential $V_{loc}(\mathbf{G})$ is

$$V_{loc}(\mathbf{G}) = \sum_{\alpha} S_\alpha(\mathbf{G}) v_{loc}^{\alpha}(\mathbf{G}), \tag{19}$$

where S_α is the structure factor for the specie α and is given by

$$S_\alpha(\mathbf{G}) = \sum_{p} exp(i\mathbf{G}.\tau_p^\alpha). \tag{20}$$

Here τ_p^α is the position of the pth atom of type α in the basis. The sum over p is on all the atoms of type α and

$$v_{loc}^{\alpha}(\mathbf{G}) = \frac{1}{\Omega_c}\int d\mathbf{r} v_{loc}^{\alpha}(r) exp(i\mathbf{G}.\mathbf{r}), \tag{21}$$

where Ω_c is the cell volume. The Hartree and the local ionic potential terms can now be transformed into real space with a fast Fourier transform to construct, together with $V_{xc}(\mathbf{r})$, the product $V(\mathbf{r})\psi(\mathbf{r})$ whose transform leads to the convolutions in Eq.(16). The non-local contribution needs separate attention as it is not in the form of a convolution. The non-local contribution to potential is

$$v_{nl}^{I}(\mathbf{G},\mathbf{G}') = 4\pi \sum_{l=0}^{l_{max}-1} (2l+1) P_l(cos(\theta_{\mathbf{GG}'})) \int dr r^2 j_l(Gr) j_l(G'r) \Delta v_l^I(r). \tag{22}$$

Here P_l is the Legendre polynomial of order l and j_l is the spherical Bessel function of the first kind of order l. $\theta_{\mathbf{GG}'}$ is the angle between $\mathbf{G}$ and $\mathbf{G}'$. Evaluation of E_{nl} would require computation of $N_s N N_p(N_p+1)/2$ integrals where N_s and N_p are respectively the number of states and the plane waves. This is a very time consuming operation due to large values of N_p. It can, however, be simplified following a suggestion from Kleinmann and Bylander[63] who transformed the semi non-local part of the potential (local in radial part but non-local in angular part) into a fully non-local operator:

$$\Delta v_{nl}^{\alpha KB} = \sum_{l,m} \frac{|\Delta v_l^\alpha(r)\Phi_{l,m}^\alpha| >< \Phi_{l,m}^\alpha|\Delta v_l^\alpha(r)|}{< \Phi_{l,m}^\alpha|\Delta v_l^\alpha|\Phi_{l,m}^\alpha >}. \tag{23}$$

Here $\Phi_{l,m}^\alpha$ is the atomic pseudo wavefunction for specie α. It can easily be varified that in the case of atom, the fully non-local and the semi non-local forms lead to the same solution. However, the same is not true when this separable form is used in other situations. In particular ghost states have been found[64] in some cases. This can ,however, be taken care of by a suitable choice of the local potential. With this choice, the Fourier components of the non-local pseudopotential can be written into a separable form such that

$$v_{nl}(\mathbf{G}, \mathbf{G}') = \sum_{l=0}^{l_{max}-1} F_l(\mathbf{G})F_l(\mathbf{G}'), \tag{24}$$

where F_l and $F_{l'}$ are functions of only one wave vector. This reduces the number of integrals to $N_s N N_p$.

2.2.2 Total energy and forces on ions

The total energy[65] can be written as

$$E = E_{KE} + E_H + E_{ps} + E_{xc} + E_{ion}. \tag{25}$$

The kinetic energy of electrons and the Hartree term are given by

$$E_{KE} = (1/2)\sum_{i,\mathbf{G}} f(i)\mathbf{G}^2 C_{\mathbf{G}}^{i\,*} C_{\mathbf{G}}^i, \tag{26}$$

and

$$E_H = \frac{1}{2}\sum_{\mathbf{G}}{}' V_H(\mathbf{G})n_e(-\mathbf{G}). \tag{27}$$

The prime indicates that the $\mathbf{G} = 0$ term is excluded in the summation. This represents the long range slowly varying part of the Coulomb potential. For a neutral system, the *total* contribution to energy from $\mathbf{G} = 0$ term is zero in the case of a purely $1/r$ potential. However, since we are using pseudopotentials, there is a non-zero contribution coming from these terms and it needs to be treated carefully. It is convinient to smear the ion core charge into a Gaussian[66] given by

$$n_c^I(\mathbf{r} - \mathbf{R}_I) = \frac{Z_I}{(R_I^c)^3}\pi^{3/2} exp(-|\mathbf{r} - \mathbf{R}_I|^2/(R_I^c)^2), \tag{28}$$

where R_I^c is the width of the Gaussian. The potential felt by an electron due to this distribution is

$$v_c(\mathbf{r} - \mathbf{R}_I) = -\frac{Z_I}{|\mathbf{r} - \mathbf{R}_I|} erf(|\mathbf{r} - \mathbf{R}_I|/R_I^c). \tag{29}$$

The total charge distribution and potential from the Gaussian smeared ion-cores are,

$$n_c(\mathbf{r}) = \sum_I n_c^I(\mathbf{r} - \mathbf{R}_I),$$

and

$$V_c(\mathbf{r}) = \sum_I v_c(\mathbf{r} - \mathbf{R}_I). \tag{30}$$

Considering the electronic and ionic charge distributions together as a neutral system, the Hartree, the local pseudopotential and the Madelung (E_{ion}) terms can be written together as

$$E_H + E_{ps}^{loc} + E_{ion} = \frac{1}{2}\int d\mathbf{r}d\mathbf{r}'\frac{n(\mathbf{r})n(\mathbf{r}')}{|\mathbf{r}-\mathbf{r}'|} + \int d\mathbf{r}n_e(\mathbf{r})\sum_I w_{loc}(|\mathbf{r}-\mathbf{R}_I|) - E_{self} + \Delta_c, \tag{31}$$

where $n = n_e + n_c$ and

$$w_{loc}(r) = [v_{l_{max}}^I(r) + \frac{Z_I}{r}erf(r/R_I^c)]. \tag{32}$$

The second term in the above equation arises from Gaussian smeared ion cores (Eq. 29) and should be subtracted as it has been included in the first term on the right hand side of Eq.(31). The latter can now be evaluated in the Fourier space and is given by,

$$\frac{1}{2}\sum_{\mathbf{G}} n(\mathbf{G})^* n(\mathbf{G})4\pi/G^2, \tag{33}$$

where,

$$n(\mathbf{G}) = n_e(\mathbf{G}) + \sum_\alpha S_\alpha(\mathbf{G})n_c^\alpha(\mathbf{G}), \tag{34}$$

and $n_c^\alpha(\mathbf{G})$ is the Fourier transform of the smeared ion-core charge of the α specie. Therefore, the explicit evaluation of the Madelung sums is avoided. w_{loc} is a short range potential as the long range parts of the two terms in Eq. (32) cancel each other. E_{self} is the self-interaction of the Gaussian smeared ion core charges which is included in the first term in Eq. (31) and should be subtracted. Δ_c is a correction to the first term in Eq. (31). It is equal to the difference in the interaction between the original point charges Z_I and the Gaussian broadened charges. These are given by,

$$E_{self} = 1/2\sum_I \int d\mathbf{r}d\mathbf{r}'\frac{n_c^I(\mathbf{r}-\mathbf{R}_I)n_c^I(\mathbf{r}'-\mathbf{R}_I)}{|\mathbf{r}-\mathbf{r}'|}$$

which from Eq.(28) simplifies to

$$\frac{1}{\sqrt{(2\pi)}}\sum_I \frac{Z_I^2}{R_I^c}, \tag{35}$$

and

$$\Delta_c = 1/2\sum_{I\neq J}[\frac{Z_I Z_J}{|\mathbf{R}_I - \mathbf{R}_J|} - \int d\mathbf{r}d\mathbf{r}'\frac{n_c^I(\mathbf{r}-\mathbf{R}_I)n_c^J(\mathbf{r}'-\mathbf{R}_J)}{|\mathbf{r}-\mathbf{r}'|}]$$

which reduces to

$$1/2\sum_{I\neq J}\frac{Z_I Z_J}{R_{IJ}}erfc(\frac{R_{IJ}}{R_{IJ}^c}), \tag{36}$$

where $R_{IJ} = |\mathbf{R}_I - \mathbf{R}_J|$ and $R_{IJ}^c = \sqrt{(R_I^{c\,2} + R_J^{c\,2})}$.

The remaining contribution to the total energy comes from the non-local part of the pseudopotential and it is given by

$$E_{nl} = \sum_i f(i)\sum_I\sum_{\mathbf{G},\mathbf{G}'} e^{-i\mathbf{G}.\mathbf{R}_I}C_{\mathbf{G}}^{i\,*}v_{nl}(\mathbf{G},\mathbf{G}')e^{i\mathbf{G}'.\mathbf{R}_I}C_{\mathbf{G}'}^i, \tag{37}$$

whereas the exchange correlation energy can be calculated from Eq. (6).

The forces on ions can be calculated from Eq. (7) following the Hellmann-Feynman theorem according to which it is sufficient to know the changes in the Hamiltonian when the ions move, the contribution from the changes in the wave-function being zero. The $\mathbf{R}_I$ dependent term in the Hamiltonian is the electron-ion interaction. The other contribution to force would come from the Madelung energy. From the above reformulation of the various contributions to the total energy, it can be noted that the contributions would come from Eq.(33), w_{loc}, Δ_c and E_{nl}. These can be written as,

$$\sum_{\mathbf{G}}[n^*(\mathbf{G})n_c^I(\mathbf{G})4\pi/G^2 + n_c^*(\mathbf{G})w_{loc}(\mathbf{G})](i\mathbf{G})e^{i\mathbf{G}.\mathbf{R}_I} + \frac{\partial\Delta_c}{\partial\mathbf{R}_I} + \frac{\partial E_{nl}}{\partial\mathbf{R}_I}, \tag{38}$$

where,

$$\frac{\partial\Delta_c}{\partial\mathbf{R}_I} = \sum_J{}'[(-)Z_I Z_J\frac{\mathbf{R}_{IJ}}{R_{IJ}^3}erfc(\frac{R_{IJ}}{R_{IJ}^c}) + \frac{Z_I Z_J}{R_{IJ}R_{IJ}^c}\frac{2}{\sqrt{\pi}}e^{-(R_{IJ}/R_{IJ}^c)^2}], \tag{39}$$

and

$$\frac{\partial E_{nl}}{\partial\mathbf{R}_I} = 2\sum_{il} f(i)\Re(\frac{\partial W_{ilI}^*}{\partial\mathbf{R}_I}W_{ilI}), \tag{40}$$

with

$$W_{ilI} = \sum_{\mathbf{G}} C_{\mathbf{G}}^i F_l(\mathbf{G})e^{i\mathbf{G}.\mathbf{R}_I}. \tag{41}$$

The prime in Eq. (39) indicates that $J = I$ term is excluded.

2.2.3 Orthogonalization of wave functions

After the forces are calculated, the set of Eqs. (9) have to be solved with the constraints (10). Starting with a set of orthonormalized ψ_i, the parameters Λ_{ij} can be calculated[46] by differentiating (10) twice and using (9a), to obtain:

$$\Lambda_{ij} =< \psi_j(\mathbf{r},t)|H|\psi_i(\mathbf{r},t) > -\mu \int d\mathbf{r}\dot{\psi}_j^*(\mathbf{r},t)\dot{\psi}_i(\mathbf{r},t). \tag{42}$$

Thus the Lagrangian multipliers depend upon the trajectories of the ψ_i's and therefore on the algorithm used to integrate the equations of motion. The solution of the equations of motion is achieved in two steps. First an unconstrained wavefunction $\bar{\psi}_i$ is obtained. In the Verlet algorithm this can be written as,

$$\bar{\psi}_i(t+\Delta t) = -\psi_i(t-\Delta t) + 2\psi_i(t) - \frac{\Delta t^2}{\mu} H\psi_i(t). \tag{43}$$

The constraint forces are then added to the wavefunctions $\bar{\psi}_i$,

$$\psi_i(t+\Delta t) = \bar{\psi}_i(t+\Delta t) + \sum_j x_{ij}^* \psi_j(t), \tag{44}$$

where $x_{ij}^* = (\Delta t^2/\mu)\Lambda_{ij}$. Now imposing the orthonormality condition on $\psi_i(t + \Delta t)$,

$$\int d\mathbf{r}\psi_i^*(\mathbf{r},t+\Delta t)\psi_j(\mathbf{r},t+\Delta t) = \delta_{ij}, \tag{45}$$

we get,

$$A_{ij} + \sum_k (x_{ik}^* B_{kj} + x_{jk} B_{ki}^* + \sum_l x_{ik} x_{jk}^*) = \delta_{ij}, \tag{46}$$

which can be written in matrix form as

$$\mathbf{A} + \mathbf{X}\mathbf{B} + \mathbf{B}^\dagger\mathbf{X}^\dagger + \mathbf{X}\mathbf{X}^\dagger = \mathbf{1}, \tag{47}$$

where, $A_{ij} =< \bar{\psi}_i(t+\Delta t)|\bar{\psi}_j(t+\Delta t) >$ and $B_{ij} =< \psi_i(t)|\bar{\psi}_j(t+\Delta t) >$. Eq. (47) can be solved iteratively. Using Eq. (43) and making a Taylor series expansion of wavefunctions at time $(t-\Delta t)$, it can be shown that $A_{ij} = \delta_{ij} + \mathcal{O}(\Delta t^2)$ and $B_{ij} = \delta_{ij} + \mathcal{O}(\Delta t)$. Therefore, from Eq. (46) within second order in (Δt) we can write

$$\mathbf{X}^o = (1/2)(\mathbf{1} - \mathbf{A}). \tag{48}$$

Then the solution to (47) can be found iteratively from

$$\mathbf{X}^{(k)} = \frac{1}{2}[\mathbf{1} - \mathbf{A} + \mathbf{X}^{(k-1)}(\mathbf{1}-\mathbf{B}) + (\mathbf{1}+\mathbf{B}^\dagger)\mathbf{X}^{(k-1)} - \mathbf{X}^{(k-1)^2}], \tag{49}$$

where k is the number of iterations. Typically it requires a few iterations to solve (47).

2.3 Simulated annealing

As the energy surface in the multidimensional configuration space is in general complicated, simulated annealing technique[67] is used to obtain the lowest energy structures. In this technique the cluster is heated upto a certain temperature T so that the ions start diffusing. This temperature is in general different for different materials and will in general also differ for different clusters of a material as the binding energy per atom may change significantly with cluster size. The cluster is then allowed to evolve at this temperature for a few pico-seconds which is a rather short time as compared to standard classical molecular dynamics simulations where the observation time is of the order of nano-seconds. Experience with different systems has, however, shown that this period is sufficient to make the cluster loose completely the memory of its starting configuration. Subsequently it is then cooled such that the system remains close to the BO surface. The time step for integration of the equations of motion is taken to be such that there is negligible transfer of energy from the ionic to the electronic degrees of freedom. As discussed above, in the case of clusters having a large HOMO-LUMO gap it has been found that the system can remain close to the BO surface even for the whole period of the simulation. The simulated annealing procedure has proved very successful in finding new structures which have lower energy than the lowest energy structures obtained by other methods in some cases. Also it provides a natural way to study the dynamical properties of clusters and the changes in the atomic and electronic structure at finite temperatures which are going to be important in understanding reactions on clusters. In the next section we discusss some of the systems to which this technique has been successfully applied.

3. Applications to clusters

3.1 Clusters of metals

In their pioneering experiments on simple metal clusters, Knight *et al*[68] observed marked stability of sodium clusters with 8, 20, 40, 58, and 92 ... atoms (Fig. 2a). Sodium has one valence electron and therefore clusters with 8, 20, 40, 58 and 92 ... valence electrons are particularly abundant. Such clusters are referred to as *magic* clusters as the abundance of clusters having one more atom is significantly less. Similar results have been obtained for clusters of other alkali metals[69] and their mixed clusters[70] as well as clusters of noble[71] metals. Clusters of divalent[72] elements also show marked stability for the same number of valence electrons. However, small clusters of divalent elements are weakly bonded due to the ns^2 closed shell configuration. Therefore, as the cluster size increases, there occurs a non-metal - metal transition which we shall discuss in detail later. Studies of the ionization

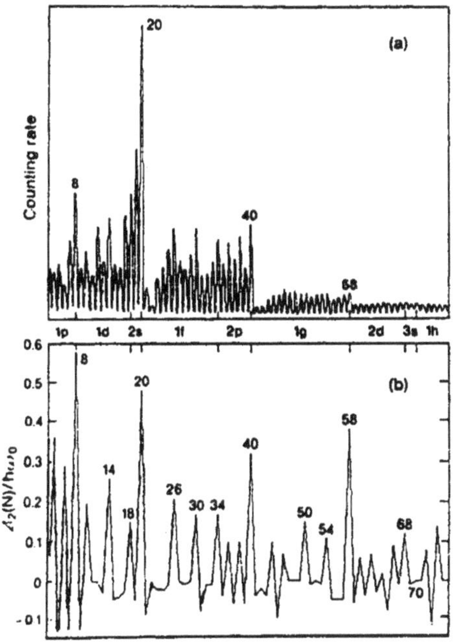

Figure 2: (a) Abundance spectrum of sodium clusters and (b) the second order derivative of energy calculated within an ellipsoidal modified harmonic oscillator model. (After Clemenger[81])

potentials[73], fragmentation[74] and polarizabilities[75] of metal clusters and reactions of some gases with them[19,20] have provided further information about their stability and electronic structure.

While experimental studies have been carried out on clusters of several metals, ab-initio MD method has been applied mainly to clusters of $s-p$ bonded metals such as Na, K, Be, Mg, Al, Sn, and binary clusters of Na-K , Na-Mg as well as doped clusters of Al. Calculations have also been performed on dimers of group IB and IIB elements[76] and clusters of semi-metals Sb[33,34] and Ga[37,38]. These studies have provided information about the low lying structures, bonding nature and dynamics of clusters. Before these calculations became possible, the spherical jellium model[22,77] (SJM) played the central role in the study of metal clusters and it is still a nice model to understand many of the results obtained by these elaborate calculations. We therefore first describe the essential results of this model.

3.1.1 The jellium model

Alkali and other simple metals are quite well described by a nearly free electron model as the effective ionic (pseudo-) potential seen by the valence electrons is nearly constant throughout the metal. The weak pseudopotential can be approximately treated as a perturbation on the free electron like behaviour of the $s-p$ conduction electrons in these metals. The same features can be expected to hold for clusters of these elements. However, as the structure of clusters is not known in most cases, one can consider a cluster to be spherical in the lowest approximation and solve the problem of electrons in a spherical potential well. Thus Knight and coworkers[68] used a simple spherical potential to study the stability of clusters as a function of the number of valence electrons. Very similar to the shell structure in atomic spectrum[78] as well as in nuclear physics[79] they correlated the stability of clusters with the filling of the electronic shells 1s, 1p, 1d, 2s, 1f, 2p, 1g, ... in the spherical potential. Later, Cohen and collaborators[22,80] used SJM in which the ionic charges were smeared out in a uniform positive spherical background such that

$$n_+(\mathbf{r}) = \begin{cases} n_0 & \text{if } r < R \\ 0 & \text{otherwise} \end{cases} \tag{50}$$

Here R is the radius of the cluster which is related to the number of atoms N in the cluster,

$$\frac{4}{3}\pi R^3 = N\Omega, \tag{51}$$

where Ω is the volume per atom in the macroscopic metal. The constant positive density n_0 is related to Ω and the number of valence electrons Z (the subscript I is dropped as only one type of atoms are considered) by $Z = n_0\Omega$. The Kohn-Sham equations (1) are then solved to obtain the eigenvalues and the total energies. Similar to the spherical potential model the energy levels of the electrons are again characterized by the radial and angular quantum numbers. The 1s, 1p, 1d, 2s, 1f, 2p, 1g, 2d, 3s, 1h, ... shells are respectively filled for the total valence electron numbers, 2, 8, 18, 20, 34, 40, 58, 68, 70, 92, and so on in the cluster. The closed shell configurations are stable due to energy gaps between electronic shells and again this simple model provides a good explanation for the marked stability observed in alkali metal clusters for 8, 20, 40, 58 and 92 atoms. The fine structure in the abundance spectrum can be understood in terms of the completion of the subshells that arise due to deformation of the spherical positive background. Clemenger[81] has studied the effects of the ellipsoidal deformations on a three dimensional modified harmonic oscillator model. As shown in Fig. 2b this deformation leads to fine structure in the second order energy difference, Δ_2 which shows peaks at $N = 10, 14, 18, 26, 30, 34, 36, 38, 44, 46, 54$ etc. corresponding to subshell filling in addition to the main peaks at $8, 20, 40, 58, \ldots$ For shell closing numbers not observed in the experiment the reason may be that only a large gap to the next shell enhances the stability of a closed shell or the SJM is not a good representation of the actual potential seen by the electrons in a cluster. The nucleation conditions can also affect the abundance spectrum. The absence of fine structure in the abundance spectrum for large clusters is due to smaller gaps for higher shells while the variation in the intensities of clusters of different elements in the same group is possibly due to the difference in the kinetic behaviour and different ionic potentials. While the general trends in several properties such as relative stability, ionization potential, polarizability etc. of metal clusters are well described[22,77] by the jellium model and it has the advantage of being applicable easily to large clusters containing several hundred atoms, it is of interest to study these aspects from a more realistic model and to understand their atomic structure and dynamical properties. In the following we review results on clusters of several systems obtained by using the ab-initio molecular dynamics method.

3.1.2 Alkali metal clusters

Several theoretical studies have been made on clusters of alkali metals but sodium is one of the most extensively studied. Ballone *et al*[28] studied Na_8, and $Na_{10}K_{10}$ clusters. More recently a very detailed LDA study on sodium clusters having upto 20 atoms has been carried out by Röthlisberger and Andreoni[29]. Several interesting observations have been made. First of all when the results of these calculations are analysed in terms of the angular character of the electronic states around the center of mass of the cluster, then the low temperature electronic structure agrees with the predictions of the shell model. However, one finds that with increasing temperature the clusters become less and less spherical and the hybridization between states of different angular momentum increases due to the decrease of energy gaps. This is in contrast to the widely held belief that the clusters tend to be spherical as the temperature increases and that the validity of the jellium model is enhanced in warm clusters. The hybridization is less pronounced for magic clusters but increases with increasing size due to the decrease in the spacing between the one-electron energy levels.

The potential energy surface is found to be rather flat and the electronic charge density, delocalized. However, there is a preference for pentagonal rings analogous to the complexes of Lennard-Jones systems. The energy difference between several low-energy structures is tiny. But even in this case the SA has proved very useful as a powerful optimization technique. The small energy differences for various cluster structures are also in agreement with the results in bulk where accurate LDA calculations[82] predict a difference of 2 meV between hcp and bcc and 3.5 meV between hcp and fcc structures.

The non-magic clusters such as Na_{10} and Na_{13} tend to have many more isomers which are nearly degenerate with the lowest energy structure, as compared to the magic clusters such as Na_8 and Na_{20}. The lowest energy structure of Na_8 is found to be dodecahedron which has the D_{2d} symmetry. This is nearly degenerate with the one having the T_d symmetry which was predicted to be the lowest energy structure from configuration interaction calculations[83].

Fig. 3 shows four structures for Na_{20} cluster. (a) and (b) have been obtained from simulated annealing and are nearly degenerate. These are based on pentagonal motif. Interestingly (b) is also the ground state and (a) a local minimum for Lennard-Jones clusters. (c) and (d) have tetrahedral symmetry. In this case the electronic states can be classified with good accuracy by their angular momentum in agreement with the shell model. However, in contrast to the Na_8 cluster, Na_{20} looses this shell model character even at as low temperatures as 200 K. Accordingly 8-atom clusters are predicted to be more stable with respect to increasing temperature. In fact the diffusion coefficient was estimated to be $\tilde{0}.5 \times 10^{-4}$ cm^2/sec for Na_{20} at 640 K from the root mean square displacement and it is only a factor of 3 lower than the value for liquid sodium at the same temperature. Further, the electronic structure is found not to be very sensitive to the atomic structure and therefore the electronic properties may not be useful as a probe to study the structure. In Fig. 4 we have shown the spherical average of the electronic charge density around the center of the cluster corresponding to four isomers shown in Fig. 3. One can see that there is very little variation from one structure to another beyond the core region. However, the vibrational spectra has been found to show a stronger dependence on structure and can be used as a probe to get some information about the structure.

For non-magic clusters such as Na_{10}, one can consider the cluster to be made up of a core which is more rigid and some cap ions which are constantly farther away from the center of the cluster. Thus individual atoms can undergo displacements larger than the typical interatomic distance. As shown in Fig. 5, the displacement of cap ions is far more than for those in the core region. However, the radial movement is much smaller than the root mean square displacement which indicates that the cap ions travel more on the surface of the cluster. While such a detailed study is not available for clusters of other alkali metals, we expect to have similar behaviour for them as also indicated from results of a few other studies and those in the bulk. In fact calculations[84] on Cs clusters do show that the general trends remain the same though the energy differences between different isomers become even smaller than in the case of Na.

Very recently Fois *et al*[85] have done a LSD study of Na clusters and have also studied the effect of the self-interaction correction (SIC). While for Na_6 the LSD calculation predicts the same structure as obtained from the LDA calculation, namely a pentagonal pyramid to be lower in energy than a 2-dimensional triangular structure, in the LSD-SIC calculation the triangular structure has a lower energy than the pentagonal pyramid. These results no doubt raise the question of the correctness of the LDA/LSD geometries. However, we expect that the trends and the information about the dynamics and the charge densities obtained from the LDA/LSD calculations are useful in understanding the general behaviour of clusters of differet sizes.

Mixed clusters of Na and K have also been studied within LDA and a ten-

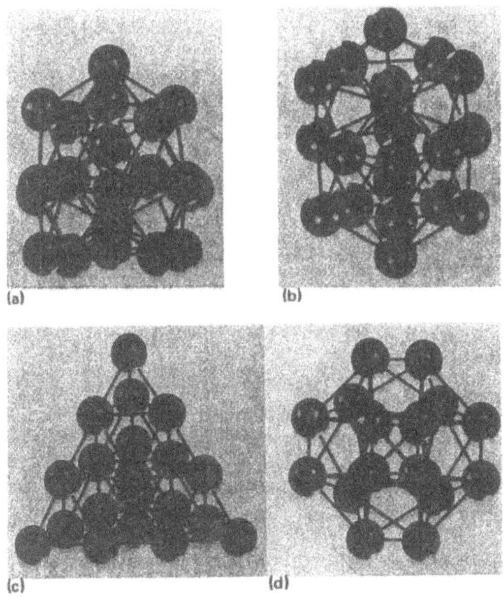

Figure 3: Isomers of Na_{20}. (a) and (b) are lowest energy structures whereas (c) and (d) are higher energy structures of T_d symmetry. (After Röthlisberger and Andreoni[29])

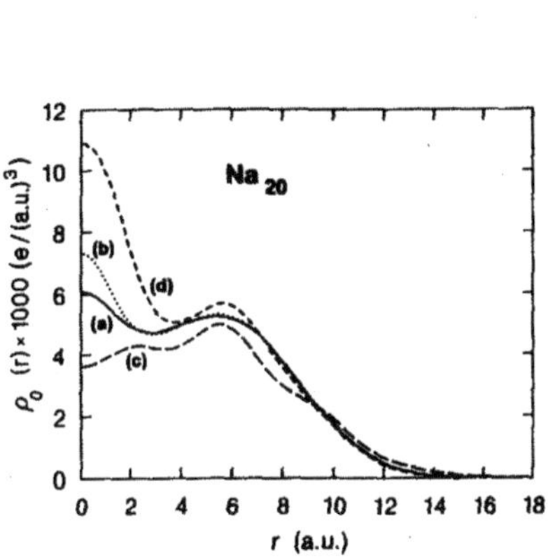

Figure 4: Spherical average of the electron density (here denoted by ρ_0) in the four structures of Fig. 3 for Na_{20}. r = 0 corresponds to the center of the cluster. (After Röthlisberger and Andreoni[29]).

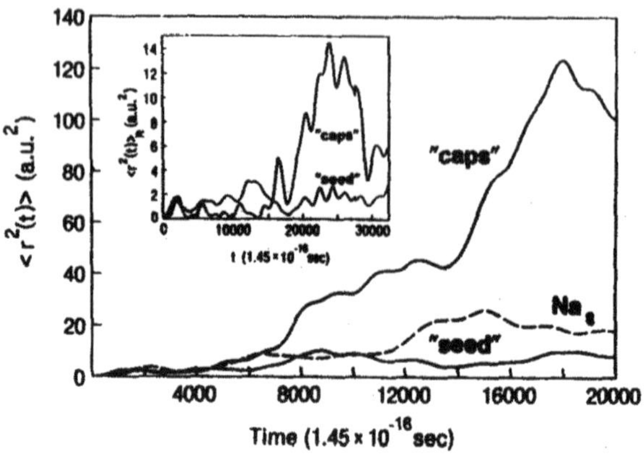

Figure 5: Time variation of the mean square displacement of 'caps' and 'seed' ions for Na_{10} cluster at 260 K. The inset shows the radial part. Note the difference between the scale of the main plot and the inset. For comparison the mean square displacement of the atoms in Na_8 around the same temperature is also shown by dashed line. (After Röthlisberger and Andreoni[29])

dency for K segregation on the surface has been found[28]. This is expected also on the basis of simple models based on broken bonds on a surface and the atomic size mismatch. According to these models, the element having the lower surface energy (larger size) tend to segregate at the surface[86]. This segregation might lead to preferential evaporation of the larger specie and favour clusters enriched with smaller atoms. The lowest energy structure of this mixed cluster is similar to that of Na_{20} and the one - electron orbitals are delocalized over the entire cluster. In agreement with the experimental observations[70], the shell model behaviour is applicable to this alloy in spite of the ionic segregation. Also for Mg doped Na_N clusters a tendency for Mg ion segregating to the surface has been obtained[44]. However, in this case due to surface segregation there is hybridization between states of different angular momentum components and the shell model is not quite applicable. On the other hand when a small size atom is doped as in the case of Na_7Al cluster, then the impurity ion prefers to sit inside the cluster and the electronic structure agrees with the predictions of the jellium model.

The finite temperature studies of clusters provide useful information about their dynamical behaviour. The vibrational spectrum (Fig. 6) of Na_7Mg cluster has a soft 'diffusive' mode as well as a high frequency mode, the latter being absent in pure sodium clusters. Even at 100 K a sodium atom which caps a pentagonal pyramid (Fig. 6) diffuses around the fivefold axis and gives rise to the soft mode. Whereas the high frquency mode corresponds to a strong bond between Na and Mg atoms which also causes Mg to have a high coordination with Na atoms.

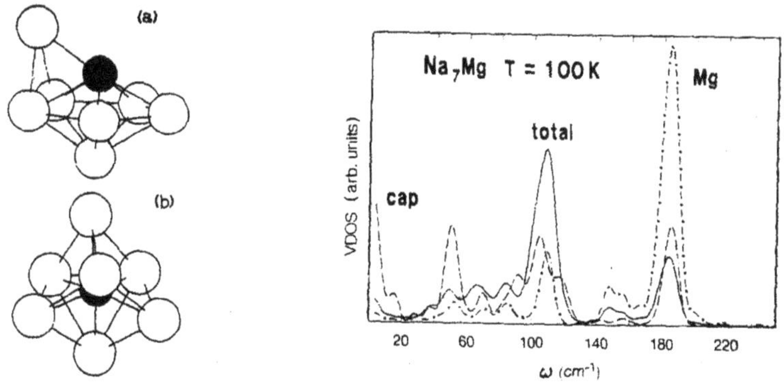

Figure 6: Structure and the vibrational spectrum of Na_7Mg cluster. (a)lowest energy structure (pentagonal bipyramid +1) and (b) tetracapped tetrahedron which also corresponds to a low energy isomer of Na_7Al. The vibrational density of states is shown for (a) at an average temperature of 100K. Total (solid line), sodium cap (dashed line) and magnesium atom (dashed-dotted line) (After Röthlisberger and Andreoni[44]).

3.1.3 Divalent metal clusters: growth and the non-metal - metal transition

For divalent metals, extensive studies have been made recently for Mg[32,87] and Be[31] clusters. The equilibrium structures of Mg clusters having upto 13 atoms have been studied by Kumar and Car[32]. The lowest energy structures with singlet states are shown in Fig. 7. Due to a closed shell atomic electronic configuration, Mg_2 is found to be very weakly bonded with a bond length 6.33 a.u.. Experimentally Mg_2 is found[88] to have a singlet ground state with a bond length 7.35 a.u. and cohesive energy 0.027 eV/atom. The larger (smaller) value of the calculated cohesive energy (bond length) is due to the use of the LDA which over- (under-) estimates the binding energy (bond length). Ballone *et al*[89] have done a calculation with a non-local exchange correlation functional and find an improvement over the LDA results. However, the results of Kumar and Car agree very well with a LSD calculation[90] which gives 0.11 eV and 6.37 a.u. respectively for the binding energy and the bond length.

The lowest energy state for Mg_3 is an equilateral triangle with bond length 5.93 a.u.. Mg_3 chain is 0.111 eV/atom higher in energy than the equilateral triangle. Thus Mg clusters favour to maximize the coordination number. This trend is continued for bigger clusters as well. The lowest energy structure for Mg_4 is a regular tetrahedron. This is a particularly stable cluster with considerably smaller bond length. Mg_5 is a slightly elongated trigonal bipyramid. These results are in agreement with a LSD calculation[90] which also finds a singlet state for all these clusters to be of lowest energy. These results were obtained by performing steepest descent calculations for a few selected geometries of the clusters. However, simulated annealing technique was used for larger clusters. For Mg_6 it leads to a structure shown in Fig. 7. This can be obtained by capping a face (3-4-5) of Mg_5. Independently Reuse *et al*[90] obtained an octahedron with a rectangular base to be of lowest energy for this cluster. A steepest descent calculation done by Kumar and Car[32] for this structure gives it only 0.004 eV/atom higher in energy than the simulated annealing structure, thus making them nearly degenerate. This is due to the fact that the pair distribution function for the two structures is nearly the same.

For Mg_7 the lowest energy structure is a pentagonal bipyramid. Its two apex atoms with coordination 6 have the shortest distance (5.3 a.u.) whereas the base atoms with coordination four have a bond length 6.04 a.u. As we shall discuss later, the short distance is significant from the point of view of non-metal - metal transition in these clusters. This structure can be obtained from Mg_5 by capping two faces and is also in agreement with the results obtained by Reuse et al[90] and Pacchioni et al[91] using the LSD and cofiguration interaction calculations respectively.

The lowest energy structure of Mg_8 can be simply obtained by capping a face (1-3-7) of the pentagonal bipyramid. For larger clusters a new structure starts. Mg_9 is a tricapped trigonal prism as shown in Fig. 7. This structure has some similarity with a fragment of the hcp lattice except that the bond lengths 1-4, 2-5 and 3-6 are shorter whereas for bulk Mg they correspond to the c-axis. Adding an atom to one of the triangular faces of the trigonal prism leads to the structure of Mg_{10} as shown in Fig. 7. The same result has also been obtained independently by Andreoni[24] and de Coulon *et al*[87]. This provides further confidence in SA strategy to obtain lowest energy structures. Among Mg clusters this is another most stable structure as it can be seen from the second order finite difference of the energy which is plotted in Fig. 8 along with the cohesive energy. $E(N)$ is the energy of a cluster with N atoms. While the cohesive energy increases with the size, the peaks in the difference spectrum in this plot can be correlated with the abundance in the mass spectrum. The marked stability of 4 and 10 atom clusters agree with the jellium model. Also a small peak and a shoulder appear for 7- and 9-atom clusters respectively which again agree with the fine structure in the calculated spectrum in the jellium model.

Mg_{11} is obtained by capping the other triangular face of the trigonal prism. While the eleventh atom caps the Mg_{10} cluster symmetrically, the simulated annealing results indicate it to be weakly bonded to the cluster and similar to the Na_{10} cluster, its mean square displacement is much larger as compared to other atoms in the cluster. This can cause evaporation of an atom and absence of the abundance of Mg_{11} clusters in the mass spectrum as it is generally noted. Mg_{12} is also a capped trigonal prism. However, instead of face capping, in this case an atom caps an edge of the trigonal prism.

13 atom clusters are interesting as these have been considered to have either

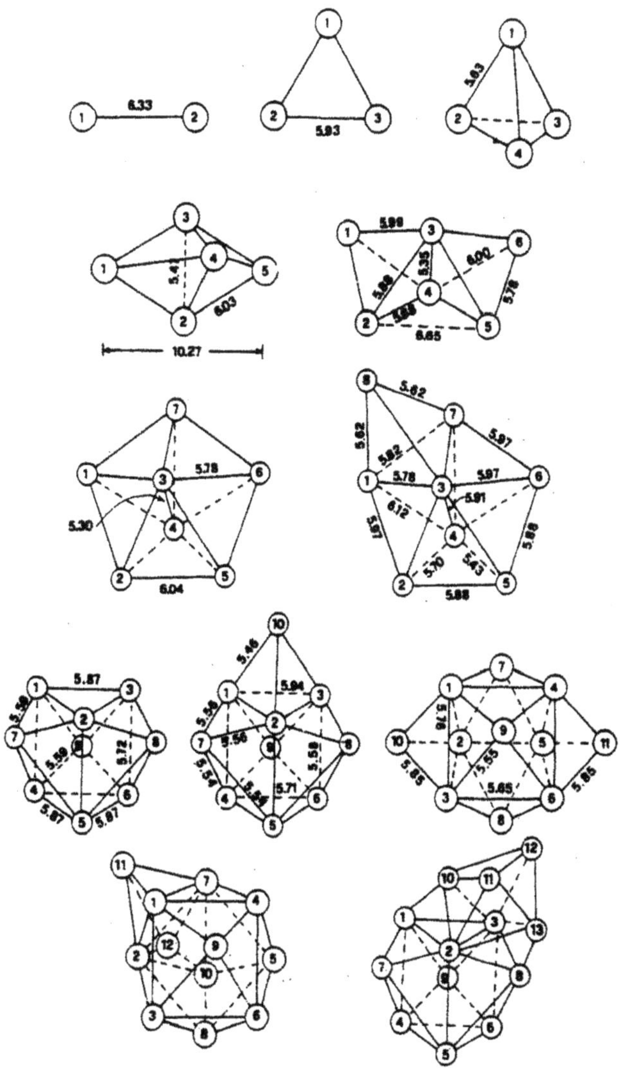

Figure 7: Lowest energy structures of Mg clusters with 2 to 13 atoms. (After Kumar and Car[32]).

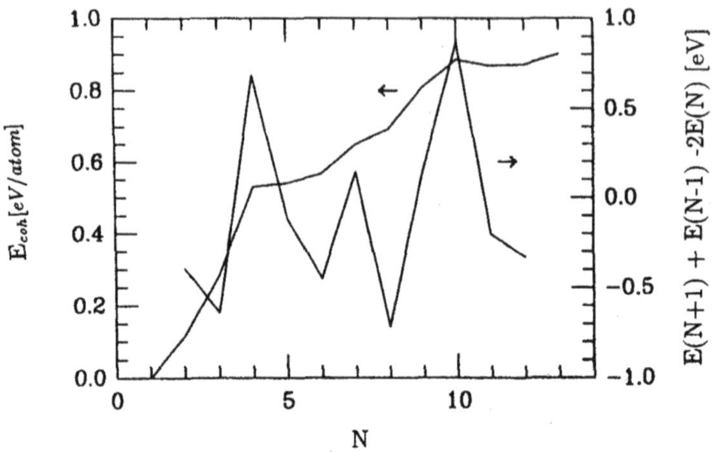

Figure 8: Plot of the cohesive energy and its second order difference spectrum as a function of the number of atoms in Mg clusters. (After Kumar and Car[32]).

an icosahedral structure which is the closest packed for 13 atoms or a cubooctahedral structure which is the closest packing in the bulk. However, Mg_{13} is neither an icosahedron nor a cuboctahedron. It can be considered to arise from the fusion of a Mg_9 and a Mg_4 cluster (atoms 10 to 13 in Fig. 7). The tetrahedron of Mg_4 opens up to form a bent distorted rhombus. Calculations[32] on Mg_4 cluster have shown that this is possible as there is no barrier for this motion. It is interesting to note that a relaxed fcc and an icosahedral structure are respectively 0.032 and 0.043 eV/atom higher in energy than the simulated annealing result. The structures themselves relax very significantly. However, relaxing a hcp cluster leads to a structure which is nearly degenerate with the one obtained from the simulated annealing. From table I it can be seen that for a 13-atom cluster the cohesive energy (see Fig. 8) is about 54% of the bulk value (1.687 eV/atom) calculated by Moruzzi *et al*[53] and much larger clusters may be needed for getting the bulk like behaviour.

An independent study of Mg clusters by de Coulon *et al*[87] also finds a tetra-capped trigonal prism for 10-atom cluster using the Car-Parrinello method. However, their results for 6-, 8-, 12- and 13- atom clusters are different from those obtained by Kumar and Car[32] and their structures lie slightly higher in energy. It may be noted that for these calculations simulated annealing was not used.

Since a non-metal - metal transition is expected for clusters of divalent elements, Kumar and Car studied the nature of bonding in these clusters by calculating the *p* character of charge density around various ions. Fig. 9 shows the *p* character obtained for different ions in a cluster. When these results are analysed together

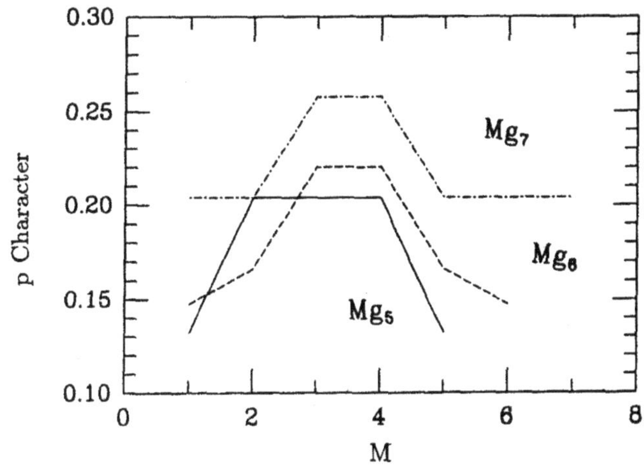

Figure 9: *p character of the electronic pseudo-charge density around different ions in a few* Mg clusters. Here M refers to the number of an ion as shown in Fig. 7. (After Kumar and Car[32]).

with the structure of the clusters, it is clear that the transition to metallic behaviour starts preferentially in some bonds such as bond 3-4 in Mg_7 and not in the whole cluster. Different atoms in a cluster have different p character and there are some short bonds which develop metallicity first. This is likely to be important for the study of reactivity of clusters. When the p character is averaged over the cluster, it shows an oscillatory behaviour and the convergence to the bulk value is slow. The charge density in these clusters indicate that bonding in these clusters changes slowly from weak chemical to covalent to metallic behaviour. Therefore the general agreement with the jellium model in such cases should be considered with caution.

The changes in the mean nearest neighbour bond length and the average coordination of an atom in the cluster are shown in Fig. 10. While the average coordination increases monotonically, the bond length shows an oscillatory behaviour which is suggested to be a feature of clusters showing a transition from van der Waals or weak chemical to metallic bonding. A simple reason for this is the fact that for van der Waals bonding the bond length is large and it should decrease till metallization occurs in the cluster. Beyond this there should be a tendency for the bond length to increase towards its bulk value. The sharp decrease in the bond length for 4- and 10-atom clusters is likely to be due to shell closing. This behaviour is different from clusters of metals like Cu for which the bond length has been reported to increase towards its bulk value monotonically[92].

Fig. 11 shows the chemisorption energy of Mg on Mg clusters and its correlation with the variation in the HOMO-LUMO gap. While both of these show an

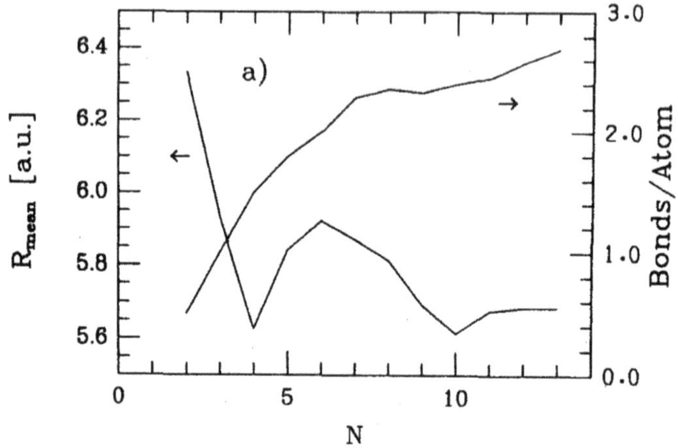

Figure 10: Mean nearest neighbour bond length (left scale) and the number of nearest neighbour bonds (right scale) as a function of the cluster size. (After Kumar and Car[32]).

oscillatory behaviour as a function of size, the magic clusters have smaller chemisorption energy and larger HOMO-LUMO gap. This is consistent with the experimental results of chemisorption of oxygen on Al clusters[20].

Kawai and Weare[31] have done a similar study for Be clusters having 2 to 20 atoms. Small clusters of Be have the same structure as for Mg. However, their calulations suggest that Be clusters tend to attain some of the features (p character) of the bulk bonding even for a 6-atom cluster whereas this is quite slow for Mg. Also Be_8 is a bicapped trigonal prism as compared to the capped pentagonal bipyramid for Mg. While Be_9 has the same structure as for Mg_9, larger clusters of Be are different and have some similarity with the bulk hcp structure. In particular, though the trigonal prism structure is retained, the capping patterns are different for the two systems. Calculations[93] for Be_{10}, a magic cluster, with the Mg_{10} structure show it to be 0.028 eV/atom higher in energy than the structure obtained by Kawai and Weare. Therefore in this size range, Be and Mg clusters have different behaviour. However, similar to Mg_{13}, for Be_{13} also, a relaxed icosahedron structure lies 0.7 eV higher in energy than the one obtained from the simulated annealing procedure. Unlike Mg clusters, there is a remarkable tendency for a two dimensional growth pattern related to hcp packing which suggests directional bonding to be very important for Be clusters and crystals. However, the orbital energies and the angular character of the wave functions for Be clusters correlate well with the predictions of the shell model described above. Also the second-order finite difference of the calculated total energy shows large maxima at N = 4, 10, and 17 which again agrees with the

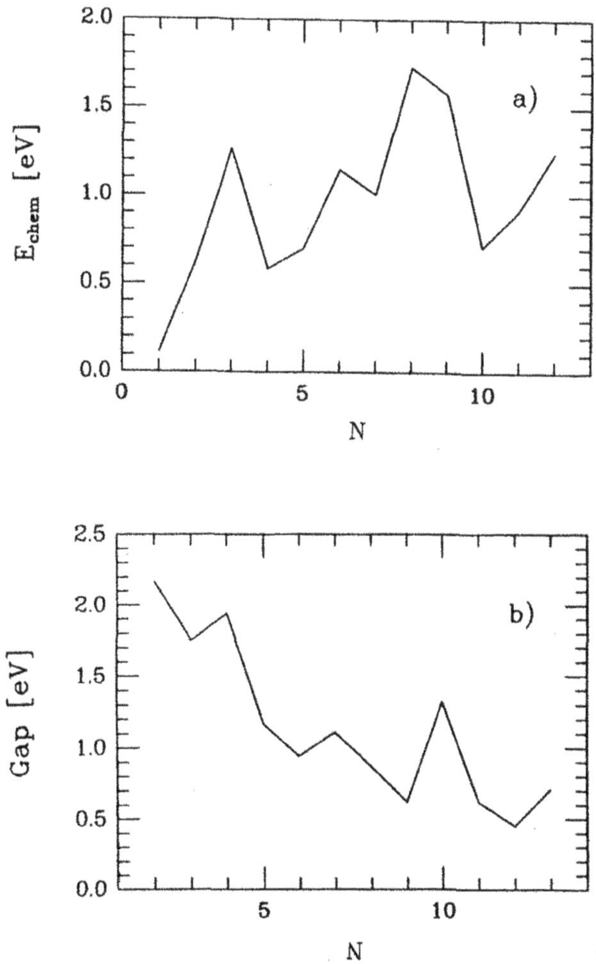

Figure 11: (a) Chemisorption energy of Mg on Mg clusters and (b) the variation of the HOMO-LUMO gap in Mg clusters. Note the correlation of small chemisorption energy with large gap. (After Kumar and Car[32]).

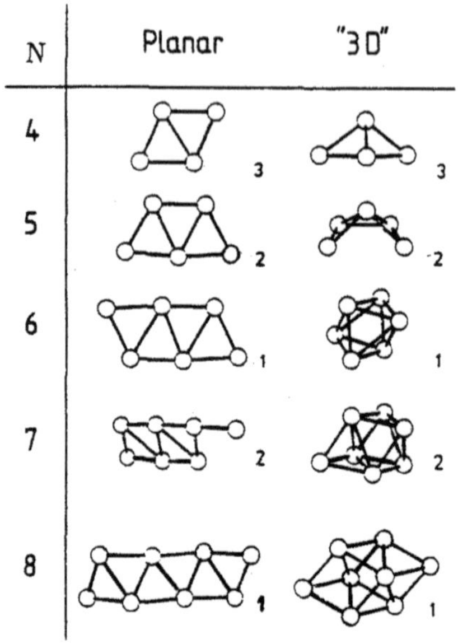

Figure 12: The structures and spin multiplicities of the most stable planar and 3-dimensional isomers of Al_N, N=4-8. *Bonds are shown if the interatomic distance is less than 5.48 a.u..* The planar structure in Al_8 corresponds to a saddle point in the energy surface and buckles readily. (After Jones[36]).

shell model.

Ballone and Galli[76] studied bonding in neutral and ionized dimers of group IB and IIB metals by solving the CP equations in cylindrical coordinates and using a large cutoff of upto about 300 Ry. This large cutoff was needed because the d electrons were treated as valence electrons. As in the case of Mg_2, the binding energy and the vibrational frequency was overestimated whereas the bond length was underestimated when compared with the experimental results. It was shown that the bonding in IIB clusters could be described reasonably well by taking only the s^2 valence electrons but by incorporating the non-linear core corrections[94] and a repulsive short-range interaction with exponential shape characteristic of the closed-shell overlap due to the d-electrons. Application of the short range repulsive interaction was shown to produce the same results as obtained with very large energy cutoff for murcury dimer.

3.1.4 Trivalent metals

In this group clusters of Al and Ga have been studied in detail. Jones[36,37] has studied Al clusters having upto 10 atoms. These calculations have been done within the LSD approximation. The lowest energy structures are shown in Fig.

12 . Small Al clusters favour planar geometries. The lowest energy structure for Al_3 is an equilateral triangle while Al_4 is a rhombus with D_{2h} symmetry and is a triplet. For larger clusters three dimensional structures are favoured. The binding energy increases monotonically towards its bulk value as the cluster size increases. For Al_5 a planar C_{2v} structure is almost degenerate with a C_s structure with similar bond lengths (both doublets). For $N \geq 6$, states with minimum spin degeneracies are favored. For Al_6, the lowest lying state is (D_{3d}) and is a singlet. This is nearly degenerate with triplet D_{2d} state. Al_7 is a nearly symmetric capping of Al_6 (D_{3d}). Al_7^+ is one of the most prominent in the mass spectrum. As it has 20 valence electrons, its stability has been correlated with the completion of the 2s shell in the spherical jellium model. For such clusters there is a large gap between the highest occupied and the lowest unoccupied orbitals. Similar behaviour was obtained by Kumar and Car for Mg_4 and Mg_{10} clusters which have 8 and 20 valence electrons respectively and are magic. Also the ionization potential of Al_7 and the dissociation energy of Al_8 are low. This is in agreement with the results for Mg_{11} cluster. In addition to these lowest energy structures, a variety of planar and buckled structures (similar to one in α - Ga) are also found to be locally stable. This is consistent with the metallic nature of Al and the fact that there are usually unoccupied orbitals near the highest occupied level and it is easy to transfer electrons between π orbitals (which dominate in the bonding in planar structures) and σ orbitals. The finding of several planar and buckled structures with arrays of triangles and having nearly the same energies can be helpful in understanding layer arrangements with triangular nets in many Al - transition metal alloys[95].

Al_{13} cluster is a classic example for studying the relative stability of icosahedral and coboctahedral structures. Negatively charged Al_{13} clusters are particularly abundant in mass spectrum[96] shown in Fig. 13. Bernholc and coworkers[36] have studied 13- atom and a few other large clusters of Al. The energy differences between the icosahedral and cuboctahedral structures for 13-, 19-, and 55- atom clusters are found to be small. For 13- atom cluster a nearly regular icosahedron is found to have the lowest energy, whereas for 55- atom cluster the structure has large distortions though its origin from an icosahedron can be discerned. While no efforts were made to make a systematic study as a function of the size, several structures were found to be close in energy for a 55- atom cluster. In the shell model of metal clusters, Al_{13} is nearly magic with 39 valence electrons and therefore a nearly perfect icosahedron has the lowest energy and a single well defined energy minimum. Adding an electron to it makes it a closed shell system which is most likely the reason for its strong abundance. On the other hand a 55- atom cluster is not magic in the shell model. The fact that several nearly degenerate structures exist for this cluster agrees also with the results of Kumar and Car[32] for Mg_{13} cluster (non-magic) in which case the structure obtained from the simulated annealing is nearly degenerate with the one obtained from relaxation of the hcp structure. However, for Al clusters the deviation from the bulk cohesive energy is not large and it was suggested that a transition from icosahedral to bulk structure may occur early.

Mixed clusters of Al have also been studied. These are interesting from

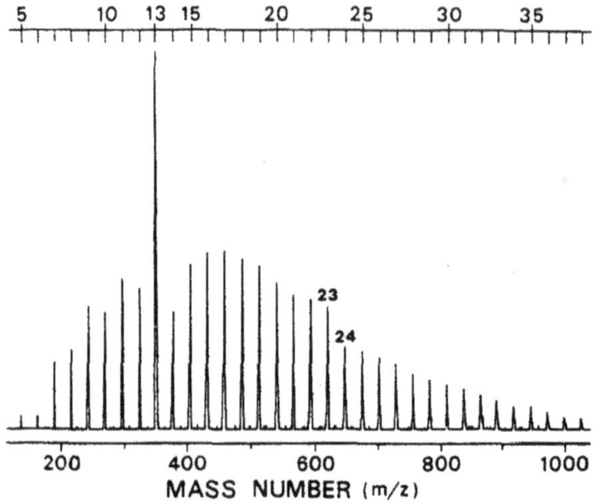

Figure 13: Time of flight mass spectrum of Al_N^- ($N = 5 - 38$) clusters. One peak at $N = 13$ and a step between $N = 23$ and 24 corresponding to magic numbers in the shell model can be observed and there are no other peakes or steps up to N = 70. (After Nakajima et al[96].)

several points of views. First of all there are several alloys of aluminium which have very complicated structures and in which an icosahedral unit (cluster) plays an important role. In addition, several aluminium alloys form quasicrystals in which again icosahedral order is prevalent. This together with the tendency of Al_{13} to form a nearly icosahedron merits study of the stability of such local units which may exist in crystalline, quasicrystalline or amorphous structures. Also the neutral clusters of Al do not satisfy the electronic shell closing condition. However, by suitable doping a cluster one can achieve the shell closing and enhance the stability of magic clusters of aluminium. Of particular interest are the 13 atom icosahedral clusters which may also act as entities to form new materials. Several binary clusters of aluminium have therefore been studied. Mixed clusters of Al-Mg were studied[97] using the steepest descent approach in the CP method. There are several phases of Al-Mg alloys. Among these, notable are the Frank-Kasper phase[98] in which icosahedral Al_xMg_{13-x}, $Al_{12}Mg_3$ and $Al_{12}Mg_5$ clusters are abundant. In the latter two, a Mg atom is surrounded by 12 Al and 2 or 4 Mg atoms respectively. 14 and 16 coordinated clusters were predicted by Frank and Kasper on the basis of geometrical arguments. Study of such clusters is interesting to understand the role of the electronic structure and the atomic sizes in complicated crystalline and quasicrystalline structures. For $Al_{12}Mg_3$ cluster which has 42 valence electrons, there is only a little distortion from the capped hexagonal antiprism structure found in the crystalline phases whereas for $Al_{12}Mg_5$ cluster which has 46 valence electrons, there are considerable John-Teller distortions from the bulk atomic packing. This suggests that the bulk structure of the 17- atom cluster is not particularly stable and there are other considerations such as those suggested by Frank and Kasper that play a more important role in the crystal packing.

Since in an icosahedron the center to vertex distance is about 5% shorter then the vertex to vertex distance, Khanna and Jena[11] suggested substitutional doping of Al_{13} clusters with a smaller atom such as C and Si at the center of an icosahedron in order to obtain a closed packed structure. This would also make this cluster a 40 valence electron system. They obtained significant gain in energy as compared to Al_{13} cluster. However, their calculated center to vertex distance was longer in the doped clusters as compared to the one in Al_{13}. Gong and Kumar[12] subsequently studied several $Al_{12}X$ (X = B, Ga, C, Si, Ge, As, and Ti) clusters and found $Al_{12}C$, $Al_{12}Si$ and $Al_{12}Ge$ to have a substantial gain in energy as shown in Fig. 14. While all these clusters except Ti have about 2 eV HOMO-LUMO gap, $Al_{12}B^-$ is the most strongly bonded in this family. This agrees with the strong abundance of $Al_{12}B^-$ clusters observed by Nakajima *et al*[99]. The main conclusion of this study has been that the chemical bonding plays the dominant role in the stability of these clusters while the atomic relaxation leads to a small gain in the binding energy when a smaller atom such as B, C, Si is sustituted at the center. For Si, a small contraction in the center to vertex distance was obtained[12] as compared to Al_{13} cluster. A similar result was also obtained by Kumar and Sundararajan[45] from CP calculations. Further, Si substitution at the center of an icosahedron was found to be more favorable than at a vertex. This is similar to the result for Na_7Al where Al was found also at the center. A large gain in energy by Si doping may also explain the improved stability of Al-Mn-Si quasicrystals as compared to Al-Mn. On the other hand doping with Ti which also has 4 valence electrons leads to a partially occupied HOMO and a small gain in binding energy as compared to Al_{13}. This is likely to be due to an increase in the Al-Ti bond length. This calculation also suggests that the jellium model may not be applicable in clusters with a transition metal element. An important point which is not resolved by these calculations is whether C at the center is most favorable. It can be noted that though the Al-C bond length is shorter than Al-Al bond, it is still much larger than in the case of Al-C dimer. The large bond length can not be accounted also from an increase in the coordination number of C in an icosahedron. In order to find this, Kumar and Sundararajan[45] have recently done an ab-initio molecular dynamics calculation for $Al_{12}C$. First of all from steepest descent calculations C at the center of an icosahedron is more favorable than C at the vertex by about 1.5 eV. Simulated annealing calculations, however, indicate that carbon tends to have a closed packed environment with 8 atoms while the remaining 4 Al atoms join the Al_8C unit. It is to be noted that $Al_{12}C$ is not a magic cluster in the mass spectrum of Al_NC^- clusters. But Al_7C^-, $Al_{14}C^-$ and $Al_{24}C^-$ are magic which respectively have 26, 47 and 77 valence electrons. None of these correspond to closing of an electronic shell in the SJM. Carbon prefers strongly directional bonding and therefore the SJM may not be appropriate for these clusters. A proper understanding of the carbon

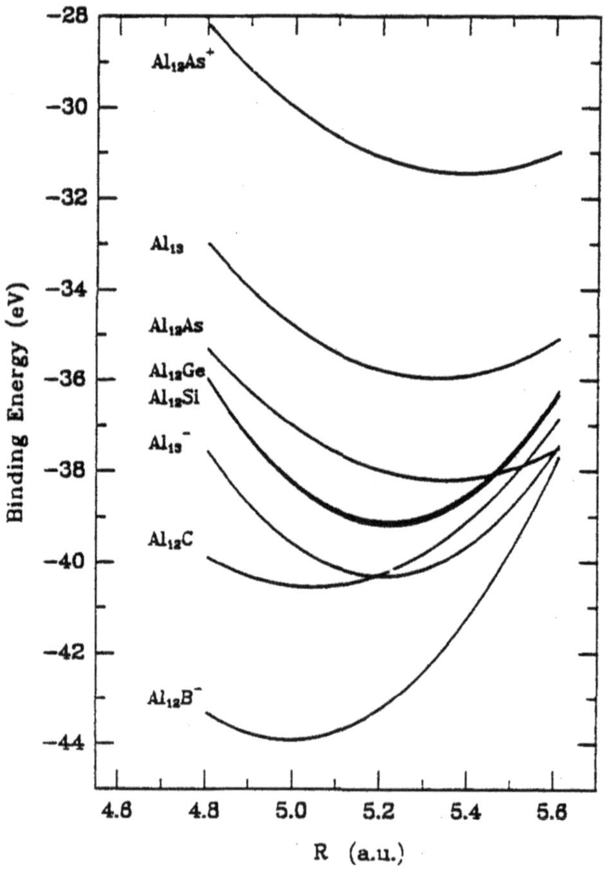

Figure 14: Binding energies of icosahedral Al_12X clusters as a function of the center to vertex distance. (After Gong and Kumar[12]).

behaviour would be useful in the study of reactions on clusters. Simulated annealing calculations for these clusters are difficult as very large cut-offs are required to represent the wavefunction but we hope that in the future such calculations can be done in order to understand the structure of these clusters better.

Ga clusters have been studied by Jones[37] and Gong and Tosatti[38]. The overall structural features are similar to those obtained for Al clusters. However, the bonding properties are slightly different due to larger $s-p$ promotion energy in Ga as compared to Al. This leads to weaker sp hybridization and consequently the bond angles in Ga clusters are closer to 90deg arising due to predominantly p bonding. Though Ga is heavier than Al, its valence orbitals are more compact than in Al. This leads to a small contraction in bond length as compared to that in Al clusters. Also though the melting temperature of Ga (29.78deg C) is much smaller than for Al, the binding energies of their clusters are very similar. This reflects the molecular nature of bonding in bulk. For both Ga and Al, there are several 2 and 3 dimensional structures with different spin multiplicities which lie close in energy. This is consistent with their metallic behaviour in bulk.

3.1.5 Sn and Sb clusters

Elements of Group IV and V are very interesting as they exihibit a variety of crystal structures and have differing bonding character in the bulk. Tin has both a metallic (white tin) and a semiconducting (grey tin) phase. Grey tin has the diamond structure, while white tin is body-centered tetragonal with a two atom basis. Kumar[35] has done a calculation for Sn_5 cluster as it has 20 valence electrons and could be expected to be a magic cluster from the point of view of the jellium model. The calculations were done for a few selected geometries and the lowest energy structure is a trigonal bipyramid. The bond length between the base atoms is large and equals 6.50 a.u. whereas the bond length between the apex and base atoms is 4.94 a.u. . This structure is similar to a 5- atom cluster of Ge or Si[24,100] (see below) and therefore its behaviour is more like the clusters of Si. It would be of interest to calculate the structure of Sn_{10} and see if it is the same as for Si_{10} cluster. Also studies on other Sn clusters would be interesting to explore the development of bonding in these clusters.

Clusters of pentavalent elements such as Sb and Bi (all of which are semimetals) have been experimentally studied in detail both using thermal evaporation and subsequent condensation[101] as well as by laser ablation[102]. For Sb clusters produced by thermal evaporation, a very important feature in the mass spectrum is the abundance of Sb_{4n} clusters (Fig. 15). Also as compared to monomer dissociation in alkali or divalent metal clusters, studies of the fragmentation of such antimony clusters produce evidence of evaporation of tetramers[103]. The abundance spectrum of the thermally evaporated Bi clusters is very different as compared to Sb. There is no particular preference to Bi_{4n} clusters (Fig. 15). However, clusters of Sb and Bi produced from laser ablation have similar features in the abundance spectrum (Fig. 16). These studies show abundance of clusters with 3, 5 or 7 atoms. Kumar[33]

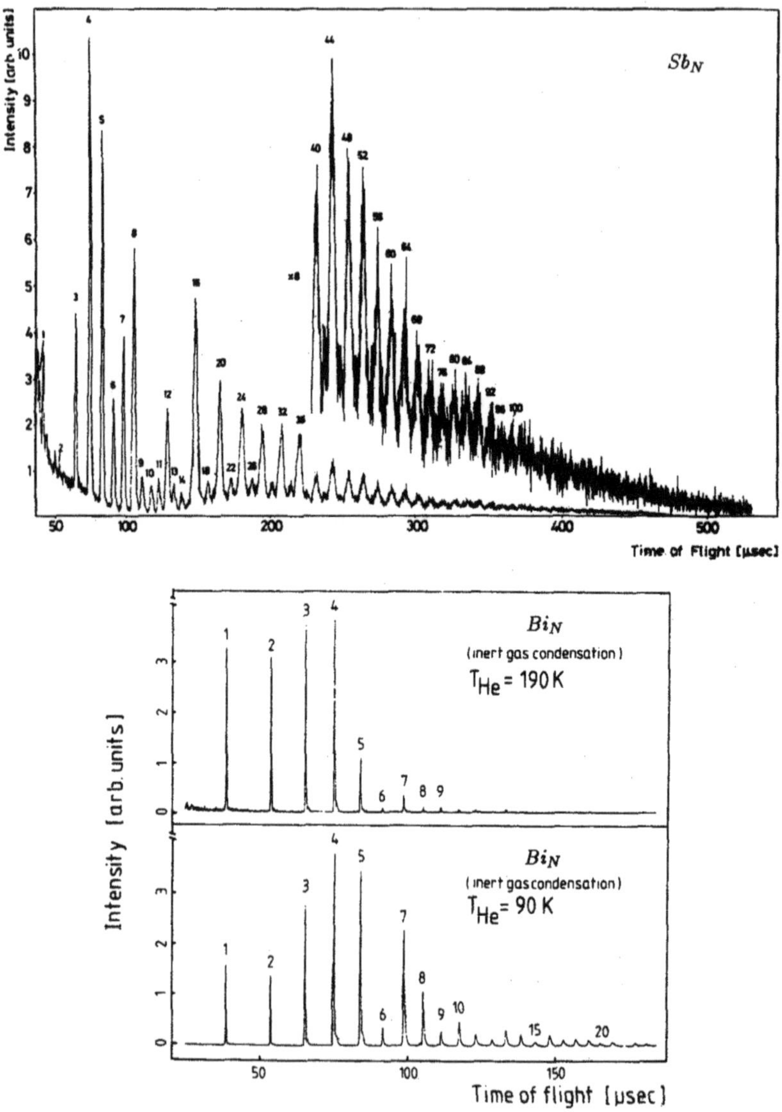

Figure 15: Time of flight mass spectrum of Sb and Bi clusters obtained from thermal evaporation. (After Sattler *et al*[101].)

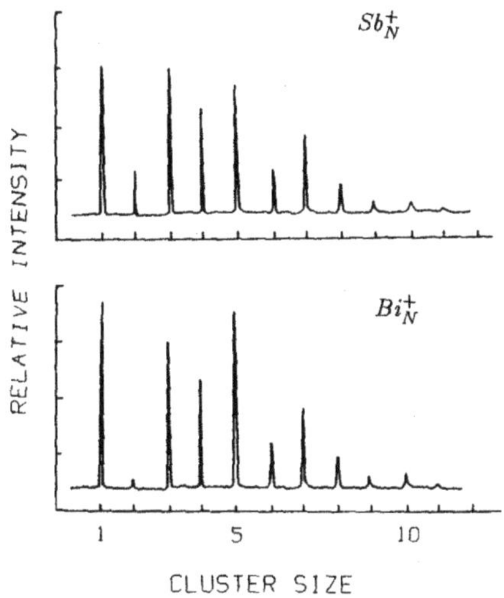

Figure 16: Mass spectrum of (a) Sb and (b) Bi clusters obtained from laser ablation. (After Geusic *et al*[102].

has studied Sb_4 and Sb_8 clusters whereas Sundararajan and Kumar[34] have recently studied some other clusters of Sb. The lowest energy structure of Sb_4 is a regular tetrahedron. A bent rhombus lies about 0.5 eV/atom higher in energy. Sb_4 has 20 valence electrons and can be expected to be a magic cluster as also observed. From this it will appear that the results of the jellium model may be applicable to Sb clusters as Sb_8 has 40 valence electrons which also corresponds to shell closing in SJM and Sb_{12} has 60 valence electrons which can correspond to the magic clusters with 58 electrons. However, recent studies[104] of the photoionization spectra of Sb clusters indicate striking differences in the behaviour of Sb_{4p} , $p > 1$ and other clusters which are not understood. Also the persistence of $4p$ type clusters for larger values of p (upto about 25) indicates a different bonding character in these clusters. Detailed calculations by Kumar[33] on Sb_8 clusters show that a cube, two fused bent rhombuses, and a capped octahedron are not even stable against two isolated tetrahedra (Fig. 17a-c). Simulated annealing calculations for this cluster result in a structure having two tetrahedra weakly interacting with each other (Fig. 17e) in confirmity with the well known result that thermal heating of Sb leads to evaporation of Sb_4 units. Another SA calculation with a different starting configuration and heat treatment leads to a different structure (Fig. 17d) which can be described as a bent rhombus interacting with a distorted tetrahedron. This structure lies only 0.117 eV higher in energy than the structure with two weakly (van der Waals type) interacting tetrahedra. It is well known that LDA is not good for describing van der Waals interactions and therefore binding energy of the two tetrahedra is

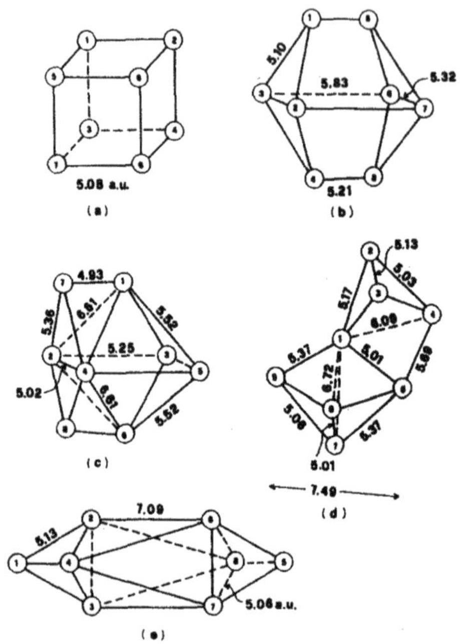

Figure 17: Different isomers of Sb_8. (a)-(c) are not stable against two isolated tetrahedra. (e) has the lowest energy and is nearly degenerate with (d). (After Kumar[33]).

likely to be overestimated. Therefore these two isomers can be treated to be nearly degenerate. These calculations indicated the importance of bent rhombus structure for larger clusters and suggested that the molecular architecture could depend upon nucleation conditions.

Results on 2- 5 atom Sb clusters are shown in Fig. 18. For Sb_5 a square pyramid has the lowest energy while for Sb_6 a prism has the lowest energy. The Kohn-Sham eigenvalues are shown in Fig. 19. It can be noted that for 3, 5 and 7 atom clusters there is a large gap between the highest and the next occupied levels. This explains the abundance of Sb_3^+, Sb_5^+ and Sb_7^+ clusters in the laser ablation experiments in which the growth is expected to be atom by atom. Simulations on 7-atom cluster show that as the cluster is cooled, it undergoes fast isostructural transformations by making and breaking bonds (Fig. 20). Such dynamical aspects may play an important role in understanding reactions on clusters. Also the finite temperature studies help to get information about other low lying structures which

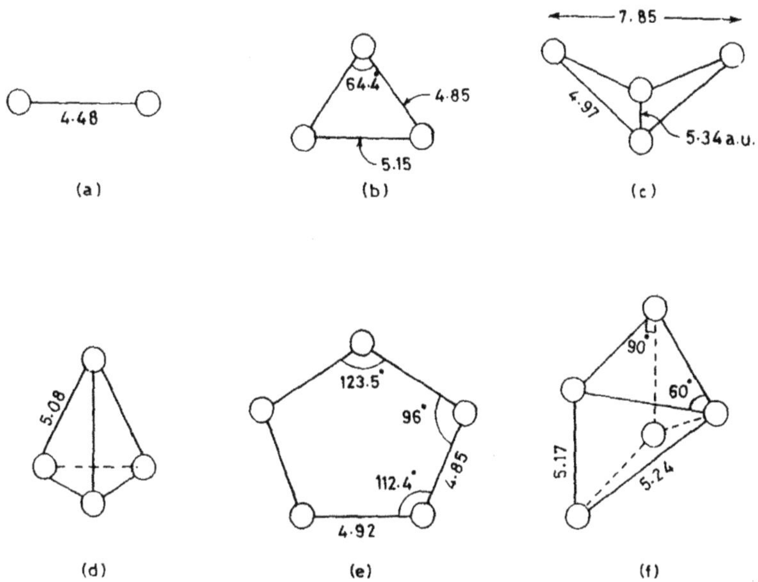

Figure 18: Low energy structures of Sb_2 - Sb_5 clusters. Tetrahedron and square pyramid are the lowest energy structures for Sb_4 and Sb_5. (After Sundararajan and Kumar[34]).

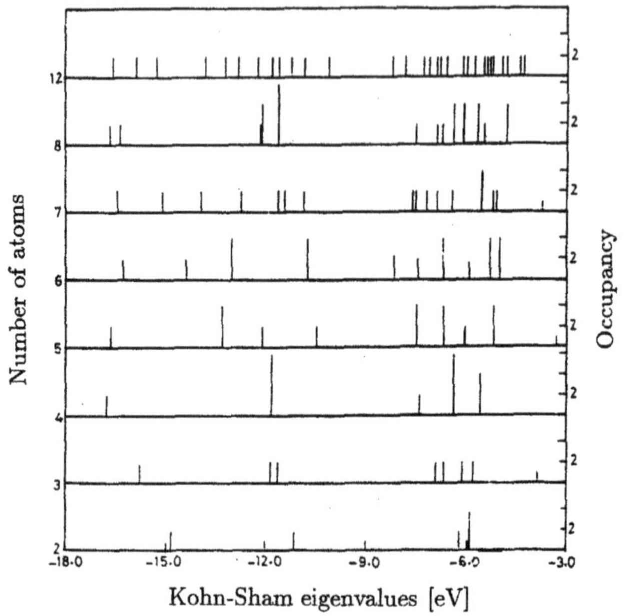

Figure 19: Kohn-Sham eigenvalues of Sb clusters. Note the large gap between the highest and the next occupied level for Sb_3, Sb_5 and Sb_7. (After Sundararajan and Kumar[34]).

the cluster may visit during the simulation. Unlike in bulk, it is not just thermal vibrations, but also much below the bulk melting temperature there could be important (dynamical) structural transformations in clusters which could be decisive in determining their properties. From Fig. 20 one can see that while above 600 K a structure with two fused bent rhombuses has a large basin of attraction, the low temperature structure is a bent rhombus interacting with a triangle.

From the above discussion it is clear that the jellium model is not applicable to Sb clusters and the bonding is strongly directional. Also unlike in other metal clusters where fragmentation is predominantly of monomers, the fragmentation channels for different clusters are different for Sb clusters[34].

3.2 Clusters of semiconductors and other materials

3.2.1 Si, Ge, GaAs, GaP and AlAs

Andreoni and coworkers[24,43,100] have studied these clusters from the CP method. As expected Si and Ge clusters have similar structures. An important aspect of the structure of these clusters is that these are closer packed as compared to more open

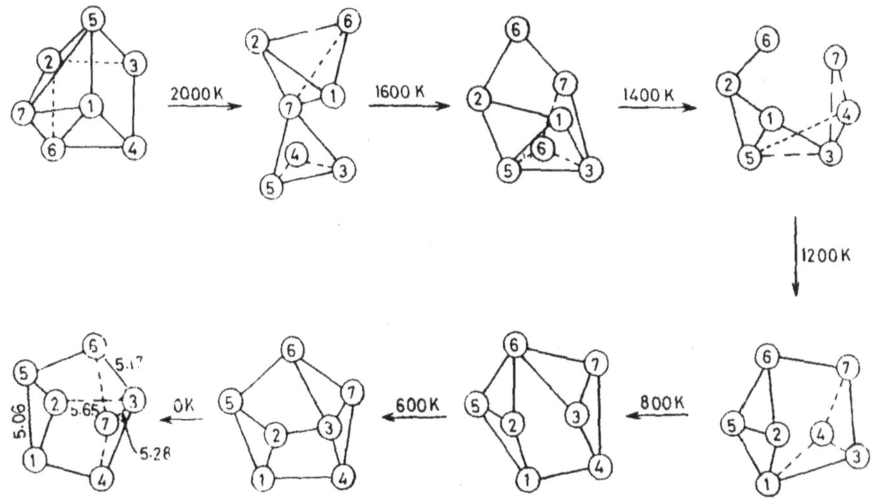

Figure 20: Snap-shots of some structures at different temperatures for Sb_7. Note the isostructural transformations during simulated annealing. (After Sundararajan and Kumar[34]).

diamond structure in the bulk. It is of considerable interest to know when directional bonding starts playing an important role in these clusters. Several efforts have been devoted to answer this question. Extensive calculations[105] for Si clusters with upto 45 atoms show no sign of bulklike features in the structure. Si was the first example where the advantage of simulated annealing became evident when an entirely new structure (tetracapped trigonal prism, same as for Mg_{10}) was found to be of lowest energy. The growth pattern of Si clusters is rather complex and no common seed can be identified. Si_5 and Ge_5 have the same structure as the one for Sn_5 which suggest directional bonding but the larger clusters seems to behave more like metal clusters. Si_7 is a pentagonal bipyramid as Mg_7. Si_8 is a bicapped octahedron (two opposite faces capped) and Si_9 is a strongly reconstructed structure and is shown in Fig. 21. Si_{10} and Ge_{10} are tetracapped trigonal prisms similar to Mg_{10} whereas Si_{13} is neither a cubooctahedron nor an icosahedron. Its lowest energy structure is also shown in Fig. 21. An important message of these calculations is that the semiempirical potentials available for Si may not be suitable for the representation of the potential energy surface of microclusters[106].

A few calculations have also been done for Ga_NAs_N, Ga_NP_N and Al_NAs_N clusters[43] containing upto 10 atoms. Some of their isomers have similarity with those of Si and Ge clusters but there are significant differences due to two species and anion-anion bonds. All the atoms in these clusters are undercoordinated as

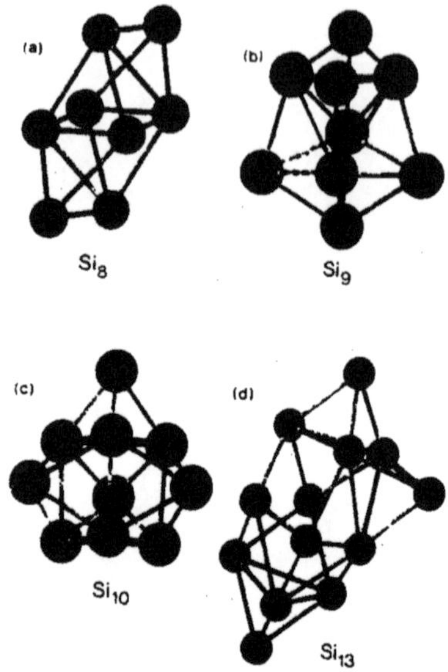

Figure 21: Lowest energy structures of 8-, 9-, 10-, and 13- atom Si clusters. (After Andreoni[24]).

compared to bulk and therefore the chemical bonding and order differs from bulk in these small aggregates. Bonds between unlike atoms are more prevalent in AlAs as the Al-As bond is more ionic than the Ga-As bond whereas in Ga-P clusters there is a preference for Ga atoms to segregate at the surface. For Ga_5As_5 cluster, though the lowest energy structure is a distorted bicapped dodecahedron, a tetracapped trigonal prism is nearly degenerate. This is an interesting result because this as well as Si_{10} and Ge_{10} clusters have 40 valence electrons each. This is a magic number in the shell model. The occurrence of the same structure for Mg_{10} cluster makes us believe that *there may be some unique structures which are either lowest or lie very close in energy with the lowest energy structures for magic clusters for which directional bonding plays a less important role.* Thus studies of low lying isomers may provide useful information about the chemical bonding in small clusters.

3.2.2 B, S, Se, P, C

Boron, a trivalent element, has a tendency for very strong directional bonding. The abundance specturm of boron clusters produced from a laser ablation source[107] shows prominent stability of positively charged 5, 10 and 13 atom clusters. This behaviour is different from clusters of aluminium and is in conformity with the observation of several crystalline phases of boron and a large variety of alloys in which atomic distribution is rather complicated[108]. One of the most prominent units in these structures is the B_{12} empty center icosahedron which has five fold rotational symmetry. However, small distortions can lead to lower symmetries compatible with crystalline order. Strong directional bonding in boron forces it to develop a complicated three dimensional crystalline order, often with large unit cells. One can then ask the question whether such icosahedral units would also be favorable for clusters. Laser ablation experiments[107] showed 12 atom cluster not to be magic. Other experiments[109] suggest that B_{13}^+ does not react with H_2 or H_2O whereas other clusters are reactive. High stability of B_{13}^+ was considered to be due to an icosahedral cage with an atom at the center. However, here it may be noted that for aluminium, negatively charged 13 atom cluster was most abundant. Kawai and Weare[42], have done ab-initio molecular dynamics study of B_{12} and B_{13} clusters. Their studies showed that an icosahedron is not even a local minimum for B_{13}. The atom at the center comes out of the cage and the lowest energy structure consists of a pentagon and a hexagon layer. The singly occupied HOMO and the next occupied level are well separated and this leads to the stability of B_{13}^+. For B_{12}, though an icosahedron is a local minimum, it has dangling bonds and one of the atoms is very loosely bonded. Simulated annealing calculations lead to an open structure which is significantly lower in energy due to removal of the dangling bonds. The existance of icosahedral clusters in bulk phases has therefore been suggested to be due to the formation of strong intericosahedral links. These calculations also indicated the inappropriateness of a SJM for these clusters. The same conclusion can be drawn from the work of kumar and Sundararajan[45] who studied carbon substitution at the center of the icosahedral B_{13} cluster. Similar to B_{13}, carbon comes out of the center and the icosahedral structure is not energetically favorable.

Jones and coworkers have made extensive calculations for S[27], Se[26] and P[30] clusters. In some of these cases it is possible to compare the results with the spectroscopic data available for the structure. These results give confidence in the simulated annealing procedure as the calculated bond lengths and angles agree remarkably with the experimental values. The atomization energies are ,however, overestimated which is due to the use of the LDA. As there is a large amount of information regarding the structure of these molecules, the reader is advised to consult the original papers for details. It would be worth to mention that similar to Sb_4, P_4 is also a regular tetrahedron. However, contrary to widespread belief, the most stable structure of P_8 is not cubic. It is a structure with C_{2v} symmetry and its energy lies 0.47 eV below that of two isolated P_4 tetrahedra. A roof-shaped tetramer (bent rhombus) is a prominent structural unit in the low-lying states of P_5, P_6, P_7 and P_8 clusters as it was also found for Sb clusters. However, the growth pattern in the two cases is different.

Some calculations have also been done[39] for C_4 and C_{10} clusters for which the structures are respectively a linear chain and a ring. Due to more localized orbitals these calculations require use of very large energy cutoff in the plane wave expansion and therefore are very expensive to do. However, due to recent excitement in fullerenes, several efforts are going on to study large clusters of carbon with 60 or

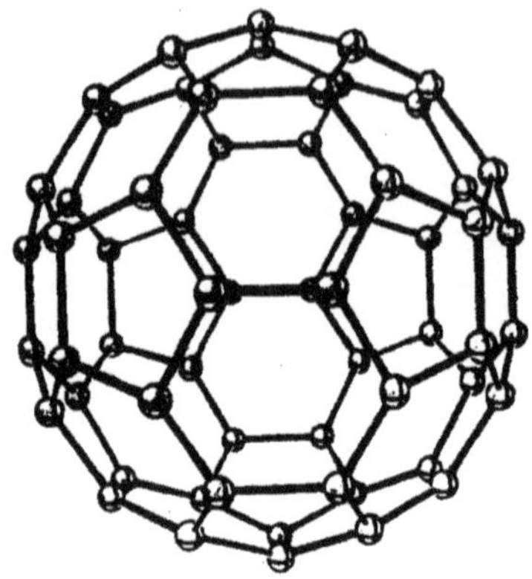

Figure 22: Truncated icosahedral structure of C_{60} cluster.

more atoms. C_{60} cluster (Fig. 22) has a football structure with 12 pentagons and 20 hexagons. There are two bond lengths (1.40 and 1.45 Å) as obtained from NMR data[110]. The bonds sharing two hexagons form double bonds and are short whereas the bonds sharing a pentagon and a hexagon are single bonds and are longer. All atoms have identical environment. Calculations by Feuston *et al*[40] produce these bond lengths to be 1.39 and 1.45 Å in very good agreement with the NMR experiments. As it is possible to do calculations at finite temperatures, they studied the structural changes and vibrational and average electronic density of states at 450 K. The fullerene structure of C_{60} was found to be very stable and the average structural parameters change by at most 0.01 Å. The calculated vibrational frequencies 530, 555, 1105 and 1345 cm^{-1} are in close agreement with 527.1, 570.3, 1169.1, 1406.9 cm^{-1} obtained from infra-red spectroscopy experiments[111]. Due to the observation of superconductivity in doped solid C_{60}, studies of the doped fullerenes are of interest. Kohanoff *et al*[41] and Laasonen *et al*[41] have studied hydrogen and La doping of fullerenes and their effects on the atomic and electronic structure of the fullerenes. Due to the difficulty in doing CP calculations with plane wave basis, efforts have been made[112,113] to use schemes based on a tight binding Hamiltonian. This has allowed the study of the relative stability of different isomers of carbon clusters upto about 100 atoms. Fig. 23 shows the heat of formation of carbon clusters for 20-94 atoms obtained by Wang *et al*[113] from tight-binding molecular dynamics calculations. It can be seen that the heat of formation slowly approaches to the value of graphite almost monotonically. However, C_{60} shows marked increase due to its high symmetry and the distribution of strain due to curvature uniformly.

Efforts are also being made to understand the properties of solids made from such large clusters. Zhang *etal*[114] have studied the structural properties of solid C_{60}

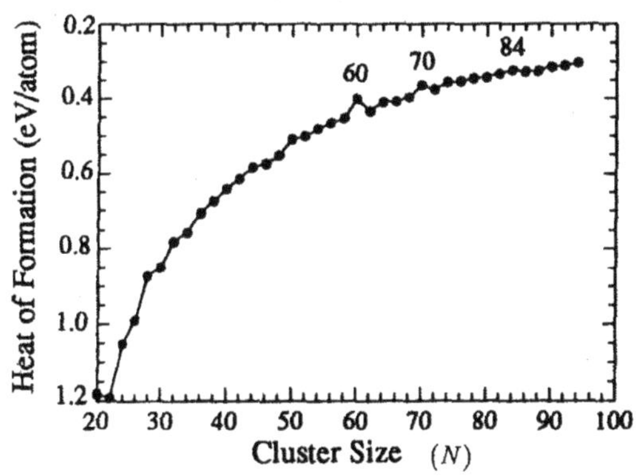

Figure 23: The heat of formation of carbon fullerenes relative to that of graphite as a function of the cluster size. (After Wang *et al*[113].)

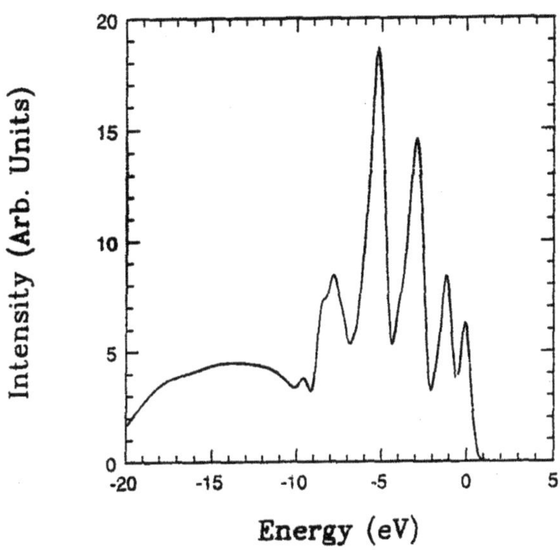

Figure 24: Gaussian broadened electron energy spectrum of solid C_{60}. (After Kumar[115].)

with a fcc structure and found the bond lengths to be 1.40 and 1.45 Å i.e. very little change from the values in the cluster. Kumar[115] has also done a similar calculation with 29 Ry. cutoff and find some anisotropy in the bond lengths due to different environments of the sites in the solid phase as compared to one in the isolated cluster. In particular there are three different environments of carbon atoms in the fcc structure. The double bonds have one bond length which is equal to 1.404 Å, whereas the single bonds have three slightly different values. These are 1.449, 1.454 and 1.462 Å. 24 atoms have their nearest neighbour bond lengths to be 1.404, 1.449 and 1.454 Å. The other 24 have 1.404, 1.449 and 1.462 Å and the remaining 12 have 1.404, 1.454 and 1.454 Å. The calculated electronic energy spectrum is shown in Fig. 24 which is good agreement with the photoemission data[118]. Experimentally[116,117] an orientational ordering of the C_{60} balls has been found below 249 K. This leads to a $Pa\bar{3}$ structure. Since the number of atoms in this structure become four times than in the fcc structure per unit cell, CP calculations becomes very difficult. However, further work on these and doped fullerenes is in progress in some laboratories.

4. Outlook

Fairly good amount of work has been carried out in recent years using the Car-Parrinello method for clusters of metals, semiconductors and other elements like B, C, S, Se, P etc. The simulated annealing technique has given a new thrust in the search for the ground state structures of clusters. The structures and the vibrational frequencies calculated with this technique are in general good agreement with experiments wherever results are available. Interesting information has been obtained on the dynamical changes in the structure at finite temperatures and the bonding properties for some systems but more work on larger clusters and other systems would be useful with the availability of better computational facilities and algorithms which scale linearly with system size.[119]. Also it would be most interesting to apply this technique to problems related to reactions on clusters and to other mixed clusters. From the point of view of applications, clusters of transition metals are very important and recent developments in pseudopotentials[120] and their implementation in the Car-Parrinello method[121] is a step forward in this direction and we hope to see applications of this to some transition metal clusters in the near future. Studies of the magnetic properties are few but hopefully these developments would lead to better understanding of magnetic properties of transition metal clusters. It would be desirable to develop more efficient algorithms for systems having d electrons or more localized orbitals such as in C or O. In the next few years we hope to see important developments in these directions.

5. Acknowledgements

I would like to express my sincere thanks to my collaborators R. Car, V. Sundararajan, X. G. Gong, I. Garzon, G. Pastore and W. Andreoni whose cooperation has led to much of the work presented here and to G. Chiarotti , G. Galli and

K. Laasonen with whom I have enjoyed several fruitful discussions.

6. References

1. *Microclusters*, Eds. S.Sugano, Y. Nishina and S. Ohnisi, Springer, Berlin (1987).
2. *Physics and Chemistry of Finite Systems: From Clusters to Crystals* , Vol. 1 and 2, Eds. P. Jena, S. N. Khanna and B.K. Rao, Kluwer Academic Publishers, Netherlands (1992).
3. *Clusters and Fullerenes*, Eds. Vijay Kumar, T.P. Martin and E. Tosatti, World Scientific (1993).
4. Articles in *Proceedings of the Fourth International Meeting on Small Particles and Inorganic Clusters*, Marseille, France (1988), published in Z. Phys. **D12**, (1989).
5. Articles in *Proceedings of the Fifth International Meeting on Small Particles and Inorganic Clusters*, Konstanz, Germany (1988), published in Z. Phys. **D19** and **20** , (1991).
6. W. Krätschmer, D.L. Lowell, K. Fostiropoulos, and D.R. Huffman, Nature **347**, 354 (1990); for C_{60} cluster see H. W. Kroto, J. R. Heath, S. C. O'Brien, R. F. Curl and R. E. Smalley, Nature, **318**, 162 (1985).
7. A.F. Hebard *et al*, Nature **350**, 600 (1991).
8. M.J. Rosseinsky *et al*, Phys. Rev. Lett. **66**, 2830 (1991); K. Holczer *et al*, Science **252**,1154 (1991).
9. K. Tanigaki *et al*, Science **352**, 222 (1991).
10. B. C. Guo, K. P. Kerns and A. W. Castleman, Jr., Science **255**, 1411 (1992); B. C. Guo, S. Wei, J. Purnell, S. Buzza and A. W. Castleman, Jr., Science **256**,515 (1992).
11. S. N. Khanna and P. Jena, Phys. Rev. Lett. **69**, 1664 (1993).
12. X. G. Gong and Vijay Kumar, Phys. Rev. Lett. **70**, 2078 (1993).
13. F. D. Weiss, J. L. Elkind, S. C. O'Brien, R. F. Curl and R. E. Smalley, J. Am. Chem. Soc., **110**, 4464 (1988); Y. Chai, T. Guo, C. Jin, R. E. Haufler, L. P. F. Chibante, J. Fure, L. Wang, J. M. Alford and R. E. Smalley, J. Phys. Chem. **95**, 7564 (1991).
14. S. Iijima, Nature (London) **354**, 56 (1991).
15. D. Ugarte in Ref. 3 pp. 231.
16. P. Ballone, W. Andreoni, R. Car and M. Parrinello, Phys. Rev. Lett. **60**, 271(1988).

17. K. Rademann, B. Kaiser, U. Even and F. Hensel, Phys. Rev. Lett. **59**, 2319 (1987); K. Rademann, Ber. Bunsenges. Phys. Chem. **93**, 653 (1989); C. Brechignac, M. Broyer, P. Cahuzac, G. Delacretaz, P. Labastie, J.P. Wolf and L. Woste, Phys. Rev. Lett. **60**,275 (1988).

18. See J. H. Sinfelt, Rev. Mod. Phys. **51**, 569 (1979).

19. R.L. Whetten *et al*, Phys. Rev. Lett. **54**, 1494 (1985).

20. R.E. Leuchtner, A.C. Harms and A.W. Castleman, Jr., J. Chem. Phys. **91**, 2753 (1989).

21. T. P. Martin, T. Bergmann, H. Göhlich and T. Lange, Chem. Phys. Lett. **176**, 343 (1991); T. P. Martin, U. Näher and H. Schaber, Chem. Phys. Lett. **199**, 470 (1992).

22. W. A. de Heer, W. D. Knight, M. Y. Chou and M. L. Cohen, Solid State Phys. Vol.**40**, Eds. H. Ehrenreich and D. Turnbull, Academic Press (1987).

23. R. Car and M. Parrinello, Phys. Rev. Lett. **55**, 2471 (1985).

24. See for a brief review W. Andreoni, in Ref. 5, **D19**, pp. 31; also a finite temperature study of a few selected clusters of Na, Mg and Si can be found in U. Röthlisberger and W. Andreoni, Z. Phys. **D20**, 243 (1991).

25. Vijay Kumar in *Atomic and Molecular Physics*, Ed. A. P. Pathak,·Narosa (1992), pp. 90.

26. D. Hohl, R. O. Jones, R. Car and M. Parrinello, Chem. Phys. Lett. **139**, 540 (1987).

27. D. Hohl, R. O. Jones, R. Car and M. Parrinello, J. Chem. Phys. **89**, 6823 (1988).

28. P. Ballone, W. Andreoni, R. Car and M. Parrinello, Europhys. Lett. **8**, 73 (1989).

29. U. Röthlisberger and W. Andreoni, J. Chem. Phys. **94**, 8129 (1991).

30. R.O. Jones and D. Hohl, J. Chem. Phys. **92**, 6710 (1990).

31. R. Kawai and J. H. Weare, Phys. Rev. Lett. **65**, 80 (1990).

32. Vijay Kumar and R. Car, in Ref.5, **D19**, pp. 177; Phys. Rev. **B44**, 8243 (1991).

33. Vijay Kumar, Phys. Rev. **B48**, 8470 (1993).

34. V. Sundararajan and Vijay Kumar, unpublished.

35. Vijay Kumar in Ref. 3 pp. 97.

36. R. O. Jones, Phys. Rev. Lett. **67**, 224 (1991); J.-Y. Yi, D. J. Oh, J. Bernholc and R. Car, Chem. Phys. Lett. **174**, 461 (1990); J.-Y. Yi, D. J. Oh and J. Bernholc, Phys. Rev. Lett. **67**, 1594 (1991).

37. R. O. Jones, J. Chem. Phys. **99**, 1194 (1993).

38. X. G. Gong and E. Tosatti, Phys. Lett. **A 43**, 369 (1992).

39. W. Andreoni, D. Scharf and P Giannozzi, Chem. Phys. Lett. **173**, 449 (1990).

40. B. Fueston, W. Andreoni, E. Clementi and M. Parrinello, Phys. Rev. **B44**, 4056 (1991); Q.-M. Zhang et al in Ref. 3, pp. 83; Vijay Kumar (unpublished).

41. J. Kohanoff, W. Andreoni and M. Parrinello, Chem. Phys. Lett. **198**, 472 (1992); K. Laasonen, W. Andreoni and M. Parrinello, Science (in press).

42. R. Kawai and J. H. Weare, J. Chem. Phys. **95**, 1151 (1991).

43. W. Andreoni, Phys. Rev. **B 45**, 4203 (1992).

44. U. Röthlisberger and W. Andreoni, Chem. Phys. Lett. **198**, 478 (1992) and in Ref. 3 pp. 91.

45. Vijay Kumar and V. Sundararajan, unpublished.

46. R. Car and M. Parrinello, in *Simple Molecular Systems at Very High Density*, Eds. A. Polian, P. Loubeyre and N. Boccara, (NATO ASI Series, Plenum, New York (1989)).

47. D.K. Remler and P.A. Madden, Mol. Phys. **70**, 921 (1990).

48. G. Galli and A. Pasquarello, preprint.

49. M. C. Payne, M. P. Teter, D. C. Allan, T. A. Arias and J. D. Joannopoulos, Rev. Mod. Phys. **64**, 1045 (1992).

50. P. Hohenberg and W. Kohn, Phys. Rev. **136**, 864B (1964).

51. W. Kohn and L.J. Sham, Phys. Rev. **140**, 1133A (1965).

52. R.O. Jones and O. Gunnarsson, Rev. Mod. Phys. **61**, 689 (1989).

53. V. L. Moruzzi, J. F. Janak and A. R. wiiliams, *Calculated Electronic Properties of Metals*, Pergamon, New York (1978).

54. H. Hellmann, *Einfuhrung in die Quantumchemie*, Deuticke, Leipzip (1937); R. P. Feynman, Phys. Rev. **56**, 340 (1939).

55. G. W. Fernando, G. X. Qian, M. Weinert and J. W. Davenport, Phys. Rev. **B40**, 7985 (1989); M. R. Pederson and K. A. Jackson, Phys. Rev. **43**, 7312 (1991).

56. S. Nose, J. Chem. Phys. **81**,511 (1984).

57. P. E. Blc̈hl and M. Parrinello, Phys. Rev. **B45**, 9413 (1992).

58. A. Williams and J. Soler, Bull. Am. Phys. Soc. **32**, 562 (1987); J. M. Soler and A. R. Williams, Phys. Rev. **B40**, 1560 (1989); ibid. B42, 9728 (1990).

59. See for a recent review, W. E. Pickett, Comput. Phys. Rep. **9**, 115 (1989).

60. D. J. Chadi and M. L. Cohen, Phys. Rev. **B8**, 5747 (1973); H. J. Monkhorst and J. D. Pack, Phys. Rev. **B13**, 5188 (1976).

61. G. B. Bachelet, D. R. Hamann and M. Schluter, Phys. Rev. **B26**,4199 (1982).

62. N. Trouillier and J. L. Martins, Phys. Rev. **B43**, 1993 (1991); G. Kerker, J. Phys. **C13**, L189 (1980); D. R. Hamann, Phys. Rev. B40, 2980 (1989); E. L. Shirley, D. C. Allan, R. M. Martin and J. D. Joannopoulos, Phys. Rev. **B40**, 3652 (1989).

63. L. Kleinman and D. M. Bylander, Phys. Rev. Lett. **48**, 1425 (1982); see also P. E. Blöchl, Phys. Rev. **B41**, 5414 (1990); M. Y. Chou, Phys. Rev. **B45**, 11465 (1992).

64. X. Gonze, P. Käckell and M. Scheffler, Phys. Rev. B41, 12264 (1990); X. Gonze, R. Stumpf and M. Scheffler, Phys. Rev. B44, 8503 (1991).

65. J. Ihm, A. Zunger and M. L. Cohen, J. Phys. **C12**, 4409 (1979); P. Bagno, L. F. Dona Dalle Rose and F. Toigo, Adv. Phys. **40**, 685 (1991).

66. In fact in the formulation of Bachelet et al (Ref. 61), the ionic pseudopotential is explicitly written as a core term coming from a Gaussian distribution and the remaining part as we have expressed in Eq. (13).

67. S. Kirkpatrick, C. D. Gelatt, and M. P. Vecchi, Science **220**, 671 (1983).

68. W. D. Knight, K. Clemenger, W. A. de Heer, W. A. Saunders, M. Y. Chou and M. L. Cohen, Phys. Rev. Lett **52**, 2141 (1984).

69. W. A. Saunders, K. Clemenger, W. A. de Heer and W. D. Knight, Phys. Rev. **B32**, 1366 (1985); M. M. Kappes, M. Schär, P. Radi and E. Schumacher, J. Chem. Phys. **84**, 1863 (1986).

70. M. M. Kappes, P. Radi, M. Schär and E. Schumacher, Chem. Phys. Lett. **119**, 11(1985); M. M. Kappes, M. Schar and E. Schumacher, J. Phys. Chem. **91**, 658 (1987).

71. I. Katakuse, I. Ichihara, Y.Fujita, T. Matsuo, T. Sakurai and H. Matsuda, Int. J. Mass Spectrom. Ion Proc. **67**, 229 (1985).

72. I. Katakuse, T. Ichihara, Y. Fujita, T. Matsuo and H. Matsuda, Int. J. Mass Spectrom. Ion Proc. **69**, 109 (1986).

73. See Refs. 22 and 69;

74. C. Brechignac, Ph. Cahuzac, F. Carlier, M. de Frutos and J. Leygnier, J. Chem. Soc. Faraday Trans. **86**, 2525 (1990).

75. W. D. Knight, K. Clemenger, W. A. de Heer and W. A. Saunders, Phys. Rev. **B31**, 2539 (1985).

76. P. Ballone and G. Galli, Phys. Rev. **B42**, 1112 (1990); ibid. **B40**, 8563(1989).

77. W. Eckert, Phys. Rev. **B29**,1558 (1984).

78. See, for example, R. M. Eisberg, *Fundamentals of Modern Physics*, p. 407, Wiley, New York, 1967.

79. See, for example H. Frauenfelder and E. M. Henley, *Subatomic Physics*, Chapter 15, Prentice-Hall, englewood Cliffs, New Jersey, 1974.

80. M.Y. Chou, A. Cleland and M. L. Cohen, Solid State Commun. **52**, 645 (1984); M. Y. Chou and M. L. Cohen, Phys. Lett. **A113**, 420 (1986).

81. K. Clemenger, Phys. Rev. **B32**, 1359 (1985).

82. M. M. Dacorogna and M. L. Cohen, Phys. Rev. **B 34**, 4996 (1986).

83. V. Bonacic-Koutecky, P. Fantucci and J. Koutecky, Phys. Rev. **B37**, 4369(1988).

84. I. Garzon, Vijay Kumar, G. Pastore and W. Andreoni, unpublished.

85. E. S. Fois, J. I. Penman and P. A. Madden, J. Chem. Phys. (1993)

86. Vijay Kumar, Phys. Rev. **B23**, 3756 (1981); Surf. Sci. **84**, L231 (1979).

87. V. de Coulon, P. Delaly, P. Ballone, J. Buttet and F. Reuse, in Ref. 5, **D19**, pp. 173.

88. K.P.Huber and G. Herzberg, *Molecular Spectra and Molecular Structure. IV. Constants of Diatomic Molecules* (Van Norstrand Reinhold, New York, 1974).

89. Ballone et al, Ref. 5.

90. F. Reuse, S.N. Khanna, U. de Coulon and J. Buttet, Phys. Rev. **B41**, 11743 (1990).

91. G. Pacchioni, W. Pewestorf and J. koutecky, Chem. Phys. **83**, 261 (1984).

92. B. Delley, D. E. Ellis, A. J. Freeman, E. J. Barends and D. Post, Phys. Rev. **B27**, 2132 (1983).

93. V. Kumar, unpublished.

94. S. G. Louie, S. Froyen and M. L. Cohen, Phys. Rev. **B26**, 1738 (1982).

95. W. B. Pearson, *The Crystal Chemistry and Physics of Metals and Alloys*, Wiley, New York 1973.

96. A. Nakajima, T. Kishi, T. Sugioka, Y. Sone and K. Kaya, Chem. Phys. Lett. **177**, 297 (1991).

97. V. Kumar, in Proc. Adriatico Annversary Conference on 'Quasicrystals', Eds. M. Jaric and S. Lundquist, World Scientific(1990).

98. F. C. Frank and J. S. Kaspar, Acta Crystallogr. **11**, 184 (1958); ibid. **12**, 483 (1959).

99. A. Nakajima, T. Kishi, T. Sugioka and K. Kaya, Chem. Phys. Lett. **187**, 239 (1991).

100. P. Ballone, W. Andreoni, R. Car and M. Parrinello, Phys. Rev. Lett. **60**, 271 (1988).

101. K. Sattler, J. Muhlbach and E. Recknagel, Phys. Rev. Lett. **45**, 821 (1980). K. Sattler in *Current Topics in Materials Science*, Vol. **20**, Ed. E. Kaldis, North Holland (1985).

102. M. E. Geusic, R. R. Freeman and M. A. Duncan, J. Chem. Phys. **89**, 223 (1988).

103. D. Rayane, P. Melinon, B. Tribollet, B. Cabaud, A. Hoareau, and M. Broyer, J. Chem. Phys. **91**, 3100 (1989).

104. C. Brechignac, M. Broyer, Ph. cahuzac, M. de Frutos, P. Labastie, and J. -Ph. Roux, Phys. Rev. Lett. **67**, 1222 (1991).

105. W. Andreoni, private communication.

106. W. Andreoni and G. Pastore, Phys. Rev. **B41**, 10243 (1990).

107. L. Hanley, J. L. Whitten and S. L. Anderson, J. Phys. Chem. **90**, 5803 (1988).

108. D. Emin, T. Aselage, C. L. Beckel, I. A. Howard and C. Wood, Eds. AIP Conf. Proc. **140**, *Boron-Rich Solids* (AIP, New York, 1986).

109. S. A. Ruatta, L. Hanley and S. L. Anderson, J. Chem. Phys. **91**, 226 (1989); P. A. Hintz, S. A. Ruatta and S. L. Anderson, J. Chem. Phys. **92**, 292 (1990).

110. C. S. Yannoni, P. P. Bernier, D. S. Bethune, G. Meijer, and J. R. Salem, J. Am. Chem. Soc. **113**, 3190 (1991).

111. C. I. Frum, R. Engleman, H. G. Hedderich, P. F. Bernath, L. D. Lamb and D. R. Huffman, Chem. Phys. Lett. **176**, 504 (1991).

112. K. Laasonen and R. M. Nieminen, J. Phys. **CM2**, 1509 (1990).

113. B. L. Zhang, C. Z. Wang, K. M. Ho, J. Chem. Phys. **96**,7183 (1992); Wang et al, in Ref. 3 pp. 249.

114. Q.-M. Zhang, J. -Y. Yi and J. Bernholc, Phys. Rev. Lett. **66**, 2633 (1991).

115. V. Kumar, unpublished.

116. P. A. Heiney, J. E. Fischer, A. R. McGhie, W. J. Romanow, A. M. Denenstein, J. P. McCauley, Jr., A. B. Smith,III and D. E. Cox, Phys. Rev. Lett. **66**, 2911 (1991).

117. W. I. F. David *et al*, Nature **353**, 147 (1991).

118. P. J. Benning, J. L. Martins, J. H. Weaver, L. P. F. Chibante and R. E. Smalley, Science **252**, 1418 (1991).

119. G. Galli and M. Parrinello, Phys. Rev. Lett. **69**, 3547 (1992).

120. D. Vanderbilt, Phys. Rev. **B41**, 7892 (1990).

121. A. Pasquarello, K. Laasonen, R. Car, C. Lee and D. Vanderbilt, Phys. Rev. Lett. **69**, 1982 (1992).

CAR-PARRINELLO MD WITH VANDERBILT'S ULTRASOFT PSEUDOPOTENTIALS

KARI LAASONEN
IBM Research Division, Zürich Research Laboratory
8803 Rüschlikon, Switzerland

ABSTRACT

A short overview of the Car-Parrinello method using the new ultrasoft pseudopotential proposed by Vanderbilt is given. The emphasis here is on the technical side of the method and few applications are only mentioned at the end.

1. Introduction

Molecular dynamics (MD) has long been an essential tool in physical chemistry and solid-state physics to simulate temperature and time dependent phenomenas [1]. Traditionally these simulations has been carried out using classical and empirical potentials, but few years ago Car and Parrinello[2] introduced an ab initio molecular dynamics scheme based on the density functional theory[3] and a plane wave expansion for the wave functions. Because of the use of plane waves as basis set one is forced to replace the core electrons with pseudopotentials[4] to achieve smoother wave functions.

The Car-Parrinello method (CP) has shown that MD can be done with almost the accuracy of LDA. The almost stands here the usually small errors due to not fully converged basis set and errors due to the pseudopotentials. With careful construction of the pseudopotentials [4,5] and large enough basis set these error can be made very small. For practical simulations the accuracy of the pseudopotentials is usually sufficient, but for some atoms, like the first-row elements and transition metals, the number of the basis functions required could be very large. The size of the basis set for these atoms will in practice make calculations larger than few atoms impossible. The underlying problem here is the norm-conserving constraint in the pseudopotential construction. In pseudopotential scheme, the pseudo wave functions matches the all-electron wave functions beyond a given cutoff radius. And in case of norm-conserving pseudopotentials the pseudo wave functions has no nodes within the cutoff radius, and are required to have the same integrals as the all-electron wave functions to ensures that both wave functions carry the same charge.

The commonly used Bachelet-Hamann-Schlüter (BHS) pseudopotentials[4] are often very steep and several improvements to reduce the basis size have been proposed [6–9], but still keeping the norm-conserving constraint. So, despite the improvements, the basis size to describe first-row elements or transitions metals is still too high to allow MD simulations of relatively extended systems.

Recently, Vanderbilt[10] has proposed a new pseudopotential scheme in which the norm-conserving condition has been relaxed. In this scheme, the pseudo wave functions are allowed to be as soft as possible within the core region yielding a dramatic reduction of the basis size. After relaxing the norm-conserving constraint the electron density is not anymore the squared modulus of the wave functions but it has to be augmented in the core regions. This augmentation leed to generalized orthonormality condition and the standard Car-Parrinello scheme has to be modified significantly. Note that the augmented part appears only in the electron density, contrary to other calculation schemes, such as LAPW, in which similar ideas have been applied to the wave functions. Also the augmented part is strictly located within the cutoff radius, which allows it to treated very effectively in the program.

In this paper we try to give an overview of the Car-Parrinello method using the Vanderbilt pseudopotential (CPV) emphasizing the differencies to the standard CP method. The standard CP code has been reviewed in several works[11–13], and we cannot repeat all the details here, thus our approach is more formal. The paper is organized as follows. In Sec 2. we give the general (density functional based) formalism of the Vanderbilt pseudopotential (PP) with electronic and ionic forces. We also describe the algorithm to obtain a Vanderbilt PP. In Sec. 3 we focus on molecular dynamics and the orthonormality constraints. And Sec. 4 is very short review of the applications of the CPV scheme.

2. Vanderbilt pseudopotential scheme

2.1. Kohn-Sham Equations

In Vanderbilt's ultrasoft pseudopotential scheme[9,14] the total energy of N_v valence electrons, described by wave functions ϕ_i, is given by the standard Kohn-Sham[15] formalism

$$\begin{aligned} E_{tot}[\{\phi_i\},\{R_I\}] &= \sum_i \langle\phi_i| -\nabla^2 + V_{NL}|\phi_i\rangle + \frac{1}{2}\iint dr\, dr' \frac{n(r)n(r')}{|r-r'|} \\ &\quad + E_{xc}[n] + \int dr\, V_{loc}^{ps}(r) n(r) + U(\{R_I\}), \end{aligned} \tag{1}$$

where $n(\mathbf{r})$ is the electron density, E_{xc} the exchange and correlation energy, $U(\{R_I\})$ the ion-ion interaction energy, and where the pseudopotential contains a local part $V_{loc}^{ps} = \sum_I V_{loc}^{ion}(r - R_I)$ and a fully non-local part given by

$$V_{NL} = \sum_{nm,I} D_{nm}^{ion} |\beta_n^I\rangle\langle\beta_m^I|, \tag{2}$$

where the functions β_n^I as well as the coefficients D_{nm}^{ion} characterize the pseudopotential and differ for different atomic species. In the following we will for simplicity consider only one atomic species. The β_n^I functions are strictly localized in the core region and they depend on the ionic positions through

$$\beta_n^I(r) = \beta_n(r - R_I). \tag{3}$$

The indices n and m in (2) run over the total number N_n of β_n functions. In the ultrasoft pseudopotential case, often two references energies are necessary for every angular momentum channel. This leads to a number N_n which is generally twice as large compared to the corresponding norm-conserving case[16].

The electron density in (1) is given by

$$n(\boldsymbol{r}) = \sum_i \Big[|\phi_i(\boldsymbol{r})|^2 + \sum_{nm,I} Q^I_{nm}(\boldsymbol{r}) \langle \phi_i | \beta^I_n \rangle \langle \beta^I_m | \phi_i \rangle \Big], \tag{4}$$

where the augmentation functions $Q^I_{nm}(\boldsymbol{r}) = Q_{nm}(\boldsymbol{r} - \boldsymbol{R}_I)$ are also provided by the pseudopotential and are strictly localized in the core regions. Thus, the electron density in (4) is separated in a *soft* delocalized contribution given by the squared modulus of the wave functions and by a *hard* one which is localized at the cores.

In the Vanderbilt scheme the pseudopotential is completely described by the quantities $V^{ion}_{loc}(\boldsymbol{r}), D^{ion}_{nm}, Q_{nm}(\boldsymbol{r})$, and $\beta_n(\boldsymbol{r})$. A more detail algorithm to obtain these quantities is reviewed in chapter 2.2.

The relaxation of the norm-conserving condition is achieved introducing a generalized orthonormality condition

$$\langle \phi_i | S(\{\boldsymbol{R}_I\}) | \phi_j \rangle = \delta_{ij}, \tag{5}$$

where S is a Hermitian overlap operator given by

$$S = 1 + \sum_{nm,I} q_{nm} |\beta^I_n\rangle \langle \beta^I_m|, \tag{6}$$

where $q_{nm} = \int d\boldsymbol{r}\, Q_{nm}(\boldsymbol{r})$. The orthonormality condition (5) is consistent with the conservation of the charge $\int d\boldsymbol{r}\, n(\boldsymbol{r}) = N_v$.

The ground-state orbitals ϕ_i are those which minimize the total energy (1) under condition (5),

$$\frac{\delta E_{tot}}{\delta \phi^*_i} = \epsilon_i S \phi_i, \tag{7}$$

where ϵ_i have been introduced as Lagrange multipliers. Because of the fact that also the augmentation part of the charge density depends on the wave functions,

$$\frac{\delta n(\boldsymbol{r}')}{\delta \phi^*_i(\boldsymbol{r})} = \phi_i(\boldsymbol{r}')\delta(\boldsymbol{r}' - \boldsymbol{r}) + \sum_{nm,I} Q^I_{nm}(\boldsymbol{r}')\beta^I_n(\boldsymbol{r})\langle \beta^I_m | \phi_i \rangle, \tag{8}$$

additional terms appear in the Kohn-Sham equations from the density dependent terms in the total energy (1). As an example, we consider the exchange and correlation energy. Using Eq. (8) we obtain:

$$\begin{aligned} \frac{\delta E_{xc}[n]}{\delta \phi^*_i(\boldsymbol{r})} &= \int d\mathbf{r}' \frac{\delta E_{xc}[n]}{\delta n(\boldsymbol{r}')} \frac{\delta n(\boldsymbol{r}')}{\delta \phi^*_i(\boldsymbol{r})} \\ &= \mu_{xc}(\boldsymbol{r})\phi_i(\boldsymbol{r}) + \sum_{nm,I} \beta^I_n(\boldsymbol{r}) \langle \beta^I_m | \phi_i \rangle \int d\mathbf{r}' \mu_{xc}(\boldsymbol{r}') Q^I_{nm}(\boldsymbol{r}'), \end{aligned} \tag{9}$$

where $\mu_{xc}(r) = \delta E_{xc}[n]/\delta n(r)$. The other terms can be calculated similarly. We obtain

$$[-\nabla^2 + V_{eff}(r)]\phi_i(r) + \sum_{nm,I} D^I_{nm}\beta^I_n(r)\langle\beta^I_m|\phi_i\rangle = \epsilon_i S\phi_i(r), \tag{10}$$

where the effective potential is

$$V_{eff}(r) = \frac{\delta E_{tot}}{\delta n(r)} = \int dr' \frac{n(r')}{|r-r'|} + V^{ps}_{loc}(r) + \mu_{xc}(r). \tag{11}$$

In Eqs. (10), all the terms arising from the augmented part of the electron density have been grouped together with the nonlocal part of the pseudopotential (2) by defining a new coefficients D^I_{nm},

$$D^I_{nm} = D^{ion}_{nm} + \int d\mathbf{r}\, V_{eff}(r) Q^I_{nm}(r). \tag{12}$$

Note, however, that the D^{ion}_{nm}'s are just parameters which characterize the pseudopotential, whereas the new D^I_{nm}'s depend trough V_{eff} (11) on the wave functions and have to be updated in the self-consistency procedure.

To obtain the correct ionic forces one should also take into account that the orthonormality constraint (eq. 5) will also depend on atomic position. This can again be done by using the Lagranges multipliers,

$$\mathbf{F}_I = -\frac{\partial E_{tot}}{\partial \mathbf{R}_I} + \sum_{ij} \Lambda_{ij}\langle\phi_i|\frac{\partial S}{\partial \mathbf{R}_I}|\phi_j\rangle\ , \tag{13}$$

where the full expression of the ionic forces due to the total energy can be obtained keeping in mind that also the electron density depends on $\mathbf{R}_I$:

$$\begin{aligned}\frac{\delta n(\mathbf{r})}{\delta R_I} &= \sum_{i,nm}\Big[Q^I_{nm}(\mathbf{r})(\langle\phi_i|\frac{d\beta^I_n}{d\mathbf{R}_I}\rangle\langle\beta^I_m|\phi_i\rangle + \langle\phi_i|\beta^I_n\rangle\langle\frac{d\beta^I_m}{d\mathbf{R}_I}|\phi_i\rangle) \\ &\quad + \frac{dQ^I_{nm}(\mathbf{r})}{d\mathbf{R}_I}\langle\phi_i|\beta^I_n\rangle\langle\beta^I_m|\phi_i\rangle\Big],\end{aligned} \tag{14}$$

causing additional terms to appear compared to the norm-conserving case. Collecting the different terms, this part of the ionic forces can be written as

$$\begin{aligned}\frac{\delta E_{tot}}{\delta R_I} &= \int dr \frac{dV^{ps}_{loc}}{dR_I}n(r) + \frac{dU}{dR_I} + \sum_{nm} D^I_{nm}\frac{\delta\rho^I_{nm}}{\delta R_I} \\ &\quad + \sum_{nm}\rho^I_{nm}\int dr V_{eff}(r)\frac{dQ^I_{nm}}{dR_I},\end{aligned} \tag{15}$$

where D^I_{nm} and V_{eff} have been defined in Eqs. (12) and (11), respectively, and where

$$\rho^I_{nm} = \sum_i \langle\phi_i|\beta^I_n\rangle\langle\beta^I_m|\phi_i\rangle. \tag{16}$$

Note that, because of the fact that the basis set consists of plane waves, the wave functions do not depend on the ionic positions and no additional Pulay-type corrections are needed.

At this stage, the difference with respect to the norm-conserving case resides in the S operator, the wave function dependent D^I_{nm} (12), and the fact that the number N_n of β^I_n functions is twice as large. The calculation of the D^I_{nm} (12) can be carried out in real space and produces a modest overhead (see Sec. 3.3). The presence of the S operator requires the calculation of $\langle\beta^I_n|\phi_i\rangle$ which are also needed for the nonlocal pseudopotential, and thus do not represent additional computation.

The number of operations to calculate one scalar product of this type scales like the number of plane waves N_w, and the number of these scalar products is given by $N_I N_n N_i$, where N_I is the number of atoms, N_n the number of β_n functions per atoms, and N_i the number of filled states, respectively. Since N_w, N_n both scale like N_I, this part scales like N_I^3, and, for large systems, it is the dominant time-consuming part. When also the ionic forces are calculated similar scalar products $\langle d\beta^I_n/dR_I|\phi_i\rangle$ are required (14). For a large number of atoms N_I, it can be assumed that the computational cost related to these scalar products is dominant, and one can deduce that the ultrasoft PP scheme is advantageous compared to the norm-conserving PP scheme, when $N_w^{NC} > 2\,N_w$, where we have taken $N_n = 2N_n^{NC}$. In terms of energy cutoff E_{cw}: $E_{cw}^{NC} > 1.6\,E_{cw}$, which is easily satisfied for first-row elements and transition metals.

Very recently King-Smith, Payne and Lin[17] have showed that it is possible to evaluate the scalar products between wave functions and β^I_n functions in real space taking advantage of the fact that the β^I_n are localized. In this way, this part of the calculation scales like $N_I N_n$, i.e. like N_I^2. Although the full calculation would still scale like N_I^3, because of the orthonormalization procedure, the cost of a previously dominant part of the computation would considerably be reduced, allowing the study of much larger systems.

2.2. *Pseudopotential generation algorithm*

In this section, we give a concise description of the specific pseudopotential generation algorithm introduced in Ref. 10.

As in other pseudopotential methods, an all-electron (AE) calculation is first carried out on a free atom in some reference configuration, leading to a screened potential $V_{\rm AE}(r)$. Then for each angular momentum l, a set of reference energies $\epsilon_{l\tau}$ is chosen ($\tau = 1, N_\tau$, typically $1 \le N_\tau \le 3$) to cover the energy range over which good scattering properties are desired (usually the range of occupied bulk valence states). At each $\epsilon_{l\tau}$, the solution of the Schördinger equation which is regular at the origin is obtained (for fixed $V_{\rm AE}$):

$$[T + V_{\rm AE}(r)]\,\psi_i(r) = \epsilon_i \psi_i(r) \ . \tag{17}$$

Here i is a composite index, $i = \{lm\tau\}$, and T is the kinetic energy operator $-\frac{1}{2}\nabla^2$. (The amplitude of ψ_i is assumed to have been fixed in some definite way, e.g., by its value at an arbitrary radius R.) Despite the fact that ψ_i is in general nonnormalizable, a bra-ket notation

$$(T + V_{\rm AE} - \epsilon_i)\,|\psi_i\rangle = 0 \tag{18}$$

is adopted as a stand-in for the previous equation. Quantities such as $\langle\psi_i|\psi_i\rangle$ are ill-defined, but the special notation $\langle\psi_i|\psi_j\rangle_R$ will be used to denote the integral of $\psi_i^*(r)\psi_j(r)$ inside the sphere of radius R.

Next, cutoff radii r_{cl} are chosen, and for each ψ_i obtained above, a pseudo wave function ϕ_i is constructed, subject only to the constraint that it join smoothly to ψ_i at r_{cl}. No norm-conservation constraint is explicitly imposed. A smooth local potential $V_{loc}(r)$ is also generated in such a way that it matches smoothly to $V_{\rm AE}(r)$ at a cutoff radius $r_c^{\rm loc}$, and a diagnostic radius R is chosen such that R is slightly larger than the maximum of the r_{cl} and $r_c^{\rm loc}$. (R must be outside the r_c's in order that scattering properties calculated there will be meaningful.) Then the orbitals

$$|\chi_i\rangle = (\epsilon_i - T - V_{loc})\,|\phi_i\rangle \tag{19}$$

are formed. These are local (they vanish at and beyond R, where $V_{loc} = V_{\rm AE}$ and $\phi_i = \psi_i$). Thus the matrix of inner products

$$B_{ij} = \langle\phi_i|\chi_j\rangle \tag{20}$$

is well defined.

We are now in a position to form the quantities V_{loc}^{ion}, D_{ij}^{ion}, $Q_{ij}(r)$, and $\beta_i(r)$ needed to specify the pseudopotential. $Q_{ij}(r)$ and $\beta_i(r)$ are given by

$$Q_{ij}(r) = \psi_i^*(r)\psi_j(r) - \phi_i^*(r)\phi_j(r) \tag{21}$$

and

$$|\beta_i\rangle = \sum_j (B^{-1})_{ji}|\chi_j\rangle \ . \tag{22}$$

The $|\beta_i\rangle$ are duals to the $|\phi_i\rangle$ and are also local; they form the "projectors" of the non-local operators. It follows from Eq.(6) that

$$q_{ij} = \langle\psi_i|\psi_j\rangle_R - \langle\phi_i|\phi_j\rangle_R \ . \tag{23}$$

It is straightforward to verify that the $|\phi_i\rangle$ obey the secular equation

$$\left(T + V_{loc} + \sum_{ij} D_{ij}|\beta_i\rangle\langle\beta_j|\right)|\phi_i\rangle = \epsilon_i\left(1 + \sum_{ij} q_{ij}|\beta_i\rangle\langle\beta_j|\right)|\phi_i\rangle \ , \tag{24}$$

where

$$D_{ij} = B_{ij} + \epsilon_j q_{ij} \ . \tag{25}$$

Finally the quantities $V_{\rm loc}^{\rm ion}(r)$ and D_{ij}^{ion} are obtained from a "descreening" procedure:

$$V_{loc}^{ion} = V_{loc}(r) - \int dr' \frac{n(r')}{|r - r'|} - \mu_{xc}(r) \ , \tag{26}$$

$$D_{ij}^{ion} = D_{ij} - \int d\mathbf{r}' V_{loc}(\mathbf{r}')n(\mathbf{r}') \ . \tag{27}$$

This pseudopotential has been generated in such a way that the following features are reproduced in the reference configuration. (Here we assume that the eigenvalues of the occupied valence orbitals in the reference configuration are included among the chosen $\epsilon_{l\tau}$; this is our standard practice.) (*i*) The pseudo eigenvalues are equal to the AE ones, and the corresponding orbitals agree exactly outside r_{cl}. (*ii*) The scattering properties are correct at each $\epsilon_{l\tau}$, in the sense that the logarithmic derivative and its energy derivative match the AE one at that energy. Thus, the transferability, as measured by scattering properties in the reference configuration, can be systematically improved by increasing the number of such energies. (*iii*) The valence charge density is precisely equal to the AE valence density in the reference configuration (except insofar as it is affected by the pseudization of the Q_{ij} as discussed in the next subsection). This helps improve the transferability with respect to changes in screening environment.

However, the main advantage of the present scheme over previous ones is that no norm conservation constraint is imposed during the pseudization $\psi_i \rightarrow \phi_i$. Thus, the pseudization can be done in such a way as to make the $\phi_i(\mathbf{r})$ as smooth as possible. This is the reason we refer to this scheme as an "ultrasoft" pseudopotential scheme.

2.3. Pseudization of the electron charge density

In norm-conserving pseudopotential schemes the electron density is defined as in Eq. (4) where only the first term is kept on the right hand side. Thus, the energy cutoff E_{cn} required to describe fully the electron density is four times the energy cutoff E_{cw} of the wave functions:

$$E_{cn} = 4E_{cw}. \tag{28}$$

This direct relationship between cutoff for electron density and wave functions does not hold in the ultrasoft pseudopotential scheme, because of the presence of the augmentation functions Q_{nm} in the electron density (4). In this scheme, it is therefore appropriate to introduce two independent energy cutoffs: one for the *soft* part of the electron density, $E_{cs} = 4E_{cw}$, and a second generally much higher one to describe the augmentation functions Q_{nm}, E_{cn}.

It is often possible to reduce the charge cutoff E_{cn} by pseudizing the functions Q_{nm} [10,18]. In this pseudization, the Q_{nm} are modified within an inner core region (defined by r_{in}). The charge density described by the pseudized Q_{nm} preserves all the charge moments, so that the electrostatic potential beyond r_{in} remains unchanged. This is achieved by decomposing the Q_{nm} according to angular momentum L

$$Q_{nm}(\mathbf{r}) = \sum_{LM} c_{LM}^{nm} Y_{LM}(\hat{\mathbf{r}}) Q_{nm}^{rad}(\mathbf{r}), \tag{29}$$

where c_{LM}^{nm} are Clebsch-Gordan coefficients, Y_{LM} are spherical harmonics, and Q_{nm}^{rad} gives the all-electron radial dependence of Q_{nm} and is independent on L and M by construction[10]. The number of possible L-channels in Eq. (29) is finite, because of the fact that a nonlocal pseudopotential is required only for the lowest angular momentum channels.

The Q_{nm}^{rad} in Eq. (29) are then replaced by L-dependent pseudized counterparts Q_{nm}^{L},

$$Q_{nm}(\mathbf{r}) = \sum_{LM} c_{LM}^{nm} Y_{LM}(\hat{r}) Q_{nm}^{L}(r), \tag{30}$$

which satisfy the condition that for each L-component the L-moment of the electron charge density be conserved:

$$\int_0^{r_{in}^L} r^2 dr\, r^L Q_{nm}^L(r) = \int_0^{r_{in}^L} r^2 dr\, r^L Q_{nm}^{rad}(r), \tag{31}$$

where L-dependent inner cutoff radii r_{in}^L have been introduced. Since the high Fourier components of Q_{nm} are mainly related to the high-L components, it is convenient, for a given cutoff E_{cn}, to use smaller r_{in}^L for low-L components. In this way, a relative better description of the lowest moments of the electron charge density is maintained.

In order to construct optimally pseudized $Q_L(r)$, we expand them in polynomials of r inside r_{in}^L as

$$Q_L(r) = r^L \rho_L(r) \qquad \text{when } r < r_{in}^L \tag{32}$$

where

$$\rho_L(r) = C_1 + C_2 r^2 + C_3 r^4 \cdots \tag{33}$$

with the number of terms insuring sufficient smoothness of the polynomial. We want $Q_L(r)$, i.e. $\rho_L(r)$ as smooth as possible. Therefore we want to have as small Fourier coefficients above a certain cutoff wavevector q_c as possible, which means we have to minimize

$$I = \int_{q_c}^{\infty} q^2\, Q_L^2(q)\, dq \tag{34}$$

where

$$Q_L(q) = \int_0^{\infty} dr\, Q_L(r)\, j_L(qr) \tag{35}$$

and j_L is the spherical Bessel function of order L. The minimization should be done subject to the constraint to eq. (30) and to the following continiuty requirements:

$$\begin{aligned} \rho(r_{in}^L) &= \rho_{AE}(r_{in}^L) \\ \rho'(r_{in}^L) &= \rho'_{AE}(r_{in}^L) \\ \rho''(r_{in}^L) &= \rho''_{AE}(r_{in}^L) \end{aligned} \tag{36}$$

This treatment gives us a very smooth charge density with a smooth and continuous first and second derivative, which are essential for the gradient corrections. It is also related to the treatment in the optimized pseudopotential by Rappe et al. [8].

When condition (28) is restored, the solution of the Kohn-Sham equations (10) can be obtained similarly to the norm-conserving case. In particular cases, such as for transition metals, the Q_{nm} require $E_{cn} > 4E_{cw}$, even after pseudization. In this case the problem of having a large number of plane waves can still be circumvented using the fact that the Q_{nm} are localized in real space, as will be discussed in Sec. 3.4.

3. Molecular dynamics

3.1. Ionic forces

In a CP simulation, the electronic wave functions and the ionic coordinates evolve according to a classical Langrangian in which the orthonormality constraints (5) have been be incorporated using Lagrange multipliers[2] Λ_{ij} :

$$\begin{aligned} \mathcal{L} &= \mu \sum_i \int dr\, |\dot{\phi}_i(r)|^2 + \sum_I \frac{1}{2} M_I \dot{R}_I^2 - E_{tot}[\phi_i, R_I] \\ &\quad + \sum_{ij} \Lambda_{ij} (\langle \phi_i | S | \phi_j \rangle - \delta_{ij}), \end{aligned} \tag{37}$$

where μ is a fictitious mass parameter for the electronic degrees of freedom and M_I is the mass of the atoms. The orthonormality constraints are holonomic, and do not cause energy dissipation during the MD run. The Euler equations associated to the Lagrangian (37) are:

$$\mu \ddot{\phi}_i = -\frac{\delta E_{tot}}{\delta \phi_i^*} + \sum_j \Lambda_{ij} S \phi_j, \tag{38}$$

$$M_I \ddot{R}_I = -\frac{\delta E_{tot}}{\delta R_I} + \sum_{ij} \Lambda_{ij} \langle \phi_i | \frac{\delta S}{\delta R_I} | \phi_j \rangle. \tag{39}$$

Eq. (38), associated to the electronic degrees of freedom, corresponds to the Kohn-Sham equations (10) encountered previously. The second equation (39) is obtained for the ionic coordinates. The two terms on the right hand side correspond to the contribution to the ionic forces due to the total energy and to the orthonormality constraints, respectively. The latter contribution appears because of the R_I-dependence of the β_n^I (3) which appear in S (6). Note that this second contribution to the ionic forces is absent in the case of norm-conserving schemes, in which the orthonormality condition does not depend in any way on the ionic positions.

3.2. Orthonormality constraints

To illustrate the consequences of the dependence of the orthonormality constraints on $\{\mathbf{R}_I\}$, we discretize the equations of motion (38) and (39) using the

Verlet algorithm. For the electronic wave functions we obtain

$$\phi_i(t+\Delta t) = 2\phi_i(t) - \phi_i(t-\Delta t) - \frac{(\Delta t)^2}{\mu}\Big(\frac{\delta E_{tot}}{\delta \phi_i^*} - \sum_j \Lambda_{ij}\, S(t)\phi_j(t)\Big), \tag{40}$$

where $S(t)$ means that the operator S is evaluated for ionic positions $\mathbf{R}_I(t)$. Similarly for the ionic coordinates:

$$\begin{aligned}\mathbf{R}_I(t+\Delta t) &= 2\mathbf{R}_I(t) - \mathbf{R}_I(t-\Delta t)\\ &\quad -\frac{(\Delta t)^2}{M_I}\Big(\frac{\delta E_{tot}}{\delta R_I} - \sum_{ij}\Lambda_{ij}\langle\phi_i(t)|\frac{\delta S(t)}{\delta R_I}|\phi_j(t)\rangle\Big).\end{aligned} \tag{41}$$

The orthonormality conditions has to be imposed at each time-step[19]:

$$\langle\phi_i(t+\Delta t)|S(t+\Delta t)|\phi_j(t+\Delta t)\rangle = \delta_{ij}\ . \tag{42}$$

Fullfilling this constraint leads to the following matrix equation for the Lagrange multipliers ($\lambda = \Delta t^2 \Lambda^*/\mu$),

$$A + \lambda B + B^\dagger \lambda^\dagger + \lambda C \lambda^\dagger = 1 \tag{43}$$

where the dagger indicates the hermitian conjugate (because of the hermiticity of S, $\lambda = \lambda^\dagger$) and where

$$\begin{aligned}A_{ij} &= \langle\bar{\phi}_i|S(t+\Delta t)|\bar{\phi}_j\rangle\\ B_{ij} &= \langle S(t)\phi_i(t)|S(t+\Delta t)|\bar{\phi}_j\rangle\\ C_{ij} &= \langle S(t)\phi_i(t)|S(t+\Delta t)|S(t)\phi_j(t)\rangle\\ \bar{\phi}_i &= 2\phi_i(t) - \phi_i(t-\Delta t) - \frac{(\Delta t)^2}{\mu}\frac{\delta E_{tot}(t)}{\delta\phi_i^*}\ .\end{aligned}$$

In norm-conserving schemes the identity operator is found at the place of S, which leads to a somewhat simpler form of eq.(43) presented in ref. 20. In the ultrasoft pseudopotential case, the solution of eq. (43) is somewhat problematic, because of the $\mathbf{R}_I(t+\Delta t)$ dependence of $S(t+\Delta t)$. But to obtain correct $\mathbf{R}_I(t+\Delta t)$ one need to know the values of Λ_{ij}. Thus it is convenient to solve this problem iteratively. First, the new $\Lambda_{ij}(t+\Delta t)$ are estimated using two previous values,

$$\Lambda_{ij}^{(0)}(t+\Delta t) = 2\Lambda_{ij}(t) - \Lambda_{ij}(t-\Delta t), \tag{44}$$

and used to find the new $\mathbf{R}_I^{(0)}(t+\Delta t)$, which are correct to $O(\Delta t^4)$. Then equation (43) is solved iteratively in a similar way (see below) as in the norm-conserving case[20], giving a new set of $\Lambda_{ij}^{(1)}(t+\Delta t)$, with which the whole procedure is repeated, and so on until convergence is achieved. Fortunately, it turns out that the ionic positions are very well determined by (44), so that it is sufficient to go just once through the orthonormalization process per time-step.

In order to solve Eq. (43), we generalize the iterative procedure used in Ref. 20, because when this procedure is applied, the result not always converges. In the norm-conserving case, matrix B converges to the identity matrix for vanishing Δt. This is not the case in the ultrasoft pseudopotential case. However, when the matrix B is decomposed in a hermitian B_h and antihermitian matrix B_a,

$$B = B_h + B_a, \tag{45}$$

it is straightforward to see that B_a vanishes in the limit of small Δt. The first-iteration $\lambda^{(0)}$ can now be obtained from

$$\lambda^{(0)} B_h + B_h \lambda^{(0)} = 1 - A \tag{46}$$

where the C-dependent term has been neglected because of higher order in Δt. Eq. (46) can be solved exactly introducing the unitary matrix U, which diagonalizes B_h, $U^\dagger B_h U = D$, where $D_{ij} = d_i \delta_{ij}$. The solution to Eq. (43) can be obtained by iterating

$$\lambda^{(n+1)} B_h + B_h \lambda^{(n+1)} = 1 - A - \lambda^{(n)} B_a - B_a^\dagger \lambda^{(n)} - \lambda^{(n)} C \lambda^{(n)}, \tag{47}$$

where at every step the new $\lambda^{(n+1)}$ are obtained in the same way as $\lambda^{(0)}$ had been obtained from Eq. (46). This improved

When imposing the orthonormality condition (42), the form of S (6) requires the calculation of an additional set of scalar products of the type $\langle \beta_i^I | \phi_n \rangle$ as compared to the norm-conserving case.

3.3. Double grid technique

As long as the electron density can be described with a cutoff $4E_{cw}$, molecular dynamics simulations can be performed without further modifications with respect to norm-conserving schemes[11,12], except for the use of appropriate forces for electronic (10) and ionic variables (15). As discussed above, there is no reason that this condition is satisfied in the ultrasoft pseudopotential scheme. In the case of Cu, for example, in which a large part of the charge of the tightly bound d-orbitals is incorporated in the Q_{ij}, E_{cn} turns out to be significantly higher than $4E_{cw}$ [18]. Also in the case of first-row elements, such as F [21], the p electronic states are so localized that condition (28) cannot be satisfied. In order to permit an optimal choice for E_{cw} and E_{cn}, it is convenient to develop a scheme in which these two parameters can be chosen independent of each other.

In the calculations, fast Fourier transforms (FFT's) are used to transform physical quantities from $\mathbf{G}$-space to $\mathbf{r}$-space and vice-versa. Some products are diagonal in $\mathbf{r}$-space, and thus should be calculated in $\mathbf{r}$-space. Products of this type occur twice: $\phi_i^*(\mathbf{r})\phi_i(\mathbf{r})$ in the calculation of the electron density (4), and $V_{eff}(\mathbf{r})\phi_n(\mathbf{r})$ which appears in (10). The FFT-grid must contain all plane waves determined by the energy cutoff $E_{cs} = 4E_{cw}$ in order to fully describe the results of those multiplications. On the other hand, in order to describe the Q_{nm}, a grid determined by the higher cutoff E_{cn} is required. A possible solution consists in using a single grid determined by the cutoff E_{cn}. However, the number of FFT's per time-step which have to be performed for the calculation of **r**-space products scales like the number of electronic states N_i, whereas only a few FFT's (independent of N_i) per time-step involve the electron density. We have therefore introduced two different FFT grids [18], a sparse and a dense grid, determined by cutoffs E_{cs} and E_{cn}, respectively. In this way, the FFT's for the **r**-space products are calculated on the sparse grid with the same numerical effort required in a norm-conserving scheme when using the same wave-function cutoff E_{cw}. The numerical cost of the few FFT's on the dense grid is negligible in a simulation of a large system. The connection between the two different grids is established in **G**-space. All the quantities defined on the sparse grid can be transferred to the denser grid, taking the plane-wave components which are absent on the sparse grid to be zero.

We illustrate this technique following the calculations stepwise. In the actual calculations, the electronic degrees of freedom are the coefficients $\phi_i(G)$ of the plane waves:

$$\phi_{j,\mathbf{k}}(r) = \frac{1}{\sqrt{\Omega}} \sum_{G}^{G_{cw}} \phi_{j,\mathbf{k}}(G) e^{-i(\mathbf{k}+G)\cdot r}, \tag{48}$$

where Ω is the volume of the simulation cell and G_{cw} the largest G-vector compatible with the condition $\frac{1}{2}|\mathbf{k}+\mathbf{G}|^2 < E_{cw}$. Due to the large simulation cell the Brillouin zone is often sampled using the only Γ-point. An advantage of this choice is that the wave functions can be taken to be real. In the following we will drop the $\mathbf{k}$ index.

The following scalar products which appear in the nonlocal pseudopotential and in the orthonormalization process are evaluated in **G**-space:

$$\langle \phi_j | \beta_n^I \rangle = \sum_{G}^{G_{cw}} \phi_j^*(G) \beta_n(G) e^{-i\mathbf{G}\cdot R_I}, \tag{49}$$

and similarly their derivatives with respect to the ionic positions:

$$\langle \phi_j | \frac{d\beta_n^I}{dR_I} \rangle = -i \sum_{G}^{G_{cw}} \mathbf{G} \phi_j^*(G) \beta_n(G) e^{-i\mathbf{G}\cdot R_I}. \tag{50}$$

The kinetic energy is diagonal in **G**-space and can directly be calculated:

$$E_{kin} = \frac{1}{2} \sum_{i,G}^{G_{cw}} G^2 |\phi_i(G)|^2. \tag{51}$$

The soft part of the electron density n^s, given by the first term on the right side of (4), is obtained in **r**-space using FFT's on the sparse grid. Then it is transformed to **G**-space using a sparse-grid FFT. The electron density is now augmented with the contribution arising from the Q_{nm}^I:

$$n(G) = n_s(G) + \sum_{i,nm,I} Q_{mn}^I(G) \langle \phi_i | \beta_n^I \rangle \langle \beta_m^I | \phi_i \rangle, \tag{52}$$

The **G**-vectors required to describe the augmented part are given by the condition $\frac{1}{2}G^2 < E_{cn}$, $\mathbf{G}_{cn}$ being the largest of such vectors. In fact, the Q^I_{mn} are first evaluated in **r**-space of the dense grid, and then transformed with a dense-grid FFT to **G**-space to be added to the soft electron density n_s in (52). The obtained $n(G)$ is used to evaluate the contribution to the total energy from the local potential

$$E_{loc} = \sum_{G}^{G_{cn}} V_{loc}^{ps}(G) n^*(G). \tag{53}$$

as well as from the Hartree energy:

$$E_H = 2\pi\Omega \sum_{G\neq 0}^{G_{cn}} \frac{n^*(G)n(G)}{G^2}. \tag{54}$$

The Ewald summation method has been used to correctly cancel the $G = 0$ component in Eq. (54). The electron density is then transformed to **r**-space of the dense grid, and the energy E_{xc} is evaluated as well as the potential μ_{xc}.

The potential V_{eff} (11) is needed on the dense grid to calculate quantities such as the D^I_{nm} (12) *and* on the sparse grid to calculate $V_{eff}\phi_i$ in the Kohn-Sham equations (10). The first two terms of V_{eff} are added in **G**-space of the dense grid. Then the result is transformed to **r**-space using a dense-grid FFT, where the μ_{xc} is added. The V_{eff} is now known in **r**-space of the dense grid were it will be used for the D^I_{nm}. The potential V_{eff} is also needed in **r**-space of the sparse grid. The connection between the two grid is in **G**-space as described previously.

In this way the dense-grid FFT is used only four times per time-step, which is negligible compared to the number of sparse-grid FFT's which scales as the number of states N_i.

3.4. Fourier interpolation scheme

We now focus on the augmentation functions Q_{nm}, which are peculiar to the ultrasoft pseudopotential scheme. These functions appear in the calculation of the electron density (52), in the integrals (12) which give the D_{nm}'s, and in similar integrals involving $dQ^I_{nm}/d\mathbf{R}_I$, which appear in the ionic forces (15). The Q_{nm} need a high cutoff E_{cn} and, if these calculations were carried in **G**-space of the dense grid, a significant increase in the computational cost would occur. It it is therefore important to take advantage of the fact that the Q_{nm} are localized in the core regions. When these integrals are calculated in **r**-space, the associated computational cost is reduced by the ratio of the volumes of the core region and of the simulation cell Ω.

The advantage of working in **G**-space is that the Q^I_{nm} can easily be evaluated for any atomic position by calculating a simple phase factor (Fourier interpolation),

$$Q^I_{nm}(G) = Q_{nm}(G)e^{-i\mathbf{G}\cdot R_I}. \tag{55}$$

When the energy and the forces are both calculated in **G**-space the forces are the analytical derivative of the expression of energy, ensuring that the constant of motion be conserved during the evolution. The accuracy of such a real-space interpolation is therefore extremely critical to guarantee an accurate and stable molecular dynamics simulation.

We have been able to combine the advantages of the locality of the Q^I_{nm} and the Fourier interpolation scheme, by introducing a small box for every ion. These boxes are taken to be large enough to contain the core regions, where the Q^I_{nm} are non vanishing. The FFT grid of the boxes locally coincides in **r**-space with the dense grid. Generally the Q^I_{nm} are obtained using phase factors similar to (55) to take into account the displacement of the ions within their boxes. But when the ion crosses one of these grid planes, the box is displaced discretely by one grid-unit to follow to motion of the ion.

The advantage of working with these small boxes is that the energy cutoff of the plane waves associated with the box-grid is the same as for the dense grid, whereas the number of plane waves is reduced by the ratio of the volumes of the box and Ω.

In the case of the electron density (52), after the $Q_{nm}(G)$ part of a single ion is calculated, it is transformed to **r**-space using a box-grid FFT, and then stored to **r**-space of the dense-grid. When these operations have been completed for all ions, the hard part of the electron density is transformed with a dense-grid FFT to its **G**-space. The integrals involving V_{eff} and density are performed transferring V_{eff} from the dense-grid **r**-space to the box-grid **r**-space, and then transforming it with box-grid FFT's to the box-grid **G**-spaces, where the integrals are carried out in the same way as in Eq. (49) and (50). The overall cost of the box-grid method is negligible.

Because of the fact that the calculation is performed in part in the Fourier space of the dense grid (e.g. Hartree energy and potential) and in part in that of the box-grid, this method does not provide ionic forces which are exact analytic derivatives of the total energy. The discrepancies appear as little jumps in the constant of motion whenever an ion crosses a grid-plane. However, these deviations can easily be eliminated by taking a large enough E_{cn}. The increase of E_{cn} affects the overall computational cost only in a modest way.

4. Applications

The CPV scheme has been allready been used for several systems containing both first-row elements and transition metals, showing the method to be both accurate and useful. One class of applications has been related to water - small water clusters[22], different phases of ice[23,24] and liquid water[25] has been studied. These studies also showed that the hydrogen bond is not described accurately within LDA. A good description of hydrogen bond is obtained if gradient corrections has been added to the xc-functional.

Transition metals copper, zinc and titanium has been studied. Of these materials copper is particularly chalenching having very localized d-state. We find that the wave functions can be accurately described with 18 Ry cutoff whereas the cutoff needed for the density is 200 Ry. In case like this, where the density cutoff is much higer than four times the one for wave functions, the double grid technique is very useful. By using the double grid technique we were able to do a simulation of liquid copper using 50 atoms[18].

The CPV method can also be used for more conventional band structure calculations, with relatively small unit cells but using several **k**-points. This type of calculations has been done for $BaTiO_3$ [26] and for ZnSe[27]. The cutoff used for both of these systems was 25 Ry, providing excellent results comparing to experiments and to all-electron calculations.

To conclude, the Vanderbilt pseudopotential makes it possible to extend firs-principle molecular dynamics simulations to systems containing first-row elements and transition metals, which could not be afforded using norm-conserving pseudopotentials.

5. Acknowledgements

This work has been done in close collaboration with Alfredo Pasquarello, Roberto Car, Changyol Lee and David Vanderbilt.

6. References

1. *Simulations od Liquid and Solids* eds. G. Ciccotti, D. Frenkel and I.R. McDonald, (North-Holland, 1986). **55** (1985) 2471.
2. R. Car and M. Parrinello, *Phys. Rev. Lett.* **55** (1985) 2471.
3. P. Hohenberg and W. Kohn, *Phys. Rev.* **136** (1964) B864.
4. D. R. Hamann, M. Schlüter and C. Chiang, *Phys. Rev. Lett.* **43** (1979) 1494.
5. G. B. Bachelet, D. R. Hamann, and M. Schlüter, *Phys. Rev.*, **B26**, 4199 (1982).
6. G. Kerker, *J. Phys.* **C13** (1980), L189.
7. D. Vanderbilt, *Phys. Rev.* **B32** (1985) 8412.
8. A. M. Rappe, K. M. Rabe, E. Kaxiras, J. D. Joannopoulos, *Phys. Rev.* **B41**(1990) 1227.
9. N. Troullier and J. L. Martins, *Phys. Rev.* **B43** (1991) 1993.
10. D. Vanderbilt, *Phys. Rev.* **B41** (1990) 7892.
11. D.K. Remler and P.A. Madden, *Mol. Phys.* **70** (1990) 921.
12. G. Galli and M. Parrinello, in *Computer Simulations In Materal Science* eds. M. Meyer and V. Pontikis (Kluwer, 1991) 283.
13. M.C. Payne, M.P. Teter, D.C. Allan, T.A. Arias, and J.D. Joannopoulos, *Rev. Mod. Phys.* **64** (1992) 1045.
14. K. Laasonen, R. Car, C. Lee, and D. Vanderbilt, *Phys. Rev.* **B43** (1991) 6796.
15. W. Kohn and L. J. Sham, *Phys. Rev.* **140** (1965) A1133.
16. L. Kleinman and D. M. Bylander, *Phys. Rev. Lett.* **48** (1982) 1425.

17. R. D. King-Smith, M. C. Payne, and J. S. Lin, *Phys. Rev.* **B44** (1991) 13063.

18. A. Pasquarello, K. Laasonen, R. Car, C. Lee, and D. Vanderbilt, *Phys.Phys.Rev. Lett.* **69** (1992) 1982.

19. J. P. Ryckaert, G. Ciccotti, and H. J. C. Berendsen, *J. Comput. Phys.* **23** (1977) 327; G. Ciccotti and J. P. Ryckaert, *Computer Physics Reports* **4** (1986) 345.

20. R. Car and M. Parrinello, in *Simple Molecular Systems at Very High Density*, eds. A. Poliani, P. Loubeyre, and N. Boccara (Plenum,New York, 1989), p. 455.

21. A. Satta, A. Pasquarello, and K. Laasonen, (unpublished).

22. K. Laasonen, M. Parrinello, R. Car, C. Lee and D. Vanderbilt, submitted to *Chem. Phys. Lett.*.

23. C. Lee, D. Vanderbilt, K. Laasonen, R. Car, and M. Parrinello, *Phys. Phys. Lett.* **69** (1992) 462.

24. C. Lee, D. Vanderbilt, K. Laasonen, and R. Car, submitted to *Phys. Rev.* B.

25. K. Laasonen, M. Parrinello, and M. Sprik, to be published.

26. R. D. King-Smith and D. Vanderbilt, *Ferroelectrics*, in press.

27. K. W. Kwak, R. D. King-Smith, and D. Vanderbilt, in ***Proceedings of the 7th Trieste Semiconductor Symposium on Wide-Band-Gap Semiconductors***, Physica B, in press.

www.ingramcontent.com/pod-product-compliance
Lightning Source LLC
LaVergne TN
LVHW051053210926
848486LV00002B/91